T0189633

Advances in Computer Vision and Pattern Recognition

More information about this series at http://www.springer.com/series/4205

Mongi A. Abidi • Andrei V. Gribok • Joonki Paik

Optimization Techniques in Computer Vision

Ill-Posed Problems and Regularization

Mongi A. Abidi
Department of Electrical and Computer Engineering
University of Tennessee
Knoxville, Tennessee
USA

Andrei V. Gribok
Department of Human Factors, Controls, and Statistics
Idaho National Laboratory
Idaho Falls, Idaho
USA

Joonki Paik
Image Processing and Intelligent Systems Laboratory
Chung-Ang University
Seoul, Korea

ISSN 2191-6586 ISSN 2191-6594 (electronic)
Advances in Computer Vision and Pattern Recognition
ISBN 978-3-319-83501-3 ISBN 978-3-319-46364-3 (eBook)
DOI 10.1007/978-3-319-46364-3

Printed on acid-free paper

This Springer imprint is published by Springer Nature
The registered company is Springer International Publishing AG
The registered company address is: Gewerbestrasse 11, 6330 Cham, Switzerland

Preface

Overview

The advent of the digital information and communication era has resulted in image processing and computer vision playing more important roles in our society today. These roles include creating, delivering, processing, visualizing, and making decisions from information in an efficient and visually pleasing manner.

The term *digital image processing* or simply *image processing* refers to comprehensively processing picture data by a digital computer. The term *computer vision* refers to computing properties of the three-dimensional world from one or more digital images.

The theoretical bases of image processing and computer vision include mathematics, statistics, signal processing, and communications theory. In a wide variety of theories for image processing and computer vision, optimization plays a major role. Although various optimization techniques are used at different levels for those problems, there has not been a sufficient amount of effort to summarize and explain optimization techniques as applied to image processing and computer vision.

The objective of this book is to present practical optimization techniques used in image processing and computer vision problems. A generally ill-posed problem is introduced and it is used to show how this type of problem is related to typical image processing and computer vision problems.

Unconstrained optimization gives the best solution based on numerical minimization of a single, scalar-valued *objective function* or *cost function*. Unconstrained optimization problems have been intensively studied, and many algorithms and tools have been developed to solve them. Most practical optimization problems, however, arise with a set of constraints. Typical examples of constraints include (a) prespecified pixel intensity range, (b) smoothness or correlation with neighboring information, (c) existence on a certain contour of lines or curves, and (d) given statistical or spectral characteristics of the solution.

Regularized optimization is a special method used to solve a class of constrained optimization problems. The term *regularization* refers to the transformation of an objective function with constraints into a different objective function, automatically reflecting constraints in the unconstrained minimization process. Because of its simplicity and efficiency, regularized optimization has many application areas, such as image restoration, image reconstruction, and optical flow estimation.

Optimization-Problem Statement

The fundamental problem of optimization is to obtain the best possible decision in any given set of circumstances. Because of its nature, problems in all areas of mathematics, applied science, engineering, economics, medicine, and statistics can be posed in terms of optimization.

The general mathematical formulation of an optimization problem may be expressed as

$$\min_{x \in R^N} f(x) \text{ subject to } x \in C, \tag{1.1}$$

where $f(x)$ represents the objective function, C the constraint set in which the solution will reside, and R^N the N-dimensional real space.

Various optimization problems can be classified based on different aspects. At first, classification based on the properties of the objective function $f(x)$ is:

1. Function of a single variable or multiple variables,
2. Quadratic function or not, and
3. Sparse or dense function.

A different classification is also possible based on the properties of the constraint C as:

1. Whether or not there are constraints,
2. Defined by equation or inequality of constraint functions, i.e., $C = \{x|c(x) = 0\}$ or $C = \{x|c(x) \geq 0\}$, where $c(x)$ is termed the constraint function, and
3. The constraint function is either linear or nonlinear.

If, for example, we want to reduce random noise from a digital image, a simple way is to minimize the extremely high-frequency component while keeping all pixel intensity values inside the range $[0, 255]$. In this case, the objective function of multiple variables represents the high-frequency component. The range of pixel intensity values plays a role in inequality constraints.

Optimization for Image Processing

Various optimization techniques for image processing can be summarized as, but are not limited to, the following:

1. Image quantization: The optimum mean square (*Lloyd*-Max) quantizer design.
2. Stochastic image models: Parameter estimation for auto-regressive (AR), moving average (MA), or auto-regressive-moving average (ARMA) model.
3. Image filtering: Optimal filter design and Wiener filtering.
4. Image restoration: Wiener restoration filter, constrained least squares (CLS) filter, and regularized iterative method.
5. Image reconstruction: Convolution/filtered back-projection algorithms.
6. Image coding: Energy compaction theory and optimum code-book design.

Optimization for Computer Vision

Some examples of optimization techniques for computer vision are summarized as:

1. Feature detection: Optimal edge enhancer, ellipse fitting, and deformable contours detection.
2. Stereopsis: Correspondence/reconstruction procedures, three-dimensional image reconstruction.
3. Motion: Motion estimation, optical flow estimation.
4. Shape from single image cue: Shape from shading, shape from texture, and shape from motion.
5. Recognition
6. Pattern matching

Organization of This Book

This book has five self-contained parts. In Part I, Chap. 1 introduces the scope and general overview of the material. Chapter 1 gives an introduction into ill-posed problems. This chapter also discusses practical reasons why many image processing and computer vision problems are formulated as ill-posed problems. Chapter 1 also presents typical examples of ill-posed problems in image processing and computer vision areas. Chapter 2 discusses different techniques to select regularization parameter.

Part II summarizes the general optimization theory that can be used to develop a new problem formulation. Practical problems are solved using the optimization formulation. Chapter 3 presents a general form of optimization problems and summarizes frequently used terminology and mathematical background. Chapters 4 and 5 describe in-depth formulation and solution methods for unconstrained and constrained optimization problems, respectively. Constrained optimization problems are more suitable for modeling real-world problems. This is true because the desired solution of the problem usually has its own constraints.

In Part III, we discuss regularized optimization, or simply regularization, that can be considered a special form of a general constrained optimization. In Chap. 6, frequency-domain regularization is discussed. Chapters 7 and 8 describe iterative type implementations of regularization and fusion-based implementation of regularization.

Part IV provides practical examples for various optimization technique applications. Chapters 9, 10, 11, and 12 give some important applications of two-dimensional image processing and three-dimensional computer vision.

Appendices summarize commonly used mathematical background.

Knoxville, TN — Mongi A. Abidi
Idaho Falls, ID — Andrei V. Gribok
Seoul, Republic of Korea — Joonki Paik

About the Authors

Mongi A. Abidi received the Principal Engineering degree in electrical engineering from the National Engineering School of Tunis, Tunisia, in 1981 and the M.S. and Ph.D. degrees in electrical engineering from the University of Tennessee, Knoxville, in 1985 and 1987, respectively. He is a Professor with the Department of Electrical and Computer Engineering, the University of Tennessee, where he directs activities in the Imaging, Robotics, and Intelligent Systems Laboratory as an Associate Department Head. He conducts research in the field of 3D imaging, specially, in the areas of scene building, scene description, and data visualization.

Andrei V. Gribok is a Principal Research Scientist at Idaho National Laboratory, Department of Human Factors, Controls, and Statistics. He received his Ph.D. in Physics from Moscow Institute of Biological Physics in 1996 and his B.S. and M.S. degrees in systems science/nuclear engineering from Moscow Institute of Physics and Engineering in 1987. Dr. Gribok worked as an instrumentation and control researcher at the Institute of Physics and Power Engineering, Russia, where he conducted research on software fault detection systems for nuclear power plants. He also was an invited research scientist at Cadarache Nuclear Research Center, France, where his research focus was on ultrasonic visualization systems for liquid metal reactors. He also holds the position of Research Associate Professor with the Department of Nuclear Engineering, University of Tennessee, Knoxville. Dr. Gribok was a member of a number of international programs including IAEA coordinated research program on acoustical signal processing for the detection of sodium boiling or sodium-water reaction in LMFRs and large-scale experiments on acoustical water-in-sodium leak detection in LMFBR. Dr. Gribok's research interests include reliability and soft computing in nuclear engineering, nuclear systems' instrumentation and control, prognostics and diagnostics for nuclear power plants, uncertainty analysis of software systems, and inverse and ill-posed problems in engineering. Dr. Gribok taught classes on soft computing at the University of Tennessee, Knoxville. Dr. Gribok is an author and coauthor of three book chapters, 37 journal papers, and numerous peer-reviewed conference papers.

Joonki Paik received his B.S. degree in control and instrumentation engineering from Seoul National University in 1984. He received the M.S. and the Ph.D. degrees in electrical engineering and computer science from Northwestern University in 1987 and 1990, respectively. From 1990 to 1993, he joined Samsung Electronics, where he designed the image stabilization chip sets for consumer camcorders. Since 1993, he has joined the faculty at Chung-Ang University, Seoul, Korea, where he is currently a Professor in the Graduate school of Advanced Imaging Science, Multimedia and Film. From 1999 to 2002, he was a visiting Professor at the Department of Electrical and Computer Engineering at the University of Tennessee, Knoxville.

Contents

Part I

Chapter 1
Ill-Posed Problems in Imaging and Computer Vision

1.1 Introduction

In many image processing problems, we need to estimate the original complete data sets, generally from incomplete and, most often, from degraded observations. One simplified example is to estimate the original pixel intensity value, which has been attenuated in the imaging system, without any correlation with neighboring pixels. If we know the nonzero attenuation factor for the imaging system, we can easily estimate the original value by multiplying by this attenuation factor. Figure 1.1 shows the corresponding attenuation and the restoration processes.

As a second example, suppose we estimate the original pixel intensity value by averaging this pixel with eight neighboring pixels as shown in Fig. 1.2. This example is frequently used in simplifying the out-of-focus blur in an imaging system. As shown in Fig. 1.2a, a pixel intensity value is distributed into the neighborhood on the imaging plane. As a result, a pixel value in the imaging plane is determined by integrating partial contributions of neighboring pixels in the object plane, as shown in Fig. 1.2b. In order to have mathematical representation of the two-dimensional (2D) signals and systems, we use the row-ordered (or lexicographically ordered) vector-matrix notation [jain89]. If we assume that both the input image contained in the object plane and the observed image contained in the imaging plane are $N \times N$, the pixel averaging process can be represented as

$$Du = f, \tag{1.1}$$

where u and f represent $N^2 \times 1$ input and output images, respectively, and D the $N^2 \times N^2$ matrix, such that

M.A. Abidi et al., *Optimization Techniques in Computer Vision*, Advances in Computer Vision and Pattern Recognition, DOI 10.1007/978-3-319-46364-3_1

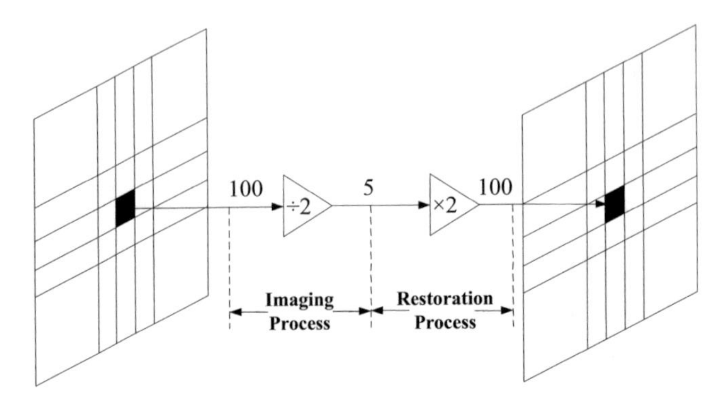

Fig. 1.1 A diagram illustrating pixel attenuation followed by restoration

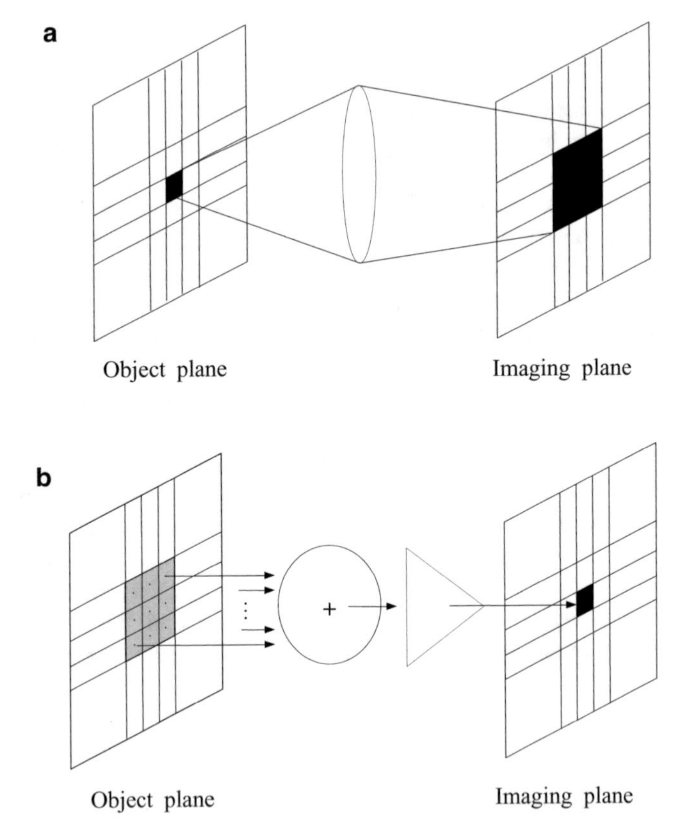

Fig. 1.2 (**a**) Simplified out-of-focus process and (**b**) the corresponding model obtained by averaging pixels in the neighborhood

$$D = \begin{bmatrix} D_1 & D_1 & 0 & \cdots & 0 \\ D_1 & \ddots & \ddots & \ddots & \vdots \\ 0 & \ddots & \ddots & \ddots & 0 \\ \vdots & \ddots & \ddots & \ddots & D_1 \\ 0 & \cdots & 0 & D_1 & D_1 \end{bmatrix}, \text{ where } D_1 = \begin{bmatrix} 1 & 1 & 0 & \cdots & 0 \\ 1 & \ddots & \ddots & \ddots & \vdots \\ 0 & \ddots & \ddots & \ddots & 0 \\ \vdots & \ddots & \ddots & \ddots & 1 \\ 0 & \cdots & 0 & 1 & 1 \end{bmatrix}. \tag{1.2}$$

See Appendix A for more details of this formulation.

To estimate u from f, we can multiply D^{-1} to the left-hand side of f, such that $\hat{u} = D^{-1}f$, where $\hat{u}$ denotes a calculated estimate of u. Keep in mind that in order to compute the inverse of a matrix, the matrix must be nonsingular. One also will recall that a nonsingular matrix has a nonzero determinant. It is desirable for the matrix D to be well conditioned. This means that a bounded perturbation in the observed data f results in a bounded error in the estimated solution $\hat{u}$. In this example, $\hat{u}$ is a good estimate, in some sense, because D is assumed to have an inverse and assumed to be well conditioned [golub96]. However, D becomes ill conditioned as the number of averaged pixels increases. Even when D has an inverse, the estimate from observed data, being possibly corrupted by noise, is not reliable if this matrix is ill conditioned. This means bounded perturbations in the observed data may result in unbounded errors in the estimated solution.

This second example deals with a simple estimation problem in which the observed data and the solution have the same dimensions. This results in the estimation process outcome being equivalent to multiplying the inverse of the distortion matrix D by the observed data f. In many engineering problems, however, the observed data and the solution generally have different dimensions and characteristics. For this reason, we need a more general description of the characteristics of inverse problems.

1.2 The Concepts of Well Posedness and Ill Posedness

J. Hadamard introduced the concept of a well-posed problem, resulting from physical mathematical models, to clarify the most natural boundary conditions for various types of differential equations in the early 1900s [tikhonov77]. As mentioned previously, a linear equation with a well-conditioned matrix is a good example of a well-posed problem. A formal definition of well posedness is now presented.

We consider a solution of any quantitative problem, where the solution u is to be estimated from given data f and the operator A, which relates u and f, such as $A: u \rightarrow f$. We shall consider u and f as elements of metric spaces U and F with metrics $\rho_U(u_1, u_2)$ for $u_1, u_2 \in U$ and $\rho_F(f_1, f_2)$ for $f_1, f_2 \in F$.

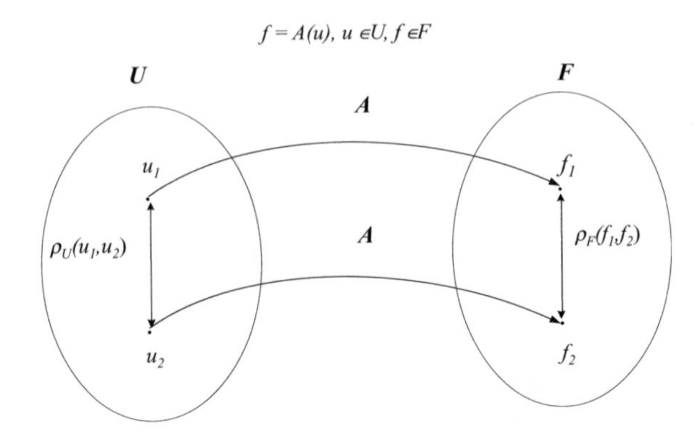

Fig. 1.3 Relationships among spaces, elements, the operator, and metrics

Figure 1.3 shows the relationships among spaces, elements, the operator, and metrics. Frequently, the Euclidean distance is used for metrics ρ_U and ρ_F. U and F are then assumed to be Hilbert spaces.

The problem of determining the solution u in the space U from the given data f in the space F is said to be well posed on the pair of metric spaces (U, F) if the following three conditions are satisfied:

1. For every element $f \in F$, there exists a solution u in the space U.
2. The solution is unique.
3. The problem is stable on the spaces (U, F).

The problem is otherwise ill posed. For a long time, it was assumed that a mathematical problem must satisfy the above conditions and that an applied problem should be formulated in the same manner [courant62]. This assumption, however, was revealed to be invalid after many physical phenomena were studied.

1.3 Ill-Posed Problems Described by Linear Equations

In general, ill-posed problems arise in a wide variety of applied physics and engineering areas such as nuclear physics, plasma physics, radiophysics, and geophysics as well as electrical, nuclear, and mineral engineering. In order to generalize the solution of the linear equation (1.1), we need the following operator equations of the first class[1]:

$$Au = f, \tag{1.3}$$

[1]If u is defined on the discrete space, each element of f in (1.3) represents a weighted sum, and we call it the linear or the first order operation. On the other hand, if u is defined on the real space, the corresponding operation is called the *first class* instead of the *first order*.

where f represents the given, incomplete, or distorted data, u the complete or original data to be estimated, and A the operator which can have various forms.

Suppose $u(t)$ is an unknown function in space U, $f(t)$ a known function in space F, and that operator A is defined by

$$Au = \int_a^b K(t,s)u(s)\,ds. \tag{1.4}$$

Equation (1.3) then becomes the Fredholm integral equation of the first kind, such that

$$\int_a^b K(t,s)u(s)\,ds = f(t), \quad \text{for } c \leq t \leq d, \tag{1.5}$$

where $[c, d]$ represents the domain on which f is defined.

Equation (1.5) is almost always an ill-posed problem. This is true because of the instability resulting from a bounded perturbation in the given data causing an unbounded value of the integration in Eq. (1.5). Detailed analysis about the existence and stability of a solution for Eq. (1.5) can be found in [tikhonov77].

In this section we will describe ill-posed problems by analyzing a discrete counterpart of Eq. (1.5). This discrete counterpart better represents digital image processing and computer vision problems.

In Eq. (1.3), suppose that u is an unknown vector, f is a known vector, and the A operator represents a square matrix D, with elements d_{ij}. Equation (1.3) then becomes a linear equation having an $N \times N$ square matrix D, such that

$$Du = f. \tag{1.6}$$

When U and F represent Hilbert spaces ($u \in U$ and $f \in F$), the problem denoted by Eq. (1.6) is well posed if there exists a unique least square solution which continuously depends on the data [nashed81].

If we consider Eq. (1.6) as the discretization of an ill-posed continuous problem given in Eq. (1.5), quantization error and noise may be added to the data, as in Eq. (1.7).

$$Du + v = f, \tag{1.7}$$

where the D matrix has a number of zero, or very small singular values. As a result, ill posedness of the continuous problem translates into an ill-conditioned matrix D, with additive quantization error v [kats91]. Various types of difficulties exist in solving an ill-conditioned linear equation as discussed below.

- Singularity check: If D is nonsingular, a unique solution vector u exists. If D is singular, a solution which is not unique only exists when a special condition is satisfied. Because we want to have a unique solution, it is generally necessary to check the singularity of D before beginning to solve the linear equation. A singularity check for the $N \times N$ square matrix can be made by investigating the determinant of D. This requires approximately N^3 operations. When the dimension of the system becomes large, a singularity check is a significant burden.

- Observation and numerical errors: In practical problems, we cannot avoid observation errors incorporated with data. With observation errors, the linear equation can be rewritten as $\widetilde{D}\widetilde{u} = \widetilde{f}$, where $\| \widetilde{D} - D \| \leq \delta$ and $\| \widetilde{f} - f \| \leq \delta$. In this case, the symbol $\sim$ is used to represent observed or measured data. Even if D is nonsingular, $\widetilde{D}$ may become singular or near singular. Furthermore, the error in the solution $\| \widetilde{u} - u \|$ may not be bounded by a sufficiently small value. In either case, or both, a stable and meaningful solution cannot be obtained.

1.4 Solving Ill-Posed Problems

It is not practical to calculate the solution of ill-posed problems by the direct method. In this section we introduce a selection method which serves as a general approach for estimating an approximate solution to the ill-posed problem. In the case of the ill-posed problem, approximate solutions are determined that are stable under small changes in the initial data based on the use of supplementary information. The concept of the selection method is to obtain multiple solutions of well-posed problems that are near the original ill-posed problem, in some sense. Next, one additional problem is formulated based on additional information from prior well-posed problem solutions.

Example 1.1 Consider the relationship

$$Du = f, \tag{1.8}$$

where $D = \begin{bmatrix} 2 & 1 \\ 2 & 2 \end{bmatrix}$ and $u = \begin{bmatrix} 1 \\ 1 \end{bmatrix}$. The operation of Du yields the vector, $f = \begin{bmatrix} 3 \\ 4 \end{bmatrix}$. Now suppose that we desire the solution u, given D and f. Since matrix D is invertible, the problem is well posed. We can solve for u directly by calculating D^{-1}, that is, $u = D^{-1}f = \begin{bmatrix} 1 & -0.5 \\ -1 & 1 \end{bmatrix} \begin{bmatrix} 3 \\ 4 \end{bmatrix} = \begin{bmatrix} 1 \\ 1 \end{bmatrix}$.

Alternatively, assume we know that the solution for u exists inside the circle of radius 2 and centered at $[1, 1]^T$. If we have three candidate solutions in the circle, $a = [0, 0]^T$, $b = [1, 0]^T$, and $c = [2, 0]^T$, then calculate $Da = [0, 0]^T$, $Db = [2, 2]^T$, and $Dc = [4, 4]^T$. Finally, we choose c as the approximate solution based on the fact that Dc is the closest to f in the Euclidean distance sense. This selection process is depicted in Fig. 1.4.

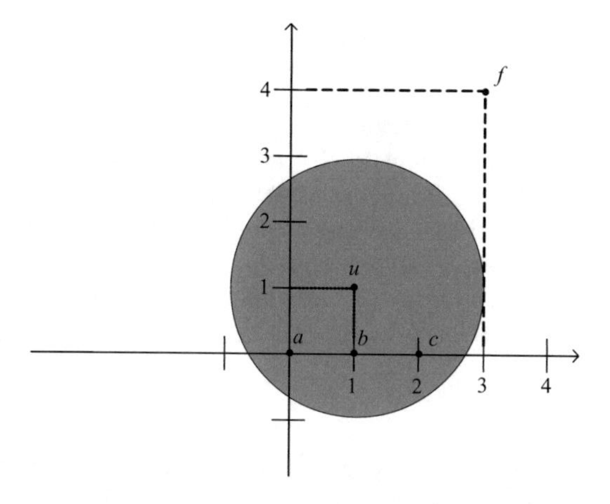

Fig. 1.4 An example of the selection method for solving an approximated linear equation with supplementary information

To describe the selection method for generally ill-posed problems, we consider an operator equation of the first kind

$$Au = f, \tag{1.9}$$

where u and f belong to metric spaces U and F, respectively, and A represents an operator that maps U onto F. Given subclass $M \subset U$, the general selection method is defined by calculating a class of Au for $u \in M$ and then taking u_0 such that

$$\rho_F(Au_0, f) = \inf_{u \in M} \rho_F(Au, f), \tag{1.10}$$

where u_0 is determined such that $\rho_F\,(Au,f)$ is minimized over $u \in M$.

The selection method can be accepted for solving a wide variety of ill-posed problems due to the use of the following proposition.

Proposition 1.1 *If, for $u_n \in M$, $\rho_f\,(Au_n,f)$ approaches zero as n becomes infinitely large, then $\rho_U(u_n, u_T) \rightarrow 0$ also approaches zero with infinitely large n, where u_T represents the exact solution. This proposition can be proved by using the following lemma.*

Lemma 1.1 *Suppose that a compact subset U of a metric space U_0 is mapped onto a subset F of a metric space F_0. If the mapping $U \rightarrow F$ is continuous and one to one, the inverse mapping $F \rightarrow U$ is also continuous.*

Proof of this lemma can be found in [tikhonov77]. An element of $\widetilde{u} \in M$ minimizing the functional $\rho_F\,(Au,f)$ on the set M is called a *quasisolution* of Eq. (1.9).

1.5 Image Restoration

Any image acquired by optical, electronic, or numerical means cannot be free from degradation due to the inherent errors produced in sensing, transmitting, and processing the image. Image restoration is concerned with estimating an image and making it as close as possible to the original image prior to degradation. The degradation may occur in the form of sensor noise, out-of-focus blur, motion blur, random atmospheric turbulence, and so forth. Most forms of image degradation can be represented, or approximated, by the linear equation

$$y = Dx + \eta, \tag{1.11}$$

where y and x represent vectors whose elements are sets of two-dimensionally distributed pixel intensity values of the observed and original images, respectively. Matrix D represents a degradation operation. Noise vector η plays a role in incorporating sensor noise and numerical errors into the degradation process [jain89, kats91]. In the previous section, Fig. 1.2b and Eqs. (1.1) and (1.2) showed a simple example for image degradation.

Accuracy of the solution of Eq. (1.11) depends on both the conditioning of the matrix D and the amount of noise η. Intuitively, the conditioning of a matrix (i.e., the accuracy of the represented information) becomes less reliable as the size and bandwidth of the matrix increases. Due to the nature of the general image degradation model, matrix D is almost always ill conditioned. In other words, Eq. (1.11), which deals with a large-size image and severe distortion, is an ill-posed problem.

Because matrix D is either singular or ill conditioned, an approximate solution of the transformed, well-posed equation is used to replace the original solution. One popular such transformation is given by

$$b = Tx, \tag{1.12}$$

where $T = D^{\mathrm{T}}D + \lambda C^{\mathrm{T}}C$ and $b = D^{T}y$. If we assume that λ is equal to zero, we can easily see that the solution for Eq. (1.12) is identical to the solution for Eq. (1.11) by ignoring noise η. However, matrix $D^{\mathrm{T}}D$ is better conditioned than D because the energy of diagonal elements increases by multiplying the matrix by itself. If λ has an appropriate value, the energy of T is further concentrated on the diagonal elements with a suitable choice of matrix C, which serves as a high-pass filter in many signal restoration problems. See Sects. 1.9, 1.10, and 1.11 for a more detailed analysis and implementation of matrix C.

In transforming the ill-posed problem into the well-posed one, supplementary or a priori information about the solution is incorporated by using the filtering term $C^{\mathrm{T}}C$. By appropriately selecting the value λ, we can make matrix T as well conditioned as desired.

Figure 1.5 shows the entire process of image degradation and the corresponding restoration. The 256×256 8-bit gray-level image is used for the original data x. As

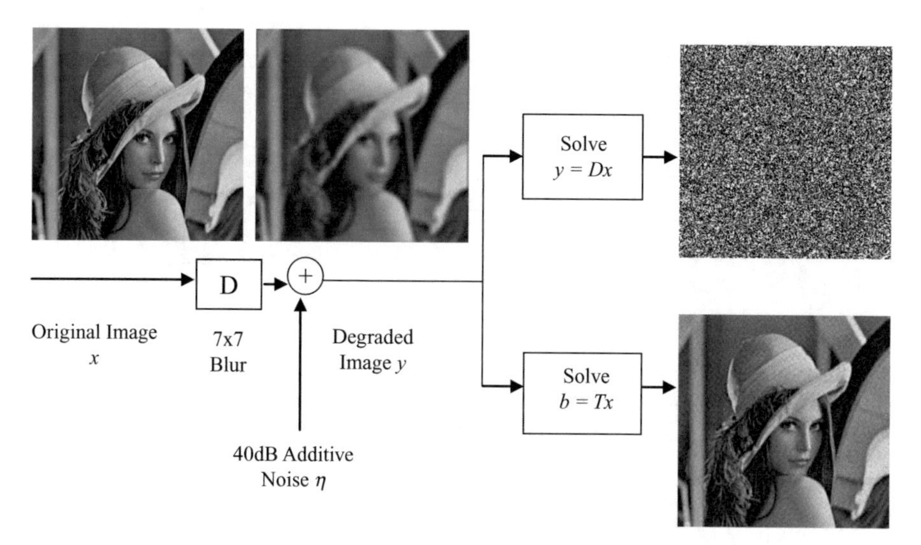

Fig. 1.5 General image degradation and restoration processes

the degradation operation, a 7×7 pixel averaging (blurring) and 40 dB additive noise are used. When we directly solve the ill-posed problem, we usually get a very poorly estimated image. On the other hand, when we solve the translated well-posed equation, we get an acceptably well-estimated image. In transforming the ill-posed problem into a well-posed one, we use supplementary smoothness information, which assumes that the original image does not have abruptly changed data patterns.

1.6 Image Interpolation

Image interpolation is used to obtain a higher-resolution image from a low-resolution image. Image interpolation is very important in high-resolution or multi-resolution image processing. More specifically, it can be used in changing the format of various types of images and videos and in increasing the resolution of images in a purely digital or numerical method.

Let $x_C\ (p,q)$ represent a two-dimensional spatially continuous image, and let $x(m,n)$ represent the corresponding digital image obtained by sampling $x_C\ (p,q)$, with sampling size $N \times N$, such that

$$x(m,n) = x_C(mT_\text{v}, nT_\text{h}), \quad \text{for } m,n = 0,1,\ldots,N-1 \tag{1.13}$$

where T_v and T_h represent the vertical and the horizontal sampling intervals, respectively. In a similar way, the image with four times lower resolution in both horizontal and vertical directions can be represented as

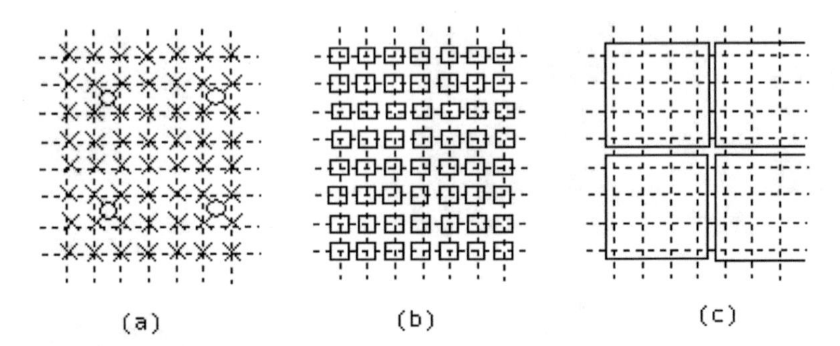

Fig. 1.6 Relationship between two images with different resolutions: (**a**) Sampling grids for two different images; "O" for $x_{1/4}$ and "X" for x, (**b**) an array structure of photodetectors for $x(m, n)$ and (**c**) an array structure of photodetectors for $x_{1/4}(m, n)$

$$x_{1/4}(m,n) = \frac{1}{16}\sum_{i=0}^{3}\sum_{j=0}^{3} x(4m+i, 4n+j), \quad \text{for } m,n = 0,1,\ldots,\frac{N}{4}-1, \quad (1.14)$$

where the subscript 1/4 represents four times downsampling.

Two different sampling grids for images in Eqs. (1.13) and (1.14) are shown in Fig. 1.6a. Array structures of photodetectors corresponding to the images in Eqs. (1.13) and (1.14) are shown in Fig. 1.6b, c, respectively.

A discrete linear space-invariant degradation model for an $\frac{N}{4} \times \frac{N}{4}$ low-resolution image, which is obtained by subsampling the original $N \times N$ high-resolution image, can be given as

$$y = Hx + \eta, \quad (1.15)$$

where the $N^2 \times 1$ vector x represents the lexicographically ordered high-resolution image and $\left(\frac{N}{4}\right)^2 \times 1$ vectors y and η the subsampled low-resolution and noise images, respectively. The $\left(\frac{N}{4}\right)^2 \times N^2$ matrix H represents the series of low-pass filtering and subsampling and can be written as

$$H = H_1 \otimes H_1, \quad (1.16)$$

where $\otimes$ represents the Kronecker product and the $\left(\frac{N}{4}\right) \times N$ matrix H_1 represents the one-dimensional low-pass filtering and subsampling by a factor of 4, such as

$$H_1 = \frac{1}{4}\begin{bmatrix} 1 & 1 & 1 & 1 & 0 & 0 & 0 & 0 & \cdots & 0 & 0 & 0 & 0 \\ 0 & 0 & 0 & 0 & 1 & 1 & 1 & 1 & \cdots & 0 & 0 & 0 & 0 \\ \vdots & \vdots & \vdots & \vdots & \vdots & \vdots & \vdots & \vdots & \ddots & \vdots & \vdots & \vdots & \vdots \\ 0 & 0 & 0 & 0 & 0 & 0 & 0 & 0 & \cdots & 1 & 1 & 1 & 1 \end{bmatrix}. \quad (1.17)$$

To estimate the original high-resolution image from the low-resolution observation given in Eq. (1.15), we need to solve the corresponding inverse problem. Because H is not a square matrix, a unique solution does not exist for the inverse problem. Instead, we may obtain the solution that minimizes

$$f(x) = ||y - Hx||^2, \tag{1.18}$$

which is clearly an ill-posed problem because the solution is not unique. To select one good solution from the set of feasible solutions, we must use additional information about the solution, such as smoothness or finite bandwidth. One typical way to incorporate the additional information into the minimization process is to minimize

$$f(x) = ||y - Hx||^2 + \lambda||Cx||^2, \tag{1.19}$$

where C represents a filter that extracts a certain kind of frequency component, and λ the scalar value that controls the utilization of additional information.

1.7 Motion Estimation

Motion estimation is one of the fundamental problems in image processing and computer vision. In approaching this problem, two-dimensional image plane motion estimation serves as a theoretical basis of motion-compensated video coding standards. Three-dimensional object motion estimation is used for reconstructing three-dimensional shapes and object tracking systems.

Consider a pair of similar images which can be obtained from either two adjacent image frames in a moving image sequence, stereo image pairs, or a pair of synthetic aperture radar (SAR) images. Those images can be depicted as in Fig. 1.7. The motion estimation problem can be posed as the estimation of *image*

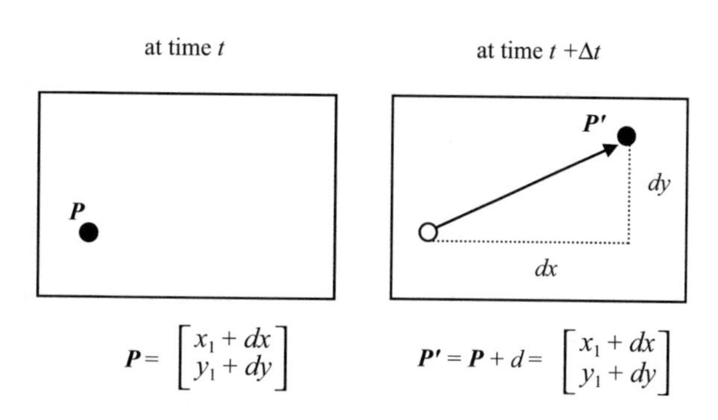

Fig. 1.7 A pair of images in which the point ***P*** in the *left image* moves onto ***P′*** in the *right image*

plane correspondence vectors or simply *motion vectors* denoted by $d = [dx, dy]^T$. Let $I(x, y, t)$ be the intensity value of the (x, y) -th pixel at time t. We can then write

$$I(x + dx, y + dy, t + \Delta t) = I(x, y, t), \quad (1.20)$$

where we assume that there is no intensity variation between two corresponding points.

Because the correspondence vector relates two different positions in each image frame, the motion estimation problem can be considered as a correspondence problem.

Among various motion vector estimation methods, the block-matching algorithm is one that minimizes the sum of absolute difference in the neighborhood of the point of interest, such as

$$\text{minimize } f(dx, dy) = \sum_{(x,y)\in S} |I(x + dx, y + dy, t + \Delta t) - I(x, y, t)|, \quad (1.21)$$

where S represents the neighborhood for the point of interest.

It is well known that motion estimation cannot be free from the following three problems [tekalp95].

1. Existence occluded problem: On the region, no correspondence can be established.
2. Uniqueness problem: If, for example, a line moves along a certain direction, many different displacements can match the two lines. It is called an aperture problem.
3. Continuity problem: Motion estimation is highly sensitive to the presence of noise. A small amount of noise may result in a large deviation in the estimated motion vector.

Because of the above-stated problems, motion estimation is an ill-posed problem. Therefore, the estimated motion vector, without proper constraints, cannot be a good approximation for the real motion.

1.8 Volumetric Models for Shape Representation

Fundamental problems in computer vision can be stated as follows:

1. Two-dimensional images are inadequate for defining the three-dimensional world.
2. More than one three-dimensional scene can produce identical two-dimensional images [ber88].

Therefore, recovery of a three-dimensional scene from two-dimensional images is an ill-posed problem. The relationship between the three-dimensional real world

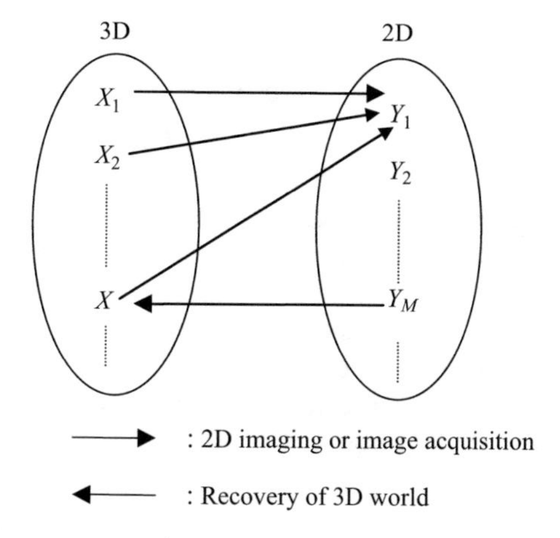

Fig. 1.8 The relationship between the real-world and two-dimensional images

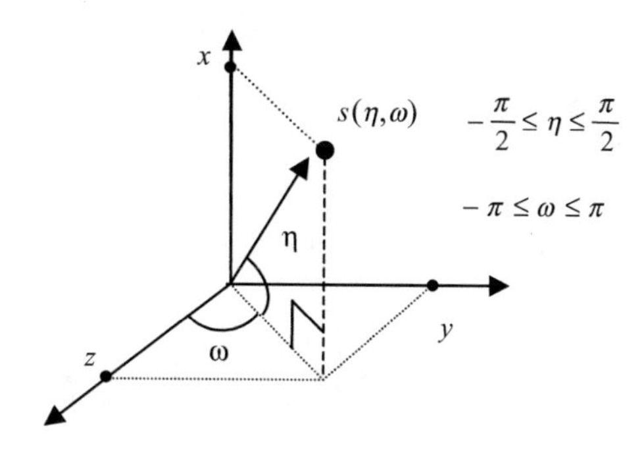

Fig. 1.9 Representation of a point in three-dimensional space

and two-dimensional images can be depicted in Fig. 1.8. Right-hand arrows represent two-dimensional image formulation or acquisition, while the left-hand arrows represent recovery of the three-dimensional world.

Due to its ill-posed nature, three-dimensional scene or shape recovery requires additional knowledge that may be incorporated in models that represent shapes. The reasons for using models for shape recovery can be summarized as follows:

1. Data compression becomes possible by using a small number of model parameters.
2. The resulting shape can be immunized against erroneous data.
3. Higher-level description is possible.
4. We can deduce occluded or missing data depending on the model.

In this section, we will introduce the superquadrics as a volumetric model for three-dimensional scene recovery [solina90]. Figure 1.9 shows a point on the superquadric surface in the three-dimensional coordinate. The point $s(\eta, w)$ can be described mathematically as

$$s(\eta, w) = \begin{bmatrix} x \\ y \\ z \end{bmatrix} = \begin{bmatrix} a_1 \cos^{\varepsilon_1}(\eta) \cos^{\varepsilon_2}(w) \\ a_2 \cos^{\varepsilon_1}(\eta) \sin^{\varepsilon_2}(w) \\ a_3 \sin^{\varepsilon_1}(\eta) \end{bmatrix}, \tag{1.22}$$

where a_1, a_2, and a_3 represent the size of the superquadric in the x, y, and z directions, respectively. ε_1 and ε_2 represent the measure of squareness in the latitude and longitude, respectively. According to the signs and magnitudes of ε_1 and ε_2, the corresponding superquadrics may have the form of a sphere, a cylinder, a parallelepiped, etc.

From Eq. (1.22), any point on the surface of a superquadric satisfies the following relationships:

$$(x/a_1)^{1/\varepsilon_2} = (\cos \eta)^{\varepsilon_1/\varepsilon_2} \cos w, \tag{1.23}$$

$$(y/a_2)^{1/\varepsilon_2} = (\cos \eta)^{\varepsilon_1/\varepsilon_2} \sin w, \tag{1.24}$$

and

$$(z/a_3)^{1/\varepsilon_1} = \sin \eta \tag{1.25}$$

On the surface of superquadric, it is also true that

$$F(x,y,z) = \left[\left\{ \left(\frac{x}{a_1}\right)^{2/\varepsilon_2} + \left(\frac{y}{a_2}\right)^{2/\varepsilon_2} \right\}^{\varepsilon_2/\varepsilon_1} + \left(\frac{z}{a_3}\right)^{2/\varepsilon_1} \right]^{\varepsilon_1} = 1, \tag{1.26}$$

where we call $F(x,y,z)$ the inside-outside function.

If the coordinate on which the superquadric is defined, rotated, and translated by (ϕ, θ, φ) and (p_x, p_y, p_z) from the three-dimensional world coordinate system, (x,y,z) can be expressed by the following homogeneous transformation:

$$\begin{bmatrix} x \\ y \\ z \end{bmatrix} = R \begin{bmatrix} x_w \\ y_w \\ z_w \end{bmatrix} + \begin{bmatrix} p_x \\ p_y \\ p_z \end{bmatrix}, \tag{1.27}$$

where R represents the rotation matrix.

Finally, the model of a superquadric surface is represented by the 11 parameters, that is, three-scale parameters (a_1, a_2, a_3), two measure of squareness parameters $(\varepsilon_1, \varepsilon_2)$, three rotation parameters (ϕ, θ, φ), and three translation parameters (p_x, p_y, p_z). For simplicity, let $\{b_1, b_2, \ldots, b_{11}\}$ represent the 11 parameters.

Given N three-dimensional surface points, such as (x_w^i, y_w^i, z_w^i), for $i = 1, \ldots, N$, the optimum superquadric that fits the data points can be obtained from the following optimization problem:

$$\text{minimize} \sum_{i=1}^{N} (a_1 a_2 a_3) \left[1 - F\left(x_w^i, y_w^i, z_w^i;\ b_1, b_2, \ldots, b_{11}\right)\right]^2, \tag{1.28}$$

where multiplication of $(a_1 a_2 a_3)$ normalizes the objective function.

In most cases, the superquadric obtained as a result of the optimization of Eq. (1.28), however, cannot completely describe a real, complicated, three-dimensional shape. It is therefore necessary to refine the superquadric for further fitting of more complicated shapes. One popular way of such refinement is using the free-form deformation [sederberg86].

Figure 1.10 shows an object that is to be deformed in a three-dimensional box. We call the set of volumetric grid points in the box control points, which links the box to the object. The box is deformed by the displacement between the control points and by estimating the new position of the deformed object. Finally, the adjusted grid points reflect deformation of the object.

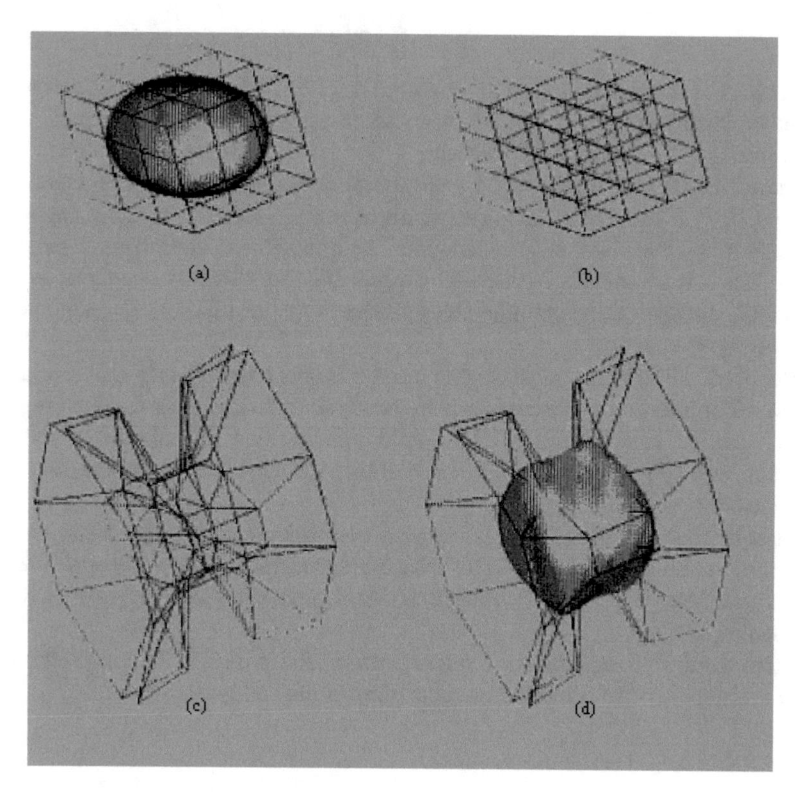

Fig. 1.10 A simple illustration of free-form deformation. In raster scanning order: (**a**) an object embedded in the initial grid points, (**b**) regular box-shaped grid points, (**c**) deformed box by displacement information, and (**d**) resulting deformed object in the deformed box [bardinet95]

Let P represent M control points regularly sampled on a three-dimensional parallelepiped box and X be the N given data points. Then, there is a matrix B that relates P and X, such that

$$X = BP. \tag{1.29}$$

The set of points on the object in Fig. 1.10a and the set of grid points in Fig. 1.10c correspond to X and P, respectively. We define ΔX as the difference between the points on the initial object and the desired shape and ΔP as the difference between the initial and desired control points. The following linear equation should then hold:

$$(X + \Delta X) = B(P + \Delta P), \tag{1.30}$$

which, from Eq. (1.27), is equivalent to

$$\Delta X = B \cdot \Delta P. \tag{1.31}$$

Since N is much larger than M, solving Eq. (1.31) for ΔP is always an overdetermined problem. We therefore must use a singular value decomposition or regularization for solving this problem.

Many practical image processing and computer vision problems can be stated, in general, in the form of constrained optimization. However, in many cases it is difficult to fit the practical problem into the constrained optimization structure. Therefore, it is reasonable to consider simpler and more intuitive methods to solve the practical constrained optimization problem at the cost of convergence rate and theoretical accuracy.

Regularization is the most widely used method to efficiently solve practical optimization problems with one or more constraints. Major advantages of regularization can be summarized as (1) simple and intuitive formulation of the cost function and (2) flexibility in incorporating one or more constraints into the optimization process.

In subsequent sections of this chapter, we will present a discussion about a typical example that explains why we use regularization and how it works. We will also show how a proper knowledge about the solution can help in finding a reasonable answer.

Mathematical formulation of regularization and a regularization method for solving linear inverse problems are described in the following sections.

1.9 Prior Knowledge About the Solution

From the material presented in previous sections, we know that many image processing and computer vision problems are ill posed. In addition, even well-posed problems tend to become ill posed as the size of the problem becomes larger, or external perturbations, such as noise or error, are involved. Although many sophisticated unconstrained optimization methods have been proposed, direct application of these methods to an ill-posed problem results in a very poor solution, which, in many cases, is much worse than the incomplete data set given initially.

One way to avoid such a disastrous result, which is caused by directly solving an ill-posed problem, is to solve a well-posed problem that is as close to the given ill-posed problem as possible, in some sense.

Consider the following linear equation.

$$\begin{bmatrix} 2 & 0 & 0 \\ 0 & 2 & 0 \\ 0 & 0 & 2\times 10^{-5} \end{bmatrix} \begin{bmatrix} x_1 \\ x_2 \\ x_3 \end{bmatrix} = \begin{bmatrix} 1 \\ 1 \\ 10^{-5} \end{bmatrix}, \tag{1.32}$$

with the solution, $x = \begin{bmatrix} 0.5 & 0.5 & 0.5 \end{bmatrix}^{\mathrm{T}}$. Suppose that there is a small perturbation in the third element of the given data, such as $\delta = \begin{bmatrix} 0 & 0 & 10^{-2} \end{bmatrix}^{\mathrm{T}}$, then the solution of the perturbed system becomes $x = \begin{bmatrix} 0.5 & 0.5 & 500.5 \end{bmatrix}^{\mathrm{T}}$.

Because the problem is almost ill posed, a small perturbation in the given data results in a significant deviation in the solution. In order to avoid such instability, we may solve

$$\begin{bmatrix} 2 & 0 & 0 \\ 0 & 2 & 0 \\ 0 & 0 & 2\times 10^{-3} \end{bmatrix} \begin{bmatrix} x_1 \\ x_2 \\ x_3 \end{bmatrix} = \begin{bmatrix} 1 \\ 1 \\ 10^{-5} \end{bmatrix}, \tag{1.33}$$

instead of Eq. (1.32). In the new problem, exact observation results in $x = \begin{bmatrix} 0.5 & 0.5 & 0.005 \end{bmatrix}^{\mathrm{T}}$, while perturbation in the third element of the given data by $\delta = \begin{bmatrix} 0 & 0 & 10^{-2} \end{bmatrix}^{\mathrm{T}}$ results in $x = \begin{bmatrix} 0.5 & 0.5 & 5.005 \end{bmatrix}^{\mathrm{T}}$. Although the solution is not exactly equal to the original one, the new system exhibits robustness against small perturbations in the given data.

The choice of a better-posed counterpart for the given ill-posed problem is very important in both obtaining the closely approximated solution and keeping the problem robust against observation error or external perturbations. The procedure of replacing the given ill-posed problem with a better-posed counterpart is called regularization. In the regularization process, a priori knowledge or information about the solution plays an important role. An example of a priori knowledge is that the intensity of a pixel in an image does not have a negative value. Another example is that there is no large difference between adjacent elements of the solution.

1.10 Mathematical Formulation of Regularization

If an ill-posed problem is given in the form of unconstrained optimization, such as

$$\text{minimize } g(x), \tag{1.34}$$

then prior knowledge can be incorporated in the form of a constraint, such as

$$x \in U, \tag{1.35}$$

where U represents a set of solutions that satisfy the prior knowledge about the solution.

We can easily see that the problem defined by Eqs. (1.34) and (1.35) is a typical form of constrained optimization. In many cases, we can represent prior knowledge about the solution given in Eq. (1.35) in the form of minimization, expressed as

$$\text{minimize } h(x). \tag{1.36}$$

In order to minimize both objective functions simultaneously, we have the following unconstrained minimization problem.

$$\text{minimize } f(x) = g(x) + \lambda h(x), \tag{1.37}$$

where $\lambda \in R$ represents a weight which controls the amount of minimization between $g(x)$ and $h(x)$. The following example shows how to formulate a regularization problem given prior knowledge.

Example 1.2 Consider the 3×3 linear system shown below.

$$\begin{bmatrix} 2 & 0 & 0 \\ 0 & 2 & 0 \\ 0 & 0 & 2 \times 10^{-5} \end{bmatrix} \begin{bmatrix} x_1 \\ x_2 \\ x_3 \end{bmatrix} = \begin{bmatrix} 1 \\ 1 \\ 10^{-5} \end{bmatrix}, \tag{1.38}$$

whose solution is equal to that of the unconstrained minimization given by

$$\text{minimize } g(x) = ||y - Dx||^2, \text{ where } y = \begin{bmatrix} 1 \\ 1 \\ 10^{-5} \end{bmatrix} \text{ and } D = \begin{bmatrix} 2 & 0 & 0 \\ 0 & 2 & 0 \\ 0 & 0 & 2 \times 10^{-5} \end{bmatrix} \tag{1.39}$$

By performing simple calculations, we find that the solution is $x = [0.5 \quad 0.5 \quad 0.5]^{\mathrm{T}}$. We note that the third diagonal element of matrix D is very small compared to its other diagonal elements and that the matrix is subject to become ill conditioned. In other words, a small perturbation in observation y, say $\delta = \begin{bmatrix} 0 & 0 & 10^{-2} \end{bmatrix}^{\mathrm{T}}$, may result in a large deviation in the solution, such as $x = [0.5 \quad 0.5 \quad 0.5 + 500]^{\mathrm{T}}$.

Suppose we are given prior knowledge of the solution indicating that each element should have the range between 0 and 1. The prior knowledge can be represented in the form of a set, such as

$$x \in \{x_i | 0 \leq x_i \leq 1, \quad i = 1, 2, 3\}. \tag{1.40}$$

Alternatively, this information can be represented in the form of minimization as

$$\text{minimize}\, h(x) = \left|\left|a\right|\right|^2, \text{ where } a_i = \begin{cases} x_i, & x_i < 0 \\ 0, & 0 \leq x_i \leq 1 \\ x_i - 1, & x_i > 1 \end{cases}. \tag{1.41}$$

Once we have a minimization-type representation for prior knowledge, we can formulate the corresponding regularization problem as

$$\text{minimize} \left|\left|y - Dx\right|\right|^2 + \lambda \left|\left|a\right|\right|^2. \tag{1.42}$$

If we set $\lambda = 0$ in Eq. (1.42), the regularization problem does not take prior knowledge into account in estimating the solution. On the other hand, if we use very large λ, the corresponding solution satisfies only prior knowledge without consideration of the given linear equation.

1.11 Regularization of Linear Inverse Problems

In the previous section, we have described that a general form of the regularization problem is to minimize

$$f(x) = g(x) + \lambda h(x), \tag{1.43}$$

where minimization of $g(x)$ involves the faithfulness of the solution to the given relationship between the original data and the observation, and minimization of $h(x)$ incorporates prior knowledge about the solution. The scalar parameter λ determines the weight of minimization between $g(x)$ and $h(x)$.

Consider the minimization of the following function:

$$\left|\left|y - Dx\right|\right|^2 = (y - Dx)^{\mathrm{T}}(y - Dx). \tag{1.44}$$

By omitting the constant term, and with an appropriate scaling, we can easily see that minimization of Eq. (1.44) is equivalent to minimization of the following quadratic function:

$$g(x) = \frac{1}{2}x^{\mathrm{T}}Dx - y^{\mathrm{T}}x. \tag{1.45}$$

Also, minimization of Eq. (1.45) is equivalent to solving the linear equation $y = Dx$, which can be easily proven by setting $\nabla_x g(x) = 0$. It seems unlikely that a minimization algorithm would be required to find the minimum of a quadratic function since it is equivalent to solving a set of linear simultaneous algebraic equations. There are, however, two important reasons for considering the minimization of a quadratic function: (1) it is the simplest type of function that can be minimized, and (2) a solution of quadratic minimization is a good indication of those requiring more complex minimization.

A solution of a linear equation with a priori knowledge can be obtained by using the regularization method, for instance, with

$$\text{minimize } f(x) = ||y - Dx||^2 + \lambda||Cx||^2, \tag{1.46}$$

where matrix C represents the counter-effect of prior knowledge about the solution. The following example explains how a typical prior knowledge of the solution can be implemented by minimization of a linear transformation.

Example 1.3 We are given a priori knowledge that the solution should be as smooth as possible. In other words, there is no abrupt change between adjacent elements in the solution. Consider a one-dimensional signal of length 8, which is represented by a vector $x = [1\ 2\ 3\ 3\ 3\ 3\ 2\ 1]^{\mathrm{T}}$ and a matrix

$$C = \begin{bmatrix} 2 & -1 & 0 & 0 & 0 & 0 & 0 & -1 \\ -1 & 2 & -1 & 0 & 0 & 0 & 0 & 0 \\ 0 & -1 & 2 & -1 & 0 & 0 & 0 & 0 \\ 0 & 0 & -1 & 2 & -1 & 0 & 0 & 0 \\ 0 & 0 & 0 & -1 & 2 & -1 & 0 & 0 \\ 0 & 0 & 0 & 0 & -1 & 2 & -1 & 0 \\ 0 & 0 & 0 & 0 & 0 & -1 & 2 & -1 \\ -1 & 0 & 0 & 0 & 0 & 0 & -1 & 2 \end{bmatrix}. \tag{1.47}$$

By multiplying C with x, we have that $Cx = [-1\ 0\ 1\ 0\ 0\ 1\ 0\ -1]^{\mathrm{T}}$. By neglecting boundary elements, it is evident that $||Cx||$ represents the amount of variance in x. Possible solutions which minimize $||Cx||^2$ include $x_1 = [1\ 1\ 1\ 1\ 1\ 1\ 1\ 1]^{\mathrm{T}}$ and $x_2 = [1\ 2\ 3\ 3\ 4\ 5\ 6\ 7]^{\mathrm{T}}$. Both x_1 and x_2 represent ultra-smooth signals that are flat or are increasing/decreasing with the same rate.

In the above example, we found that prior knowledge can be incorporated into the solution by minimizing the component with counter-characteristics.

In summary, for this chapter we have given a brief introduction to the concept of ill-posed and inverse problems in imaging and computer vision as well as to the method of regularization and how it can be used in more practical problems rather than a full-blown version of constrained optimization.

References

[courant62] R. Courant, D. Hilbert, *Methods of Mathematical Physics, II, Partial Differential Equations* (Interscience, New York, 1962)

[golub96] G.H. Golub, C.F. Van Loan, *Matrix Computations* (Johns Hopkins University Press, Baltimore, 1996)

[jain89] A.K. Jain, *Fundamentals of Digital Image Processing* (Prentice-Hall, Upper Saddle River, 1989)

[kats91] A.K. Katsaggelos (ed.), *Digital Image Restoration* (Springer, Heidelberg, 1991)

[nashed81] M.Z. Nashed, IEEE Trans. Antennas Propag. **AP-29**, 220–231 (1981)

[tikhonov77] A.N. Tikhonov, V.Y. Arsenin, *Solutions of Ill-Posed Problems* (Winston and Sons, Washington, DC, 1977)

[bardinet95] E. Bardinet, L.D. Cohen, N. Ayache, A parametric deformable model to fit unstructured 3D data. Project Report, Programme 4-Robotique, image et vision, Utite de rescherche INRIA Sophia-Antipolis (1995)

[sederberg86] T.W. Sederberg, S.R. Parry, Free-form deformation of solid geometric models, in *Proc. 1986 SIGGRAPH*, vol. 20, pp. 151–160, 1986

[solina90] F. Solina, R. Bajcsy, Recovery of compact volumetric models for shape representation of single-part objects. *IEEE Trans. Pattern Anal. Mach. Intell.* **PAMI-12**(2), 131–147 (1990)

[tekalp95] A.M. Tekalp, *Digital Video Processing* (Prentice-Hall, Upper Saddle River, 1995)

[ber88] M. Bertero, T.A. Poggio, V.Torre, Ill-posed problems in early vision. *Proc. IEEE* **76**(8), 869–889 (1988)

Additional References and Further Readings

[eng00] W.H. Engl, M. Hanke, A. Neubauer, *Regularization of Inverse Problems* (Kluwer Academic Publishers, Dordrecht, 2000)

[gro93] C.W. Groetsch, *Inverse Problems in the Mathematical Sciences* (Vieweg, Bruanschweig, Wiesbaden, 1993)

[han98] Per Christian Hansen, *Rank Deficient and Discrete Ill-Posed Problem* (SIAM, Philadelphia, 1998)

[ber98] M. Bertero, P. Boccacci, *Introduction to Inverse Problems in Imaging* (IOP Publishing, Bristol, 1998)

[vog02] C. Vogel, *Computational Methods for Inverse Problems* (SIAM, Philadelphia, 2002)

[kir96] A. Kirsch, *An Introduction to the Mathematical Theory of Inverse Problems* (Springer, New York, 1996)

[sab87] P.C. Sabatier (ed.), *Inverse Problems: An Interdisciplinary Study* (Academic, London, 1987)

[two96] S. Twomey, *Introduction to the Mathematics of Inversion in Remote Sensing and Indirect Measurements* (Dover Publications, New York, 1996)

[tikh98] A.N. Tikhonov, A.S. Leonov, A.G. Yagola, *Nonlinear Ill-Posed Problems*, vols. 1, 2 (Chapman and Hall, London, 1998)

[mor84] V.A. Morozov, *Methods for Solving Incorrectly Posed Problems* (Springer, New York, 1984)

[lav61] M.M. Lavrent'ev, *Ill-Posed Problems in Mathematical Physics* (Nauka, Novosibirsk (in - Russian), 1961). English translation in Springer, Berlin-Heidelberg

[win91] G. Milton Wing, *A Primer on Integeral Equations of the First Kind* (SIAM, Philadelphia, 1991)

[bau87] J. Baumeister, *Stable Solution of Inverse Problems* (Vieweg, Braunschweig, 1987)

[hh93] Martin Hanke, Per Christian Hansen, Regularization methods for large scale problems. *Surv. Math. Ind.* **3**, 253–315 (1993)

[eng93] W.H. Engl, Regularization methods for the stable solution of inverse problems. Surv. Math. Ind. **3**, 71–143 (1993)

[col00] D. Colton, H.W. Engl, A.K. Louis, J. McLaughlin, W. Rundell, *Surveys on Solution Methods for Inverse Problems* (Springer, Wien/New York Herausgeber, 2000)

[tar87] A. Tarantola, *Inverse Problem Theory: Methods for Data Fitting and Model Parameter Estimation* (Elsevier, New York, 1987)

[hen91] E. Hensel, *Inverse Theory and Applications for Engineers* (Prentice-Hall, Englewood Cliffs, 1991)

[tru97] D.M. Trujillo, H.R. Busby, *Practical Inverse Analysis in Engineering* (CRC Press, Boca Raton, 1997)

[tikh95] A.N. Tikhonov, A.V. Goncharsky, V.V. Stepanov, A.G. Yagola, *Numerical Methods for the Solution of Ill-Posed-Problems* (Kluwer Academic Press, Dordrecht, 1995)

[ber85] M. Bertero, C. De Mol, E.R. Pikes, Linear inverse problems with discrete data. I: General formulation and singular system analysis. *Inverse Prob.* **1**, 301–330 (1985)

[ber88a] M. Bertero, C. De Mol, E.R. Pikes, Linear inverse problems with discrete data. II: Stability and Regularization. *Inverse Prob.* **4**, 573–594 (1988)

[par77] R.L. Parker, Understanding inverse theory. Annu. Rev. Earth Planet. Sci. **5**, 35–64 (1977)

[iva62a] V.K. Ivanov, Integral equations of the first kind and the approximate solution of an inverse potential problem. *Dokl. Acad. Nauk. SSSR* **142**, 997–1000 (in Russian). English translation in *Soviet. Math. Dokl.* (1962)

[iva62b] V.K. Ivanov, On linear ill-posed problems. *Dokl. Acad. Nauk. SSSR* **145**, 270–272 (in Russian). English translation in *Soviet. Math. Dokl.* (1962)

[ph62] D.L. Phillips, A technique for the numerical solution of certain integral equations of the first kind. JACM **9**, 84–97 (1962)

[two63] S. Twomey, On the numerical solution of Fredholm integral equations of the first kind by the inversion of the linear system produced by quadrature. JACM **10**, 97–101 (1963)

[tikh63] A.N. Tikhonov, Solution of incorrectly formulated problems and the regularization method. Soviet Math. Dokl. **4**, 1035–1038 (1963)

[mil70] K. Miller, Least squares methods for ill-posed problems with prescribed bound. SIAM J. Math. Anal. **1**, 52–74 (1970)

[ber88b] M. Bertero, P. Brianzi, E.R. Pike, L. Rebolia, Linear regularizing algorithms for positive solutions of linear inverse problems. Proc. R. Soc. Lond. A **415**, 257–275 (1988)

[ber81a] M. Bertero, V. Dovi, Regularized and positive-constrained inverse methods in the problem of object restoration. *Opt. Acta* **28**, 1635–1649 (1981)

[ber81b] M. Bertero, C. De Mol, Ill-posedness, regularization and number of degrees of freedom. Atti della Fondazione Giorgio Ronchi **36**, 619–632 (1981)

[ber92] M. Bertero, E.R. Pike (eds.), *Inverse Problems in Scattering and Imaging* (Adam Hilger, Bristol, 1992)

[ber89] M. Bertero, Linear inverse and ill-posed problems, in *Advances in Electronics and Electron Physics*, ed. by P.W. Hawkes, vol 75 (Academic Press, New York, 1989), pp. 1–120
[ber86] M. Bertero, Regularization methods for linear inverse problems, in *Inverse Problems*, ed. by G. Talenti. Lecture Notes in Mathematics, vol 1225 (Springer, Berlin, 1986), pp. 52–112
[ber80] M. Bertero, C. De Mol, G.A. Viano, The stability of inverse problems, in *Inverse Scattering Problems in Optics*, ed. by H.P. Baltes. Topics in Current Physics, vol 20 (Springer, Berlin, 1980), pp. 161–214
[rog96] M.C. Roggeman, B.M. Welsh, *Imaging Through Turbulence* (CRC Press, Boca Raton, 1996)
[good96] J.W. Goodman, *Introduction to Fourier Optics*, 2nd edn. (McGraw-Hill, New York, 1996)
[har98] J.W. Hardy, *Adaptive Optics for Astronomical Telescopes* (Oxford University Press, Oxford, UK, 1998)
[louis92] A.K. Louis, Medical imaging: state of the art and future development. Inverse Prob. **8**, 709–738 (1992)
[arr97] S.R. Arridge, J.C. Hebden, Optical imaging in medicine: II. Modelling and reconstruction. Phys. Med. Biol. **42**, 841–853 (1997)
[arr99] S. Arridge, Optical tomography in medical imaging. Inverse Prob. **15**(2), R41–R93 (1999)
[bar81] H.H. Barrett, W. Swindell, *Radiological Imaging: The Theory of Image Formation, Detection, and Processing* (Academic, New York, 1981)
[her79] G.T. Herman, *Image Reconstruction from Projections* (Springer, Berlin, 1979)
[her80] G.T. Herman, *The Fundamentals of Computerized Tomography. Image Reconstruction from Projections* (Academic, New York, 1980)
[kak87] A.C. Kak, M. Slaney, *Principles of Computerized Tomographic Imaging* (IEEE Press, New York, 1987)
[mac83] A. Macovski, *Medical Imaging Systems*. Prentice-Hall Information and System Sciences Series (Prentice-Hall, Upper Saddle River, 1983)
[nat96] National Research Council Institute of Medicine, National Academy Press, *Mathematics and Physics of Emerging Biomedical Imaging* (National Research Council Institute of Medicine, National Academy Press, Washington, DC, 1996)
[nat86] F. Natterer, *The Mathematics of Computerized Tomography* (B.G. Teubner and Wiley, Chichester, 1986)
[cra86] I.J.D. Craig, J.C. Brown, *Inverse Problems in Astronomy: A Guide to Inversion Strategies for Remotely Sensed Data* (Adam Hilger, Boston, 1986)
[cast96] K.R. Castleman, *Digital Image Processing* (Prentice Hall, Toronto, 1996)
[myr92] J. Myrheim, H. Rue, New algorithms for maximum entropy image restoration. CVGIP: Graph. Models Image Process. **54**(3), 223–238 (1992)
[bur83] S.F. Burch, S.F. Gull, J. Skilling, Image restoration by a powerful maximum entropy method. Comput. Vis. Graph. Image Process. **23**(2), 113–128 (1983)
[gon92] R.C. Gonzalez, R.E. Woods, *Digital Image Processing* (Addison-Wesley, New York, 1992)
[arc95] G. Archer, D.M. Titterington, On some bayesian/regularization methods for image restoration. *IEEE Trans. Image Process.* **4**(7), 989–995 (1995). AV94
[acar94] R. Acar, C.R. Vogel, Analysis of total variation penalty methods. Inverse Prob. **10**, 1217–1229 (1994)
[buck91] B. Buck, V.A. Macaulay (eds.), *Maximum Entropy in Action* (Clarendon Press, Oxford, 1991)
[craig86] I.J.D. Craig, J.C. Brown, *Inverse Problems in Astronomy* (Adam Hilger, Bristol, 1986). Chapter 6.2, CB90
[chou90] P.B. Chou, C.M. Brown, The theory and practice of Bayesian image modeling. Int. J. Comput. Vis. **4**, 185–210 (1990)

[dem89] G. Demoment, Image reconstruction and restoration: overview of common estimation structures and problems. IEEE Trans. Acoust. Speech Signal Process. **37**(12), 2024–2036 (1989)

[hum87] R.A. Hummel, B. Kimia, S. Zucker, Deblurring gaussian blur. Comput. Vis. Graph. Image Process. **38**, 66–80 (1987)

[mar87] J.L. Marroquin, S. Mitter, T. Poggio, Probabilistic solution of ill-posed problems in computer vision. J. Am. Stat. Assoc. **82**, 76–89 (1987)

[pog85a] T. Poggio, V. Torre, C. Koch, Computational vision and regularization theory. Nature **317**, 314–319 (1985)

[pog85b] T. Poggio, C. Koch, Ill-posed problems in early vision: from computational theory to analogue networks. Proc. R. Soc. Lond. B. Biol. Sci. **226**, 303–323 (1985)

[leary94] D.P. O'Leary, Regularization of ill-posed problems in image restoration, in *Proceedings of the Fifth SIAM Conference on Applied Linear Algebra*, ed. by J.G. Lewis. (SIAM Press, Philadelphia, 1994), pp. 102–105

[terzopoulos86] D. Terzopoulos, Regularization of inverse visual problems involving discontinuities. IEEE Trans Pattern Anal. Mach. Intell. **8**(4), 413–424 (1986)

[blo97] Peter Blomgren, Tony F. Chan, Pep Mulet, C.K. Wong, Total variation image restoration: numerical methods and extensions, in *IEEE Image Proc. Proceedings of the 1997 I.E. International Conference on Image Processing*, October 1997

[dob96] D. Dobson, O. Scherzer, Analysis of regularized total variation penalty methods for denoising. Inverse Prob. **12**, 601–617 (1996)

[rud92] L. Rudin, S. Osher, E. Fatemi, Nonlinear total variation based noise removal algorithms. Phys. D **60**, 259–268 (1992)

[ban97] M.R. Banham, A.K. Katsaggelos, Digital image restoration. *IEEE Signal Process. Mag.* **14**, 24–41 (1997)

[kun96] D. Kundur, D. Hatzinakos, Blind image deconvolution. IEEE Signal Process. Mag. **13**(3), 43–64 (1996)

[kat91a] A.K. Katsaggelos, *Digital Image Restoration* (Springer, New York, 1991)

[han00] P.C. Hansen, *Numerical Aspects of Deconvolutions, Lectures Notes* (Technical University of Denmark, Department of Mathematical Modelling, Denmark, 2000)

[bie84a] J. Biemond, A.K. Katsaggelos, A new iterative restoration scheme for noisy blurred images, in *Proc. Conf. on Math. Methods in Signal Processing*, pp. 74–76, Aachen, W. Germany, September 1984

[bie84b] J. Biemond, A.K. Katsaggelos, Iterative restoration of noisy blurred images, in *Proc. 5th Inf. Theory Symposium in the Benelux*, pp. 11–20, Aalten, The Netherlands, May 1984

[kat86] A.K. Katsaggelos, A general formulation of adaptive iterative image restoration algorithms, in *Proc. 1986 Conf. Inf. Sciences and Systems*, pp. 42–47, Princeton, NJ, March 1986

[kat84] A.K. Katsaggelos, J. Biemond, R.M. Mersereau, R.W. Schafer, An iterative method for restoring noisy blurred images, in *Proc. 1984 Int. Conf. Acoust., Speech, Signal Processing*, pp. 37.2.1–37.2.4, San Diego, CA, March 1984

[and77] H.C. Andrews, B.R. Hunt, *Digital Image Restoration* (Prentice-Hall, Englewood Cliffs, 1977)

[gal92] N.P. Galatsanos, A.K. Katsaggelos, Methods for choosing the regularization parameter and estimating the noise variance in image restoration and their relation. IEEE Trans. Image Process. **1**, 322–336 (1992)

[and76] G.L. Anderlson, A.N. Natravali, Image restoration based on subjective criterion. *IEEE Trans. Syst., Man, Cyber.* **SMC-6**, 845–853 (1976)

[lim90] J.S. Lim, *Two-Dimensional Signal and Image Processing* (Prentice-Hall, Englewood Cliffs, 1990)

[kar90] N.B. Karayiannis, A.N. Venetsanopoulos, Regularization theory in image restoration—the stabilizing functional approach. *IEEE Trans. Acoust. Speech Signal Process.* **38**(7), 1155 (1990)

[kang92] M.G. Kang, A.K. Katsaggelos, Simultaneous iterative image restoration and evaluation of the regularization parameter. IEEE Trans. Signal Process. **40**(9), 2329–2334 (1992)

[kat91b] A.K. Katsaggelos, J. Biemond, R.W. Schafer, R.M. Mersereau, A regularized iterative image restoration algorithm. IEEE Trans. Signal Process. **39**(4), 914–929 (1991)

[reeves90] S.J. Reeves, R.M. Mersereau, Optimal estimation of the regularization parameter and stabilizing functional for regularized image restoration. Opt. Eng. **29**, 446–454 (1990)

[reeves92] S.J. Reeves, R.M. Mersereau, Blur identification by the method of generalized cross-validation. IEEE Trans. Image Process. **1**, 301–311 (1992)

[ber79] M. Bertero, C. De Mol, G.A. Viano, On the problems of object restoration and image extrapolation in optics. J. Math. Phys. **20**, 509–521 (1979)

[ber78a] M. Bertero, G.A. Viano, On probabilistic methods for the solution of improperly posed problems. *Bollettino U.M.I.,* **15-B**, 453–508 (1978)

[ber78b] M. Bertero, C. De Mol, G.A. Viano, Restoration of optical objects using regularization. Opt. Lett. **3**, 51–53 (1978)

[ber78c] M. Bertero, C. De Mol, G.A. Viano, On the regularization of linear inverse problems in Fourier optics, in *Applied Inverse Problems*, ed. by P.C. Sabatier. Lecture Notes in Physics, vol 85 (Sprinter, Berlin, 1978), pp. 180–199

Chapter 2
Selection of the Regularization Parameter

2.1 General Considerations

The success of all currently available regularization techniques relies heavily on the proper choice of the regularization parameter. Although many regularization parameter selection methods (RPSMs) have been proposed, very few of them are used in engineering practice. This is due to the fact that theoretically justified methods often require unrealistic assumptions, while empirical methods do not guarantee a good regularization parameter for any set of data. Among the methods that found their way into engineering applications, the most common are Morozov's Discrepancy Principle (abbreviated as MDP) [morozov84, phillips62], Mallows' CL [mallows73], generalized cross validation (abbreviated as GCV) [wahba90], and the L-curve method [hansen98]. A high sensitivity of CL and MDP to an underestimation of the noise level has limited their application to cases in which the noise level can be estimated with high fidelity [hansen98]. On the other hand, noise-estimate-free GCV occasionally fails, presumably due to the presence of correlated noise [wahba90]. The L-curve method is widely used; however, this method is nonconvergent [leonov97, vogel96]. An example of image restoration using different values of regularization parameters is shown in Figs. 2.1, 2.2, 2.3, 2.4, and 2.5. The Matlab code for this example was provided by Dr. Curt Vogel of Montana State University in a personal communication. The original image is presented in Fig. 2.1, and the observed blurred image is in Fig. 2.2.

Figures 2.3 and 2.4 represent reconstructed images with too small and too large regularization parameters, respectively. These two images illustrate the importance of the regularization parameter selection for proper image restoration. For comparison, Fig. 2.5 represents the image reconstructed using a "good" value of regularization parameter.

M.A. Abidi et al., *Optimization Techniques in Computer Vision*, Advances in Computer Vision and Pattern Recognition, DOI 10.1007/978-3-319-46364-3_2

Fig. 2.1 Original image

Fig. 2.2 Observed blurred image

Fig. 2.3 Reconstructed image with a very small regularization parameter ($\lambda = 10$–20)

2.2 Discrepancy Principle

The discrepancy principle is the most widely used method which requires a priori knowledge of some of the noise properties such as the power of the noise. The regularization parameter value is chosen as a solution of the equation

Fig. 2.4 Reconstructed image with a very large regularization parameter ($\lambda = 10$)

Fig. 2.5 Reconstructed image with a good value of regularization parameter ($\lambda = 0.0007$)

$$\|Y - Dx_{\lambda}\|_2 = \varepsilon, \quad \text{where} \quad \|\eta\|_2 \leq \varepsilon. \tag{2.1}$$

The ε is the upper bound on the variance of the noise.

The regularization parameter λ is chosen such that the corresponding residual (left-hand side) of Eq. (2.1) is less than or equal to the a priori specified bound (right-hand side) for the noise level in the response. Since a smaller λ corresponds to less stable solutions, the λ for which the residual equals the specified noise level is chosen. There is no reason to expect a residual less than the noise level. In modeling from data, a residual less than the noise level in the response corresponds to overfitting, which is a term for learning noise in the training data. The regularization method with λ chosen according to the discrepancy principle in Eq. (2.1) is convergent and of optimal order [engl00, morozov84]. Application of the discrepancy principle requires solving the following nonlinear equation with respect to λ as shown in [golub97].

$$\left\| Y - D\left(D^{\mathrm{T}}D + \lambda I\right)^{-1} D^{\mathrm{T}}Y \right\|_2 = \varepsilon. \tag{2.2}$$

For $\lambda > 0$, the identity

$$I - D(D^{T}D + \lambda I)^{-1}D^{T} = \lambda(DD^{T} + \lambda I)^{-1} \tag{2.3}$$

holds. Hence, Eq. (2.2) can be rewritten as

$$\left\| Y\lambda(DD^{T} + \lambda I)^{-1} \right\|_{2} = \varepsilon \tag{2.4}$$

or

$$\left[Y\lambda(DD^{T} + \lambda I)^{-1}\right]^{T} \cdot \left[Y\lambda(DD^{T} + \lambda I)^{-1}\right] = \varepsilon, \tag{2.5}$$

and after elementary matrix algebra, we arrive at the following nonlinear equation for λ.

$$\lambda^{2}Y^{T}(DD^{T} + \lambda I)^{-2}Y = \varepsilon. \tag{2.6}$$

Since the derivative of the left-hand side is equal to

$$2\lambda Y^{T}D(D^{T}D + \lambda I)^{-3}D^{T}Y, \tag{2.7}$$

the function $\lambda^{2}Y^{T}(DD^{T} + \lambda I)^{-2}Y$ is strictly increasing for $\lambda > 0$, and Eq. (2.6) has a unique positive solution.

A very important property of the discrepancy principle is its convergence or regularity, which means that as error $\|\eta\|_2$ in the data goes to zero, the λ selected by MDP goes to zero; hence, the approximated regularized solution x_λ converges to the exact solution or true image x_{exact}. Normally, the literature on inverse problems analyzes the rates of convergence of x_λ to x_{exact}. The faster the method converges, the better its behavior.

Statistical literature on the selection of a regularization parameter is more concerned about asymptotic behavior of different methods as the number of samples N goes to infinity.

To apply MDP, we must have a priori knowledge about the noise level in the response. Since the noise level is usually unknown, we use an estimate of the noise level. One of the methods for noise estimation is described in [wahba90] and consists in monitoring the function

$$\hat{\sigma}_{\eta}^{2} = \frac{\|Y - Dx_\lambda\|_2^2}{\operatorname{trace}\left(I - D(D^{T}D + \lambda I)^{-1}D\right)}. \tag{2.8}$$

The plateau of this function can serve as a good estimate of the noise variance. However, in practice the variance of the residuals of the least square solution is often used as a quick estimate of the noise level.

Unfortunately, MDP is very sensitive to an underestimation of the noise level. This limits its application to cases in which the noise level can be estimated with high fidelity [hansen98]. The MDP belongs to a posteriori methods for the selection of a regularization parameter. A posteriori RPSM requires the noise level to be either known or reliably estimated. Such an estimate of the noise level can be hard to obtain.

2.3 L-Curve

An alternative approach to regularization parameter selection uses noise-level-free RPSMs. Noise-level-free RPSMs are also referred to as heuristic RPSMs. We are now going to consider some of these methods. The method which attracted the attention of researchers recently is the L-curve method [hansen93]. The method is based on the plot of the logarithm of the solution norm x_λ versus the logarithm of the norm of the residuals. In many cases, such a curve has a characteristic L shape. It is then argued that the optimal regularization parameter has to be selected at the point of maximum curvature of the curve or its "elbow." Mathematically, the L-curve criterion seeks to maximize the curvature

$$C_{\mathrm{L}}(\lambda) = \frac{\rho'\eta'' - \rho''\eta'}{\left((\rho')^2 + (\eta')^2\right)^{3/2}} = \max, \text{ where} \tag{2.9}$$

$$\rho(\lambda) = \log\left\|Y - D\left(D^{\mathrm{T}}D + \lambda I\right)^{-1}DY\right\|_2 = \log\left\|\lambda\left(DD^{\mathrm{T}} + \lambda I\right)^{-1}Y\right\|_2 \text{ and} \tag{2.10}$$

$$\eta(\lambda) = \log\left\|\left(D^{\mathrm{T}}D + \lambda I\right)^{-1}D^{\mathrm{T}}Y\right\|_2. \tag{2.11}$$

The differentiation is with respect to λ. A typical L-curve is presented in Fig. 2.6. To obtain the curve, we used Hansen's regularization toolbox [hansen94].

The L-curve method recently suffered major theoretical as well as practical setbacks. It was shown in [leonov97, vogel96] that the method is generally not convergent, and in practice the L-curve may not have an "elbow" or have several ones.

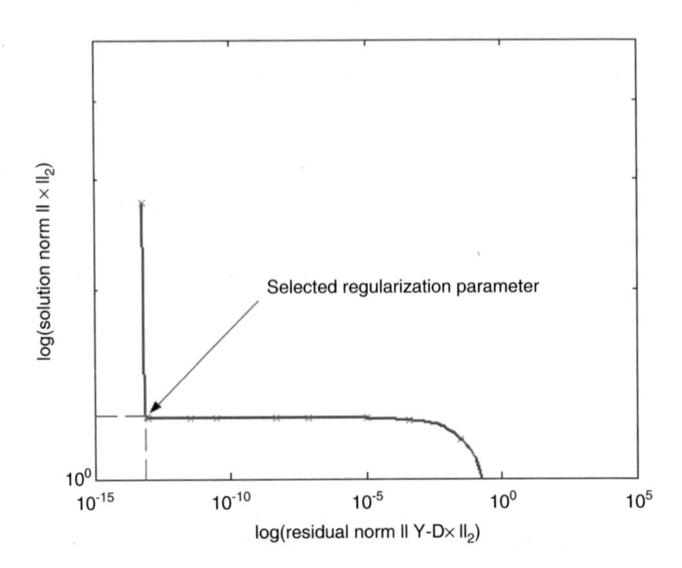

Fig. 2.6 The generic form of the L-curve

2.4 Mallow's C_L

The other heuristic method which was proposed in statistical literature is the Mallow's C_L or unbiased prediction risk estimation method. In the following derivations, we closely follow [vogel02]. Let's call the difference between the regularized image and true image an estimation error

$$\varepsilon_\lambda = x_\lambda - x_{\text{true}}. \tag{2.12}$$

Obviously, this quantity is unknown and not computable due to the unavailability of the true image x_{true}. The observed image represents convolution of the true image with point spread function plus some additive noise as in

$$Y = Dx_{\text{true}} + \eta. \tag{2.13}$$

Let's define the predictive error as the difference between two quantities

$$p_\lambda = Dx_\lambda - Dx_{\text{true}}. \tag{2.14}$$

We can express the regularized image x_λ as

$$x_\lambda = \left(D^{\mathrm{T}}D + \lambda I\right)^{-1} D^{\mathrm{T}} Y \tag{2.15}$$

or

$$x_{\lambda} = \left(D^{\mathrm{T}}D + \lambda I\right)^{-1} D^{\mathrm{T}}(Dx_{\mathrm{true}} + \eta), \tag{2.16}$$

and, hence, the predictive error can be expressed as

$$p_{\lambda} = (H - I)Dx_{\mathrm{true}} + H\eta, \tag{2.17}$$

where H is the hat or influence matrix

$$H = D\left(D^{\mathrm{T}}D + \lambda I\right)^{-1} D^{\mathrm{T}}. \tag{2.18}$$

As shown in [wahba90], the mean value of the mean squared predictive error can be written as

$$E\left(\frac{1}{n}\|p_{\lambda}\|^2\right) = \frac{1}{n}\|(H - I)Df_{\mathrm{true}}\|^2 + \frac{\sigma^2}{n}\mathrm{trace}(H). \tag{2.19}$$

Notice that this value is not computable either; however, we can introduce the training error as

$$r_{\lambda} = Df_{\lambda} - Y. \tag{2.20}$$

As shown in [vogel02],

$$E\left(\frac{1}{n}\|r_{\lambda}\|^2\right) = E\left(\frac{1}{n}\|p_{\lambda}\|^2\right) - 2\frac{\sigma^2}{n}\mathrm{trace}(H) + \sigma^2; \tag{2.21}$$

hence,

$$E\left(\frac{1}{n}\|p_{\lambda}\|^2\right) = E\left(\frac{1}{n}\|r_{\lambda}\|^2\right) + 2\frac{\sigma^2}{n}\mathrm{trace}(H) - \sigma^2. \tag{2.22}$$

The C_{L} function is given as

$$C_{\mathrm{L}} = \frac{1}{n}\|r_{\lambda}\|^2 + 2\frac{\sigma^2}{n}\mathrm{trace}(H) - \sigma^2; \tag{2.23}$$

hence, the C_{L} function is an unbiased estimator for the mean squared predictive error

$$E(C_{\mathrm{L}}) = E\left(\frac{1}{n}\|p_{\lambda}\|^2\right). \tag{2.24}$$

Fig. 2.7 Reconstructed image with regularization parameter selected with CL ($\lambda = 0.0007$)

In the case of a correctly specified point spread function and Gaussian noise, C_{L} is theoretically optimal. It should be noted that C_{L} requires the estimation of the noise variance

$$\sigma^2 = \mathrm{var}(\eta), \tag{2.25}$$

and, what is even more important, the C_{L} performance depends heavily on the accuracy of this estimate. The image reconstructed using C_{L} as parameter selection method is shown in Fig. 2.7.

2.5 Generalized Cross Validation

To overcome the inconvenience of noise estimation, Wahba [wahba90] suggested a noise-free method for the selection of a regularization parameter which is currently widely used—generalized cross validation (GCV). GCV is a rotation-invariant version of ordinary leave-one-out cross validation. The ordinary cross validation is known not to be invariant under data transformations, and GCV fixes this problem. The GCV seeks for a minimum of the following function:

$$\mathrm{GCV}(\lambda) = \frac{\frac{1}{n}\|Y - Dx_\lambda\|_2^2}{\left[\frac{1}{n}\|\mathrm{trace}(I - H)\|\right]^2}. \tag{2.26}$$

We can see that the GCV function is the ratio of two functions—the mean sum of squares $\frac{1}{n}\|Y - Dx_\lambda\|_2^2$ and a penalty function $\left[\frac{1}{n}\|\mathrm{trace}(I - H)\|\right]^2$ which is often called the effective degrees of freedom and is used to quantify the amount of information in ill-posed problems. The GCV function has a number of nice properties such as convergence to the optimal regularization parameter as n goes to infinity, and convergence rates for this method are also available. GCV is derived under the assumption of white Gaussian noise, and, if this condition fails to hold,

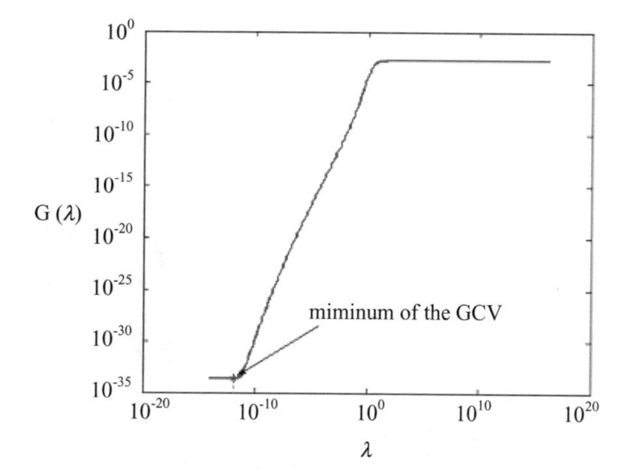

Fig. 2.8 A typical GCV function

Fig. 2.9 Reconstructed image using GCV for the selection of regularization parameter ($\lambda = 0.0007$)

the GCV may produce grossly under regularized solutions. The GCV function itself can also have very flat minima, thus leading to numerical difficulties in determining a unique value of regularization parameter. A typical GCV function is plotted in Fig. 2.8. An image reconstructed using GCV for the selection of the regularization parameter is shown in Fig. 2.9.

2.6 Information Approach to the Selection of a Regularization Parameter

There are two potential problems with all the methods that we considered so far. First of all, if the true relationships between the observed image Y and original image X are not linear or if the response function is not known exactly, then we have what is called functional misspecification. The second problem is the assumption of white Gaussian noise in the data which is rarely a valid assumption in image

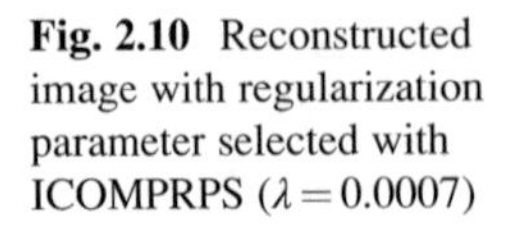

Fig. 2.10 Reconstructed image with regularization parameter selected with ICOMPRPS ($\lambda = 0.0007$)

processing applications. This second type of misspecification is called distributional misspecification. Both these misspecifications affect the estimation of the covariance matrix of the restored image which is implicitly used by many methods to select the regularization parameter. We now consider the information approach to regularization parameter selection, which is robust to the model misspecification and also robust to the underestimation of the regularization parameter. However, first, we have to consider some theoretical preliminaries such as Kullback-Leibler (abbreviated KL) distance (Fig. 2.10).

When the parameters of a specified model $f(X_i, Y_i;\ b)$ are estimated by the maximum penalized likelihood (MPL) method (see Appendix D), each particular choice of the penalty operator and regularization parameter yields some approximating density $\hat{f}_\lambda \equiv f\left(X_i, Y_i;\ \hat{b}_\lambda\right)$. The closeness of this approximating density $\hat{f}_\lambda$ to the unknown true density $g(X_i, Y_i)$, assuming such exists, can be evaluated by the Kullback-Leibler [kullback51] (abbreviated as KL) information (or distance) that measures the divergence between the densities

$$\mathrm{KL}\left(\hat{f}_\lambda;\ g\right) \equiv E_{W,Z}\left\{\log \frac{g}{\hat{f}_\lambda}\right\} = \int \cdots \int \log \frac{g(w,z)}{f\left(w,z;\ \hat{b}_\lambda\right)} \cdot g(w,z)\ dw_1, dw_2, \ldots, dw_m dz. \tag{2.27}$$

The regularization parameter can be selected to minimize the mean KL distance. The mean KL distance is the KL distance averaged over all possible data sets D which can be used to obtain the approximating density $\hat{f}_\lambda$.

$$\hat{\lambda}_{\mathrm{KL}} = \arg\min_{\lambda} \left\{E_D \mathrm{KL}\left(\hat{f}_\lambda;\ g\right)\right\}. \tag{2.28}$$

Such a choice guarantees that, on the average, the corresponding approximating density will be closest among those considered in the sense of the minimum KL distance. We can decompose the mean KL distance into a "systematic error" and a "random error":

$$
\begin{aligned}
E_D\mathrm{KL}(\hat{f}_\lambda;\ g) &= E_D\left\{E_{W,Z}\log\frac{g}{\hat{f}_\lambda}\right\} \\
&= E_D\left\{E_{W,Z}\log\frac{g}{f^*}\frac{f^*}{f_\lambda^*}\frac{f_\lambda^*}{\hat{f}_\lambda}\right\} \\
&= \underbrace{E_{W,Z}\log\frac{g}{f^*} + E_{W,Z}\log\frac{f^*}{f_\lambda^*}}_{\text{Systematic Error}} + \underbrace{E_D\left\{E_{W,Z}\log\frac{f_\lambda^*}{\hat{f}_\lambda}\right\}}_{\text{Random Error}},
\end{aligned}
\tag{2.29}
$$

where $f^* \equiv f(W,Z;\ b^*)$, and b^* is a solution of

$$
E_{W,Z}\left\{\frac{\partial}{\partial b}\mathrm{LL}(W,Z|b)\right\} = 0, \tag{2.30}
$$

or the limiting value of the maximum likelihood (ML) estimator $f_\lambda^* \equiv f(W,Z;\ b_\lambda^*)$, and b_λ^* is a solution of

$$
E_{W,Z}\left\{\frac{\partial}{\partial b}\mathrm{PLL}(W,Z|b)\right\} = 0, \tag{2.31}
$$

or the limiting value of the MPL estimator.

The systematic error, which can also be termed the bias, consists of two terms. The first term represents the error of modeling and vanishes when the model is correctly specified. The second term represents the error due to using a penalization and vanishes when the maximum likelihood method of estimation is used. The random error, also called the variance, arises due to the inaccuracy of the model's parameter estimation because of a limited number of observations. When the model is correctly specified and the ML method is used, only the variance term contributes to the mean KL distance. However, as we know, the variance in a case of ill-conditioned data sets can be very large and make the approximating density useless. Although penalization introduces a bias, it also drastically reduces the variance, allowing for a trade-off which may reduce the mean KL distance. This means that, on the average, with a properly chosen regularization parameter, the penalized model can be closer to the true model.

From the definition of the KL distance, it can be seen that, since $E_D\{E_{W,Z}\log g\}$ does not depend on the model $\hat{f}_\lambda$, minimization of the mean KL distance is equivalent to maximization of the mean expected log likelihood (abbreviated as MELL) which is defined as

$$
\mathrm{MELL}(\lambda) \equiv E_D\{E_{W,Z}\log\hat{f}_\lambda\}, \tag{2.32}
$$

where, as before, W and Z have the same joint distribution as X_i and Y_i and are independent of them. That is why the mean expected log likelihood is extensively used in statistical model selection as a powerful tool for evaluating the model performance and for choosing one model from the competing models. In a pioneering work, [akaike73] introduced the MELL as a model selection method and justified the use of ML for parameter estimation.

In the Gaussian case (when $Z|W$ is normally distributed) and with a correctly specified model, maximization of the mean expected log likelihood is equivalent to minimization of the mean predictive error (abbreviated as MPE). As with MPE, the mean expected log likelihood is not computable because of the unknown true distribution, but it can be estimated by plugging the empirical distribution into Eq. (2.32). By this means, the so-called average log likelihood (abbreviated as ALL) is obtained as follows:

$$\mathrm{ALL}\left(\hat{b}_{\lambda}\right) = \frac{1}{n}\sum_{i=1}^{n} \log f\left(Y_i|X_i;\ \hat{b}_{\lambda}\right). \tag{2.33}$$

Despite the fact that $\mathrm{ALL}(b) \rightarrow \mathrm{ELL}(b)$ as $n \rightarrow \infty$, due to the law of large numbers, the ALL, evaluated at MPLE $\hat{b}_{\lambda}$, is a biased estimator of the MELL of the MPL model, i.e., $E_D\ \mathrm{ALL}\left(\hat{b}_{\lambda}\right) \neq \mathrm{MELL}(\lambda)$. This bias should be corrected when we use MELL as an RPSM. In the next section, one of the methods for bias correction is presented. This method is usually used for deriving information model selection criteria as in [akaike73, sakamoto86, bozdogan87, konishi96, shibata89].

An information-based RPSM is given as the maximization of the mean expected log likelihood (Eq. 2.32) of maximum penalized likelihood models

$$\hat{\lambda}_{\mathrm{MELL}} = \arg\max_{\lambda} \{\mathrm{MELL}(\lambda)\}. \tag{2.34}$$

As already mentioned, the MELL is not computable and can be estimated by the ALL. The ALL, evaluated at the MPLE, is a biased estimator of MELL. To quantify the bias of ALL in estimating the MELL, we first define the expected penalized log likelihood (abbreviated as EPLL) as

$$\mathrm{EPLL}(b) \equiv E_{W,Z}\mathrm{PLL}\left(W, Z|b\right) \tag{2.35}$$

and expand it in a Taylor series at $\hat{b}_{\lambda}$ around b_{λ}^{*}, which is the limiting value of the MPLE $\hat{b}_{\lambda}$ as $n \rightarrow \infty$.

$$\begin{aligned}\mathrm{EPLL}(\hat{b}_\lambda) &\approx \mathrm{EPLL}(b_\lambda^*) + \left\{\tfrac{\partial}{\partial b}\mathrm{EPLL}(b_\lambda^*)\right\}^{\mathrm{T}}(\hat{b}_\lambda - b_\lambda^*) \\ &\quad + \frac{1}{2}(\hat{b}_\lambda - b_\lambda^*)^{\mathrm{T}}\left\{\frac{\partial^2}{\partial b \partial b^{\mathrm{T}}}\mathrm{EPLL}(b_\lambda^*)\right\}(\hat{b}_\lambda - b_\lambda^*) \\ &= \mathrm{EPLL}(b_\lambda^*) - \frac{1}{2}(\hat{b}_\lambda - b_\lambda^*)^{\mathrm{T}} J (\hat{b}_\lambda - b_\lambda^*),\end{aligned} \tag{2.36}$$

where

$$J \equiv -\frac{\partial^2}{\partial b \partial b^{\mathrm{T}}}\mathrm{EPLL}(b_\lambda^*). \tag{2.37}$$

Next, we expand the average penalized log likelihood (abbreviated as APLL) defined as

$$\mathrm{APLL}(b) \equiv \frac{1}{n}\sum_{i=1}^{n}\mathrm{LL}(X_i, Y_i | b) - \lambda p(b) \tag{2.38}$$

in a Taylor series at b_λ^* around $\hat{b}_\lambda$ as

$$\begin{aligned}\mathrm{APLL}(b_\lambda^*) &\approx \mathrm{APLL}(\hat{b}_\lambda) + \left\{\tfrac{\partial}{\partial b}\mathrm{APLL}(\hat{b}_\lambda)\right\}^{\mathrm{T}}(b_\lambda^* - \hat{b}_\lambda) \\ &\quad + \frac{1}{2}(b_\lambda^* - \hat{b}_\lambda)^{\mathrm{T}}\left\{\frac{\partial^2}{\partial b \partial b^{\mathrm{T}}}\mathrm{APLL}(\hat{b}_\lambda)\right\}(b_\lambda^* - \hat{b}_\lambda) \\ &\approx \mathrm{APLL}(\hat{b}_\lambda) - \frac{1}{2}(b_\lambda^* - \hat{b}_\lambda)^{\mathrm{T}} J (b_\lambda^* - \hat{b}_\lambda)\end{aligned} \tag{2.39}$$

We used the fact that

$$\frac{\partial}{\partial b}\mathrm{APLL}(\hat{b}_\lambda) = 0 \tag{2.40}$$

and that, by the law of large numbers, as $n \to \infty$

$$\left\{\frac{\partial^2}{\partial b \partial b^{\mathrm{T}}}\mathrm{APLL}(b_\lambda^*)\right\} \to \left\{\frac{\partial^2}{\partial b \partial b^{\mathrm{T}}}\mathrm{EPLL}(b_\lambda^*)\right\}, \tag{2.41}$$

and, since $\hat{b}_\lambda \to b_\lambda^*$ as $n \to \infty$, we have

$$\left\{\frac{\partial^2}{\partial b \partial b^{\mathrm{T}}}\mathrm{APLL}(\hat{b}_\lambda)\right\} \to \left\{\frac{\partial^2}{\partial b \partial b^{\mathrm{T}}}\mathrm{EPLL}(b_\lambda^*)\right\}. \tag{2.42}$$

Using $E_D\text{EPLL}(b_\lambda^*) = E_D\text{APLL}(b_\lambda^*)$ and combining Eqs. (2.36) and (2.39), we obtain

$$E_D\text{EPLL}(\hat{b}_\lambda) \approx E_D\text{APLL}(\hat{b}_\lambda) - E_D\left\{(b_\lambda^* - \hat{b}_\lambda)^{\mathrm{T}} J (b_\lambda^* - \hat{b}_\lambda)\right\}. \tag{2.43}$$

Since

$$E_D\,\text{EPLL}(\hat{b}_\lambda) = E_D\,\text{ELL}(\hat{b}_\lambda) - \lambda E_D p(\hat{b}_\lambda) \quad \text{and} \tag{2.44}$$

$$E_D\text{APLL}(\hat{b}_\lambda) = E_D\text{ALL}(\hat{b}_\lambda) - \lambda E_D p(\hat{b}_\lambda), \tag{2.45}$$

we have

$$\begin{aligned} E_D\text{ELL}(\hat{b}_\lambda) &\approx E_D\text{ALL}(\hat{b}_\lambda) - E_D\left\{(b_\lambda^* - \hat{b}_\lambda)^{\mathrm{T}} J (b_\lambda^* - \hat{b}_\lambda)\right\} \\ &\approx E_D\text{ALL}(\hat{b}_\lambda) - \frac{1}{n}\text{trace}(IJ^{-1}), \end{aligned} \tag{2.46}$$

where we use the asymptotic normality of the maximum penalized likelihood estimator and the trace result from Appendix D to obtain

$$E_D\left\{(b_\lambda^* - \hat{b}_\lambda)^{\mathrm{T}} J (b_\lambda^* - \hat{b}_\lambda)\right\} = \frac{1}{n}\text{trace}(IJ^{-1}). \tag{2.47}$$

Therefore, an unbiased estimator of the mean expected log likelihood is defined as

$$T_{\text{MELL}}(\hat{b}_\lambda) \equiv \text{ALL}(\hat{b}_\lambda) - \frac{1}{n}\text{trace}\left(\hat{I}\hat{J}^{-1}\right), \tag{2.48}$$

where

$$\hat{I} = \frac{1}{n}\sum_{i=1}^{n} \frac{\partial}{\partial b}\text{PLL}(X_i, Y_i|\hat{b}_\lambda) \frac{\partial}{\partial b^{\mathrm{T}}}\text{PLL}(X_i, Y_i|\hat{b}_\lambda) \quad \text{and} \tag{2.49}$$

$$\hat{J} = -\frac{1}{n}\sum_{i=1}^{n} \frac{\partial^2}{\partial b \partial b^{\mathrm{T}}}\text{PLL}(X_i, Y_i|\hat{b}_\lambda), \tag{2.50}$$

and the corresponding RPSM is

$$\hat{\lambda}_{\text{MELL}} = \underset{\lambda}{\text{argmax}}\left\{\text{ALL}(\hat{b}_\lambda) - \frac{1}{n}\text{trace}\left(\hat{I}\hat{J}^{-1}\right)\right\}. \tag{2.51}$$

A number of RPSMs can follow from this. When the model is Gaussian, correctly specified, and X is fixed, the well-known Mallows' (1973) CL method is obtained.

$$\hat{\lambda}_{\mathrm{CL}} = \underset{\lambda}{\operatorname{argmin}} \left\{ \frac{1}{n} \left\| Y - X\hat{b}_{\lambda} \right\|^2 + \frac{2\sigma^2}{n} \operatorname{trace}\left(X^{\mathrm{T}} X \left(X^{\mathrm{T}} X + n\lambda I_m \right)^{-1} \right) \right\}. \tag{2.52}$$

When the model is Gaussian and σ^2 is treated as a nuisance parameter, and J and I are estimated as

$$\hat{J} = -\frac{1}{\sigma^2} \left(\frac{1}{n} \sum_{i=1}^{n} X_i X_i^{\mathrm{T}} + \lambda I_m \right) \quad \text{and} \tag{2.53}$$

$$\hat{I} = \frac{1}{n\sigma^4} \sum_{i=1}^{n} r_{\mathrm{ols}\,i}^2 X_i X_i^{\mathrm{T}}, \tag{2.54}$$

Shibata's (1989) regularization information criterion (abbreviated as RIC) is obtained, and the corresponding RPSM is

$$\hat{\lambda}_{\mathrm{RIC}} = \underset{\lambda}{\operatorname{argmin}} \left\{ \frac{1}{n} \left\| Y - X\hat{b}_{\lambda} \right\|^2 + \frac{2\sigma^2}{n} \sum_{i=1}^{n} \frac{r_{\mathrm{ols}\,i}^2}{\sigma^2} H_{ii} \right\}, \tag{2.55}$$

where $H = X \left(X^{\mathrm{T}} X + n\lambda I_m \right)^{-1} X^{\mathrm{T}}$ and $r_{\mathrm{ols}\,i} = Y_i - X_i^{\mathrm{T}} \hat{b}$.

When $\hat{b}_{\lambda}$ is an M-estimator [huber81], Konishi and Kitagawa [konishi96] propose an information criterion for choosing the regularization parameter which is similar to RIC (Eq. 2.55).

We also suggest a RPSM that uses [bozdogan96a, bozdogan96b] an informational complexity framework to account for interdependencies between parameter estimates when evaluating the bias of ALL in estimating the MELL. The resulting method, by means of a more severe penalization of the inaccuracy of estimation, produces slightly overestimated regularization parameter values as compared to that given by CL or RIC. Overestimation, however, is in a safe direction and is shown to be beneficial in situations with a limited number of observations.

Despite its simplicity, the Gaussian correctly specified case is very important, especially for the numerical solution of integral equations with a method of regularization, because X is fixed and there is no functional misspecification. In the Gaussian correctly specified case, the information RPSM (Eq. 2.51) becomes similar to CL.

The MELL RPSM (Eq. 2.51) reduces to Mallows' CL under the following conditions: the approximating distribution (model) belongs to the Gaussian family, i.e.,

$$W \sim N_m(\mu, A) \quad \text{and} \tag{2.56}$$

$$Z | W \sim N\left(m(W), \sigma^2 \right), \tag{2.57}$$

and the model is correctly specified, meaning that there exists b_0, referred to as the true regression coefficients (or the true solution), such that

$$f(W,Z;\ b_0) = g(W,Z), \tag{2.58}$$

where $g(W,Z)$ is the actual (true) data-generating distribution and where σ^2, the conditional variance of the output (or noise variance), is treated as a nuisance parameter. In particular, correct specification implies that

$$E_{Z|W}\{Z - W^{\mathrm{T}}b^*\} = 0 \quad \text{and} \tag{2.59}$$

$$E_{Z|W}\left\{(Z - W^{\mathrm{T}}b^*)(Z - W^{\mathrm{T}}b^*)^{\mathrm{T}}\right\} = \sigma^2. \tag{2.60}$$

The log likelihood in this case is

$$\begin{aligned}\log f(Z|W;b) &= \log\frac{1}{\sqrt{2\pi\sigma^2}}\exp\left(-\frac{1}{2\sigma^2}(Z - W^{\mathrm{T}}b)^{\mathrm{T}}(Z - W^{\mathrm{T}}b)\right)\\ &= \log\frac{1}{\sqrt{2\pi\sigma^2}} - \frac{1}{2\sigma^2}(Z - W^{\mathrm{T}}b)^{\mathrm{T}}(Z - W^{\mathrm{T}}b).\end{aligned} \tag{2.61}$$

Its derivatives with respect to b are

$$\frac{\partial}{\partial b}\log f(Z|W;\ b) = \frac{1}{\sigma^2}W(Z - W^{\mathrm{T}}b), \tag{2.62}$$

$$\frac{\partial}{\partial b^{\mathrm{T}}}\log f(Z|W;\ b) = \frac{1}{\sigma^2}(Z - W^{\mathrm{T}}b)^{\mathrm{T}}W^{\mathrm{T}}, \quad \text{and} \tag{2.63}$$

$$\frac{\partial^2}{\partial b\partial b^{\mathrm{T}}}\log f(Z|W;\ b) = -\frac{1}{\sigma^2}WW^{\mathrm{T}}. \tag{2.64}$$

Using the quadratic penalty, matrix J becomes

$$\begin{aligned}J &= -\frac{\partial^2}{\partial b\partial b^{\mathrm{T}}}E_{W,Z}\{\log f_\lambda^* - \lambda p(b_\lambda^*)\}\\ &= E_W\left\{\frac{1}{\sigma^2}WW^{\mathrm{T}} + \lambda p'(b_\lambda^*)p'(b_\lambda^*)^{\mathrm{T}}\right\}\\ &= \frac{1}{\sigma^2}E_W\{WW^{\mathrm{T}}\} + \lambda p'(b_\lambda^*)p'(b_\lambda^*)^T = \frac{1}{\sigma^2}(E_W\{WW^{\mathrm{T}}\} + \lambda I_m)\end{aligned} \tag{2.65}$$

and can be estimated as

$$\hat{J} = \frac{1}{\sigma^2}\left(\frac{1}{n}\sum_{i=1}^{n}X_iX_i^{\mathrm{T}} + \lambda I_m\right) = \frac{1}{n\sigma^2}(X^{\mathrm{T}}X + n\lambda I_m). \tag{2.66}$$

Matrix I becomes

$$\begin{aligned} I &= E_{W,Z}\left\{\frac{\partial}{\partial b}\left(\log f_\lambda^* - \lambda p(b_\lambda^*)\right)\frac{\partial}{\partial b^{\mathrm{T}}}\left(\log f_\lambda^* - \lambda p(b_\lambda^*)\right)\right\} \\ &= E_{W,Z}\left\{\frac{\partial}{\partial b}\log f_\lambda^* \frac{\partial}{\partial b^{\mathrm{T}}}\log f_\lambda^*\right\} - E_{W,Z}\left\{\frac{\partial}{\partial b^{\mathrm{T}}}\log f_\lambda^*\right\}E_{W,Z}\left\{\frac{\partial}{\partial b^{\mathrm{T}}}\log f_\lambda^*\right\} \\ &= \frac{1}{\sigma^2}E_W\{WW^{\mathrm{T}}\} + \frac{1}{\sigma^4}E_W\left\{WW^{\mathrm{T}}(b^* - b_\lambda^*)(b^* - b_\lambda^*)^{\mathrm{T}}WW^{\mathrm{T}}\right\} \\ &\quad - \frac{1}{\sigma^4}E_W\{WW^{\mathrm{T}}\}(b^* - b_\lambda^*)(b^* - b_\lambda^*)^{\mathrm{T}}E_W\{WW^{\mathrm{T}}\}, \end{aligned} \tag{2.67}$$

and, for a large n, it can be estimated as

$$\hat{I} = \frac{1}{n\sigma^2}\sum_{i=1}^{n} X_i X_i^{\mathrm{T}} = \frac{1}{n\sigma^2}X^{\mathrm{T}}X. \tag{2.68}$$

The trace term becomes

$$\begin{aligned} \mathrm{trace}\left(\hat{I}\hat{J}^{-1}\right) &= \mathrm{trace}\left(\frac{1}{n\sigma^2}X^{\mathrm{T}}X \cdot n\sigma^2\left(X^{\mathrm{T}}X + n\lambda I_m\right)^{-1}\right) \\ &= \mathrm{trace}\left(X^{\mathrm{T}}X\left(X^{\mathrm{T}}X + n\lambda I_m\right)^{-1}\right) \\ &= \mathrm{trace}(H), \end{aligned} \tag{2.69}$$

where the hat matrix is defined as $H \equiv X\left(X^{\mathrm{T}}X + n\lambda I_m\right)^{-1}X^{\mathrm{T}}$.

The RPSM becomes

$$\hat{\lambda}_{\mathrm{MELL}} = \arg\min_{\lambda}\left\{\frac{1}{2\sigma^2}\frac{1}{n}\sum_{i=1}^{n}\left(Y_i - X_i^{\mathrm{T}}\hat{b}_\lambda\right)^2 + \frac{1}{n}\mathrm{trace}(H)\right\} \quad \text{or} \tag{2.70}$$

$$\hat{\lambda}_{\mathrm{MELL}} = \arg\min_{\lambda}\left\{\frac{1}{n}\left\|Y - X\hat{b}_\lambda\right\|^2 + \frac{2\sigma^2}{n}\mathrm{trace}(H)\right\}. \tag{2.71}$$

This is exactly CL. Therefore, CL can be viewed as an information RPSM when the model is correctly specified and is Gaussian with fixed X.

Dropping the assumption of correct model specification and using the Gaussian approximating distribution as in the previous case, a similar expression for J is obtained as

$$J = \frac{1}{\sigma^2}\left(E_W\{WW^{\mathrm{T}}\} + \lambda I_m\right) \tag{2.72}$$

and estimated as

$$\hat{J} = \frac{1}{\sigma^2}\left(\frac{1}{n}\sum_{i=1}^{n} X_i X_i^{\mathrm{T}} + \lambda I_m\right) = \frac{1}{n\sigma^2}\left(X^{\mathrm{T}}X + n\lambda I_m\right). \tag{2.73}$$

Matrix I becomes

$$\begin{aligned} I &= E_{W,Z}\left\{\frac{\partial}{\partial b}\log f_\lambda^* \frac{\partial}{\partial b^{\mathrm{T}}}\log f_\lambda^*\right\} - E_{W,Z}\left\{\frac{\partial}{\partial b^{\mathrm{T}}}\log f_\lambda^*\right\}E_{W,Z}\left\{\frac{\partial}{\partial b^{\mathrm{T}}}\log f_\lambda^*\right\} \\ &= \frac{1}{\sigma^4}E_{W,Z}\left\{W\left(Z - W^{\mathrm{T}}b^*\right)^2 W^{\mathrm{T}}\right\} \end{aligned} \tag{2.74}$$

and is estimated as

$$\hat{I} = \frac{1}{\sigma^4 n}\sum_{i=1}^{n} X_i\left(Y_i - X_i^{\mathrm{T}}\hat{b}\right)^2 X_i^{\mathrm{T}}. \tag{2.75}$$

The RPSM becomes

$$\hat{\lambda}_{\mathrm{MELL}} = \arg\min_{\lambda}\left\{\frac{1}{n}\left\|Y - X\hat{b}_\lambda\right\|^2 + \frac{2\sigma^2}{n}\mathrm{trace}\left(\hat{I}\hat{J}^{-1}\right)\right\}. \tag{2.76}$$

This RPSM uses the Gaussian model but does not assume that the conditional mean is correctly specified. That means the choice of the regularization parameter value remains consistent even if a functional misspecification is present, i.e., when $m(x) \equiv E\{Y_i|X_i = x\} \neq x^{\mathrm{T}}b$ for any parameter $b \in R^m$.

As mentioned already, distributional misspecification does not affect the estimation of the location parameter b. However, when an estimate of the covariance matrix of the MLE or MPLE is needed, an estimator that is consistent under distributional misspecification must be used because the usual covariance matrix estimators are not consistent under distributional misspecification. To account for possible distributional misspecifications, the estimation of σ^2, treated so far as a nuisance parameter, must be considered. This allows one to account for a nonzero skewness and kurtosis in the response variable $Z|W$.

With a limited number of observations, the inaccuracy penalization in Eq. (2.76) becomes inadequate, and further refinement is needed. Starting from Eq. (2.76) and using Bozdogan's [bozdogan96a, bozdogan96b] refinement argument, we obtain an information complexity regularization parameter selection (abbreviated as ICOMPRPS) method that behaves favorably for a limited number of observations.

Notice that the term $\mathrm{trace}\left(\hat{I}\hat{J}^{-1}\right)$ in Eq. (2.76) can be interpreted as the effective number of parameters of a possibly misspecified model. ICOMPRPS also penalizes the interdependency between the parameter estimates. ICOMPRPS imposes a more severe penalization of estimation inaccuracy caused by the fact that the data-generating distribution is unknown.

For the MPLE method, the ICOMPRPS has the form [urmanov02]

$$\mathrm{ICOMPRPS}(\lambda) \equiv \mathrm{ALL}(\hat{b}_\lambda) - \frac{1}{n}\mathrm{trace}\left(\hat{I}\hat{J}^{-1}\right) - \frac{1}{n}C_1\left(\hat{J}^{-1}\right), \tag{2.77}$$

and the corresponding RPSM is

$$\hat{\lambda}_{\mathrm{ICOMPRPS}} = \arg\max_{\lambda}\left\{\mathrm{ALL}(\hat{b}_\lambda) - \frac{1}{n}\mathrm{trace}\left(\hat{I}\hat{J}^{-1}\right) - \frac{1}{n}C_1\left(\hat{J}^{-1}\right)\right\}, \tag{2.78}$$

where C_1 is the maximal covariance complexity index proposed by van Emden [emden71] to measure the degree of interdependency between parameter estimates. C_1 is a function of a covariance matrix and is computed as in Eq. (2.79) using the eigenvalues of the covariance matrix. Notice that the more ill conditioned the data matrix X, the more dependent the parameter estimates become; therefore, the covariance complexity can be used to quantify ill conditioning.

Under the assumption that the vector of parameter estimates $\hat{b}_\lambda$ is approximately normally distributed, the maximal covariance complexity reduces to

$$C_1\left(\hat{J}^{-1}\right) = \frac{m}{2}\log\frac{\overline{v}_a}{\overline{v}_g}, \tag{2.79}$$

where $\overline{v}_a = \frac{1}{m}\sum_{j=1}^{m} v_j$, $\overline{v}_g = \left(\prod_{j=1}^{m} v_j\right)^{\frac{1}{m}}$, and ν_j are the eigenvalues of $\hat{J}^{-1}$.

In the Gaussian case, ICOMPRPS for correctly specified models (abbreviated as ICOMPRPSCM) becomes

$$\mathrm{ICOMPRPSCM}(\lambda) = \frac{1}{n}\left\|Y - X\hat{b}_\lambda\right\|^2 + \frac{2\sigma^2}{n}\left(\mathrm{trace}(H) + C_1\left(\hat{J}^{-1}\right)\right), \tag{2.80}$$

and the corresponding RPSM is

$$\hat{\lambda}_{\mathrm{ICOMPRPSCM}} = \arg\min_{\lambda}\left\{\frac{1}{n}\left\|Y - X\hat{b}_\lambda\right\|^2 + \frac{2\sigma^2}{n}\left(\mathrm{trace}(H) + C_1\left(\hat{J}^{-1}\right)\right)\right\}, \tag{2.81}$$

where

$$\hat{J} = X^{\mathrm{T}}X + n\lambda I_m \quad \text{and} \tag{2.82}$$

$$H = X\left(X^{\mathrm{T}}X + n\lambda I_m\right)^{-1}X^{\mathrm{T}}. \tag{2.83}$$

There is a strong bond between the RPSMs based on maximizing the mean expected log likelihood and minimizing the mean predictive error. Namely, if the parametric family of approximating distributions (the model) is Gaussian,

$$f(Y_i|X_i;\ b) \equiv N(X_i^T b, \sigma^2), \tag{2.84}$$

then maximizing the MELL is equivalent to minimizing the MPE. This fact allows us to write an MPE analog of the information criterion (Eq. 2.48). Indeed, using the Gaussian model, the ALL can be written as the sum of the training error (abbreviated TE) and a constant term

$$\begin{aligned}
\mathrm{ALL}(\hat{b}_\lambda) &= \frac{1}{n}\sum_{i=1}^{n} \log f(X_i, Y_i|\hat{b}_\lambda) \\
&= \frac{1}{n}\sum_{i=1}^{n} \log \frac{1}{\sqrt{2\pi\sigma^2}} \exp\left(-\frac{1}{2\sigma^2}(Y_i - X_i^{\mathrm{T}}\hat{b}_\lambda)^2\right) \\
&= \log \frac{1}{\sqrt{2\pi\sigma^2}} - \frac{1}{n2\sigma^2}\sum_{i=1}^{n}(Y_i - X_i^{\mathrm{T}}\hat{b}_\lambda)^2 \\
&= \log \frac{1}{\sqrt{2\pi\sigma^2}} - \frac{1}{2\sigma^2}\mathrm{TE}(\hat{b}_\lambda),
\end{aligned} \tag{2.85}$$

where the training error is defined as

$$\mathrm{TE}(\hat{b}_\lambda) \equiv \frac{1}{n}\sum_{i=1}^{n}(Y_i - X_i^T\hat{b}_\lambda)^2. \tag{2.86}$$

The expected log likelihood for the Gaussian model is

$$\begin{aligned}
\mathrm{ELL}(\hat{b}_\lambda) &= E_{W,Z}\log f(W, Z|\hat{b}_\lambda) = E_{W,Z}\left\{\log \frac{1}{\sqrt{2\pi\sigma^2}} \exp\left(-\frac{1}{2\sigma^2}(Z - W^{\mathrm{T}}\hat{b}_\lambda)^2\right)\right\} \\
&= \log \frac{1}{\sqrt{2\pi\sigma^2}} - \frac{1}{2\sigma^2}E_{W,Z}\left\{(Z - W^{\mathrm{T}}\hat{b}_\lambda)^2\right\} \\
&= \log \frac{1}{\sqrt{2\pi\sigma^2}} - \frac{1}{2\sigma^2}E_{W,Z}\left\{(Z - m(W))^2\right\} \\
&\quad - \frac{1}{2\sigma^2}E_W\left\{(m(W) - W^{\mathrm{T}}\hat{b}_\lambda)^T(m(W) - W^{\mathrm{T}}\hat{b}_\lambda)\right\} \\
&= \log \frac{1}{\sqrt{2\pi\sigma^2}} - \frac{1}{2} - \frac{1}{2\sigma^2}\mathrm{PE}(\hat{b}_\lambda),
\end{aligned} \tag{2.87}$$

where the predictive error is defined as

$$\mathrm{PE}(\hat{b}_\lambda) \equiv E_W\left\{(m(W) - W^{\mathrm{T}}\hat{b}_\lambda)^{\mathrm{T}}(m(W) - W^{\mathrm{T}}\hat{b}_\lambda)\right\}. \tag{2.88}$$

Plugging these representations into Eq. (2.48), an MPE analog of the information RPSM is obtained. The mean predictive error is approximated as

$$E_D\mathrm{PE}(\hat{b}_\lambda) \approx E_D\mathrm{TE}(\hat{b}_\lambda) + \frac{2\sigma^2}{n}\mathrm{trace}\left(\hat{I}\hat{J}^{-1}\right) - \sigma^2. \tag{2.89}$$

Therefore, an unbiased estimator of the MPE is given by

$$T_{\text{MPE}}(\lambda) \equiv \text{TE}\left(\hat{b}_{\lambda}\right) + \frac{2\sigma^2}{n}\text{trace}\left(\hat{I}\hat{J}^{-1}\right) - \sigma^2, \tag{2.90}$$

and the corresponding RPSM is

$$\hat{\lambda}_{\text{MPE}} = \arg\min_{\lambda}\left\{\text{TE}\left(\hat{b}_{\lambda}\right) + \frac{2\sigma^2}{n}\text{trace}\left(\hat{I}\hat{J}^{-1}\right) - \sigma^2\right\}. \tag{2.91}$$

Therefore, when the Gaussian model is used, the MELL and MPE have the same minimizer. When the model is correctly specified, $\text{trace}\left(\hat{I}\hat{J}^{-1}\right) = \text{trace}(H)$, and the CL method follows as

$$\text{CL}(\lambda) = \text{TE}\left(\hat{b}_{\lambda}\right) + \frac{2\sigma^2}{n}\text{trace}(H) - \sigma^2, \tag{2.92}$$

with the corresponding RPSM

$$\hat{\lambda}_{\text{CL}} = \arg\min_{\lambda}\left\{\text{TE}\left(\hat{b}_{\lambda}\right) + \frac{2\sigma^2}{n}\text{trace}(H) - \sigma^2\right\}. \tag{2.93}$$

We now present an example of an image reconstructed using the information approach to the selection of regularization parameter. Notice that in this example of a correctly specified model, the C_{L} and ICOMPPRS selected identical parameters as expected from theoretical derivations.

References

[akaike73] H. Akaike, Information theory and an extension of the maximum likelihood principle, in *2nd International Symposium on Information Theory*, ed. by B.N. Petrov, F. Csaki (Akademiai Kiado, Budapest, 1973), pp. 267–281

[bozdogan87] H. Bozdogan, Model selection and Akaike's information criterion (AIC): the general theory and its analytical extensions. Psychometrika **52**(3), 345–370 (1987)

[bozdogan88] H. Bozdogan, ICOMP: a new model selection criterion, in *Classification and Related Methods of Data Analysis*, ed. by Hans H. Bock (Elsevier Science Publishers B.V. (North-Holland), Amsterdam, 1988), pp. 599–608

[bozdogan96a] H. Bozdogan, A new informational complexity criterion for model selection: the general theory and its applications, in *Information Theoretic Models and Inference (INFORMS)*, Washington D.C., 5–8 May 1996

[bozdogan96b] H. Bozdogan, Informational complexity criteria for regression models, in *Information Theory and Statistics Section on Bayesian Stat. Science*, ASA Annual Meeting, Chicago, IL, 4–8 August 1996

[engl00] H.W. Engl, M. Hanke, A. Neubauer, *Regularization of Inverse Problems* (Kluwer Academic, Dordrecht, 2000)

[golub97] G.H. Golub, U. von Matt, Tikhonov regularization for large scale problems. Technical report SCCM-97-03, Stanford University, 1997

[hansen93] P.C. Hansen, D.P. O'Leary, The use of the L-curve in the regularization of discrete ill-posed problems. SIAM J. Sci. Comput. **14**, 1487–1503 (1993)

[hansen94] P.C. Hansen, Regularization tools: a Matlab package for analysis and solution of discrete ill-posed problems. Numer. Algorithms **6**, 1–35 (1994)

[hansen98] P.C. Hansen, *Rank-Deficient and Discrete Ill-Posed Problems*. SIAM Monographs on Mathematical Modeling and Computation (SIAM, Philadelphia, 1998)

[huber81] P.J. Huber, *Robust Statistics* (Wiley, New York, 1981)

[konishi96] S. Konishi, G. Kitagawa, Generalized information criteria in model selection. Biometrika **83**(4), 875–890 (1996)

[kullback51] S. Kullback, R.A. Leibler, On information and sufficiency. Ann. Math. Stat. **22**, 79–86 (1951)

[leonov97] A.S. Leonov, A.G. Yagola, The L-curve method always introduces a non removable systematic error. Mosc. Univ. Phys. Bull. **52**(6), 20–23 (1997)

[mallows73] C.L. Mallows, Some comments on CP. Technometrics **15**(4), 661–675 (1973)

[morozov84] V.A. Morozov, *Methods for Solving Incorrectly Posed Problems* (Springer, New York, 1984)

[phillips62] D.L. Phillips, A technique for the numerical solution of certain integral equations of the first kind. JACM **9**, 84–97 (1962)

[sakamoto86] Y. Sakamoto, *Akaike Information Criterion Statistics* (KTK Scientific publishers, Tokyo, 1986)

[shibata89] R. Shibata, Statistical aspects of model selection, in *From Data to Model*, ed. by J.C. Willems (Springer, New York, 1989), pp. 215–240

[urmanov02] A.M. Urmanov, A.V. Gribok, H. Bozdogan, J.W. Hines, R.E. Uhrig, Information complexity-based regularization parameter selection for solution of ill-conditioned inverse problems. Inverse Prob. **18**, L1–L9 (2002)

[emden71] M.H. van Emden, An analysis of complexity, in *Mathematical Centre Tracts*, vol. 35 (Mathematisch Centrum, Amsterdam, 1971)

[vogel96] C.R. Vogel, Non-convergence of the L-curve regularization parameter selection method. Inverse Prob. **12**, 535–547 (1996)

[vogel02] C.R. Vogel, *Computational Methods for Inverse Problems*. SIAM, Frontiers in Applied Mathematics Series, vol 23 (SIAM, Philadelphia, 2002)

[wahba90] G. Wahba, *Spline Models for Observational Data* (Society for Industrial and Applied Mathematics, Philadelphia, PA, 1990)

Part II

Chapter 3
Introduction to Optimization

Abstract Many engineering problems, particularly in image processing, can be expressed as optimization problems. Often, we must make approximations of our mathematical models in order to cast the problems into optimization form. This chapter considers basic optimization theory and application as related to image processing. The focus presented in the examples will be to functions of one dimension, or line functions. Later chapters will give emphasis to the multidimensional case.

3.1 Optimization Problems

Suppose we want to find a set of data $\{x_1, x_2, \ldots, x_N\}$ that minimizes an objective function $f(x_1, x_2, \ldots, x_N)$. If there is a constraint on the data, the optimization problem can be described as

$$\text{minimize } f(x), \quad \text{subject to } x \in U, \tag{3.1}$$

where $x = [x_1, x_2, \ldots, x_N]^T, f(x) : R^N \to R$, and $U \subset R^N$ [chong96].

For example, we consider that x represents an image with N pixels, each of which has a continuous intensity value in the range $[0, 255]$. If the image is observed as y and is obtained by the relationship

$$y = Dx, \tag{3.2}$$

then the original image can be estimated by

$$\text{minimizing } \|y - Dx\|^2 \text{ subject to } x \in \{x_i | 0 \leq x_i \leq 255, \; i = 1, \ldots, N\}. \tag{3.3}$$

The problem described in Eq. (3.3) can fit into the general optimization structure given in Eq. (3.1) if $f(x) = \|y - Dx\|^2$ and $U = \{x_i | 0 \leq x_i \leq 255, \; i = 1, \ldots, N\}$.

As another example, consider a 3×5 array of edges with gradient magnitude, as shown in Fig. 3.1a. By using the gradient magnitude table, we want to link edge

M.A. Abidi et al., *Optimization Techniques in Computer Vision*, Advances in Computer Vision and Pattern Recognition, DOI 10.1007/978-3-319-46364-3_3

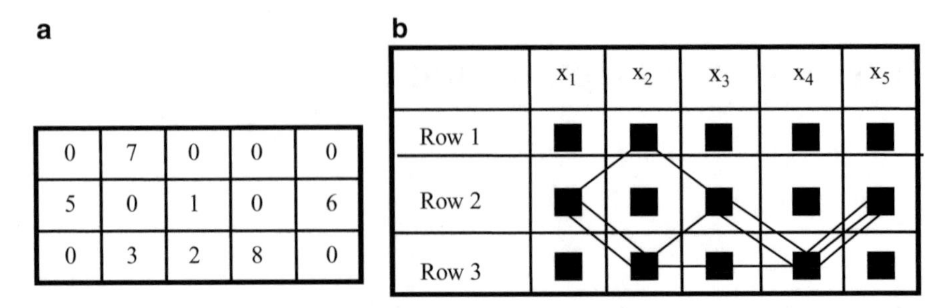

0	7	0	0	0
5	0	1	0	6
0	3	2	8	0

Fig. 3.1 (**a**) Table for gradient magnitudes of a 3×5 array of edges and (**b**) three possible contours that link edges with nonzero gradient magnitude

points for boundary extraction. Note that we did not use gradient directions for simplicity, which should be considered for more practical edge linking problems [jain89]. If we assume that the contour should pass an edge of nonzero gradient magnitude from left to right, then three possible contours are shown in Fig. 3.1b.

Let x_i represent the row position of the ith column in Fig. 3.1a. Then the vector for contour A in Fig. 3.1b can be described as $x^A = (2, 1, 2, 3, 2)$. Likewise, contours B and C, respectively, can be described as $x^B = (2, 3, 2, 3, 2)$ and $x^C = (2, 3, 3, 3, 2)$. To choose the best contour, we define the objective function as the sum of cumulative gradient magnitudes. We then have $g(x^A) = 5 + 7 + 1 + 8 + 6 = 27$, $g(x^B) = 23$, and $g\left(x^C\right) = 24$. Thus, to maximize $g(x)$, the best contour is x^A, which gives the maximum cumulative gradient magnitudes. This example can be described as an optimization problem for which

$$\text{maximize } g(x), \quad \text{subject to} \, x \in \left\{x^A, x^B, x^C\right\}. \tag{3.4}$$

The problem described in Eq. (3.3) becomes equivalent to Eq. (3.1) if we define $f(x) = -g(x)$ and $U = \left\{x^A, x^B, x^C\right\}$.

The problems discussed above have the general form of a constrained optimization problem. This is true since the variables are constrained to be in the constraint set U. If $U = R^N$, where N is the number of variables, we refer to the problem as an unconstrained optimization problem. Generally, most practical problems have constraints on the solution. However, as a logical progression, starting in Chap. 4, we first discuss the basic unconstrained optimization problem. This approach is taken because a good mathematical description and analysis for unconstrained optimization problems can serve as a basic approach to solve the more general constrained optimization problem. Constrained optimization problems will be discussed in Chap. 5. Many useful methods for solving constrained problems have been developed. However, to help simplify the numerical work, it is not uncommon to first transform the constrained optimization problem to an appropriate unconstrained optimization problem. Then the approximate solution is obtained by solving the unconstrained problem. This approach, called the regularization method, will be presented in Chap. 6.

3.2 Definition of Minimum Points

A minimization problem can be defined as one in which a solution is found that minimizes a given function. If the function has only one variable, the solution constitutes a scalar value. If the function is defined on the N-dimensional integer space, then the solution is an $N \times 1$ vector with integer elements. According to the characteristics of the minimization problem, the solution space may have various types. We will use the N-dimensional real space, denoted by R^N, as a generalized solution space in this book. A general unconstrained optimization problem can then be described as

$$\text{minimize} f(x) \quad \text{for} x \in R^N, \quad \text{given} f : R^N \to R. \tag{3.5}$$

The first task in understanding unconstrained optimization problems is to define various types of solutions [murray72]. As a background, we consider the following definitions.

Definition 3.1: Strong Local Minimum A point x^* is a strong local minimum of $f(x)$ over R^N if there exists ε such that $f(x^*) < f(x)$, for all $x \in R^N$ and $\|x - x^*\| < \varepsilon$.

The terminology "minimum" stands for a point or a solution giving the minimum value of the given function. Alternatively, "minimizer" or "minimum point" is used to express the same meaning.

Definition 3.2: Weak Local Minimum A point x^* is a weak local minimum of $f(x)$ over U if there exists ε such that $f(x^*) \leq f(x)$, for all $x \in U$ and $\|x - x^*\| < \varepsilon$.

In Definition 3.2 we have replaced the less than sign, "$<$" in Definition 3.1, with the less than or equal sign, "$\leq$," in order to have a weak local minimum. Similar inequalities apply for strong and weak global minima, as defined in Definitions 3.3 and 3.4.

Definition 3.3: Strong Global Minimum A point x^* is a strong global minimum of $f(x)$ over U if $f(x^*) < f(x)$, for all $x \in U$.

Definition 3.4: Weak Global Minimum A point x^* is a weak global minimum of $f(x)$ over U, if $f(x^*) \leq f(x)$, for all $x \in U$.

There are a number of alternative expressions used to describe the different types of minima. The more common of these are listed in Table 3.1. In addition, Fig. 3.2 shows various types of minima for a single-variable function.

Suppose we have a continuous and convex function, $f(x_1, x_2) = x_1^2 + x_2^2$, as shown in Fig. 3.3. Intuitively, we can guess that the minimum of $f(x)$ occurs where the sign of its first derivative changes from negative to positive. For a more rigorous analysis, the first- and the second-order derivatives and associated terminologies are defined in the following definitions and theorems.

Table 3.1 Alternative terminology that describes the different types of minima

Type of minima	Alternative expressions
Strong local minimum	Strict local (relative) minimum
	Proper local (relative) minimum
Weak local minimum	Improper local (relative) minimum
Strong global minimum	Strict global (absolute) minimum
	Proper global (absolute) minimum
Strong global minimum	Improper global (absolute) minimum

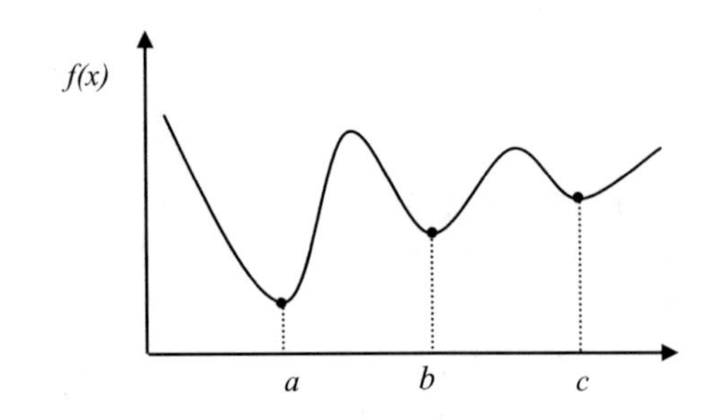

Fig. 3.2 A function of a single variable and the corresponding minima: *a* global minimum, *b* strong local minimum, and *c* weak local minimum

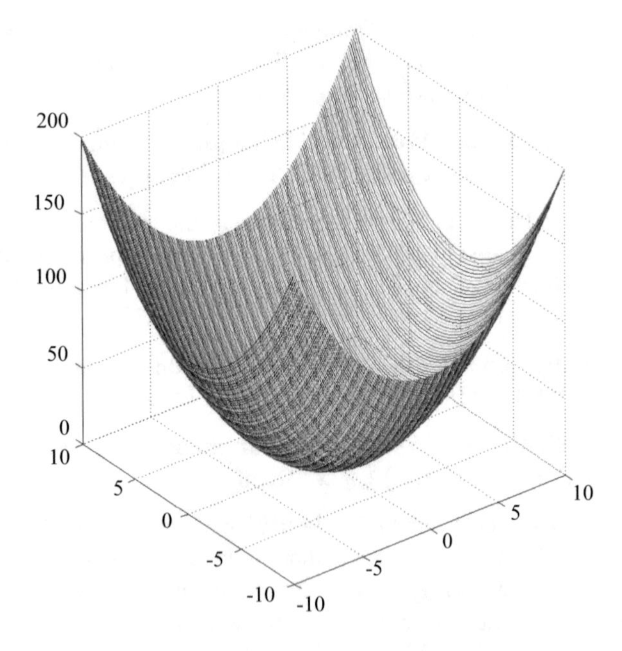

Fig. 3.3 A continuous and convex function, $f(x_1, x_2) = x_1^2 + x_2^2$

Definition 3.5: Gradient Vector The gradient of $f(x) : R^N \to R$ is a vector whose ith element is equal to the first derivative of $f(x)$ with respect to x_i, that is,

$$\nabla f(x) = \left[\frac{\partial f}{\partial x_1}, \frac{\partial f}{\partial x_2}, \ldots, \frac{\partial f}{\partial x_N} \right]^T. \tag{3.6}$$

The ith partial derivative is defined as

$$\frac{\partial f(x)}{\partial x_i} = \lim_{h \to 0} \frac{f(x + he_i) - f(x)}{h}, \tag{3.7}$$

where e_i represents the ith column of the $N \times N$ identity matrix.

The gradient of a function provides important information about functional increase or decrease. The following theorem summarizes an important characteristic of the gradient.

Theorem 3.1: Rate of Increase *If d is a unit vector, that is, $\|d\| = 1$, the inner product of $\nabla f(x)$ and* d, *denoted by $\nabla f(x)^T d$ or $< \nabla f(x)^T d >$, is the rate of increase of* f(x) *at the point* x *in the direction* d.

Proof Let $f(x) : R^N \to R$ be a real-valued function and d be a vector in R^N.[1] The directional derivative of $f(x)$ in the direction d, denoted by $\frac{\partial f(x)}{\partial d}$, is the real-valued function defined by

$$\frac{\partial f(x)}{\partial d} = \lim_{\alpha \to 0} \frac{f(x + \alpha d) - f(x)}{h}. \tag{3.8}$$

If $\|d\| = 1$, then $\frac{\partial f(x)}{\partial d}$ is the rate of increase of $f(x)$ at x in the direction d. Given x and d, the directional derivative can be computed as a function of α,

$$\frac{\partial f(x)}{\partial d} = \left. \frac{\partial f(x + ad)}{\alpha} \right|_{\alpha=0}. \tag{3.9}$$

Applying the chain rule yields

$$\frac{\partial f(x)}{\partial d} = \left. \frac{\partial f(x + ad)}{\alpha} \right|_{\alpha=0} = \nabla f(x)^{\mathrm{T}} d = \langle \nabla f(x), d \rangle = d^{\mathrm{T}} f(x). \tag{3.10}$$

□

Definition 3.6: Hessian Matrix The Hessian of $f(x) : R^N \to R$ is a matrix whose (i, j)th element is equal to the second derivative of $f(x)$ with respect to x_i and x_j, that is,

[1]This assumption is made for an unconstrained optimization problem. When we deal with a constrained optimization problem, it is replaced by the condition that d is a feasible direction at $x \in \Omega$, where Ω represents the constraint set. Definitions of the feasible direction and the constraint set will be given in Chap. 5 for constrained optimization.

$$H(x)=\begin{bmatrix}\frac{\partial^2 f(x)}{\partial x_1^2} & \cdots & \frac{\partial^2 f(x)}{\partial x_N \partial x_1}\\ \vdots & \ddots & \vdots \\ \frac{\partial^2 f(x)}{\partial x_1 \partial x_N} & \cdots & \frac{\partial^2 f(x)}{\partial x_N^2}\end{bmatrix}. \tag{3.11}$$

Using Definitions 3.5 and 3.6, we can have the Taylor series expansion of a function as follows:

Theorem 3.2: Taylor Series Expansion *Let* $f(x): R^N \rightarrow R$ *and* $d \in R^N$. *Then the Taylor series expansion of f(x) at* $x+\alpha d$ *is expressed, for* $\alpha > 0$, *as*

$$f(x+\alpha d)=f(x)+\alpha \nabla f(x)^{\mathrm{T}} d+\frac{\alpha^2}{2} d^{\mathrm{T}} H(x) d+O(\alpha^3). \tag{3.12}$$

The necessary and sufficient condition for a function to be convex is that its Hessian matrix is positive definite. The definition of positive definiteness is given below.

Definition 3.7: Positive Definiteness An $N \times N$ matrix A is positive definite, if $x^{\mathrm{T}} A x > 0$, for $x \neq 0$, $x \in R^N$.

The function shown in Fig. 3.3 has the gradient $\nabla f(x)=[2x_1, 2x_2]^{\mathrm{T}}$ and the Hessian $H(x)=\begin{bmatrix}2 & 0\\0 & 2\end{bmatrix}$, which is positive definite.

According to Definition 3.5, we can guess that there is no change in *f*(*x*) in the neighborhood of x^*, which satisfies $\nabla f(x^*)=0$. From this result, we have the following definition.

Definition 3.8: Stationary Point A point $x \in R^N$ is said to be a stationary point of *f*(*x*) if $\nabla f(x)=0$.

Although a stationary point is a good candidate for a local or global minimum, there is an exception that is called a saddle point.

Definition 3.9: Saddle Point A point *x* is a saddle point of *f*(*x*) if *x* is a stationary point but not a local minimum or maximum.

Example 3.1: A Saddle Point Consider a two-variable, quadratic function

$$f(x_1, x_2)=x_1^2-x_2^2. \tag{3.13}$$

Figure 3.4 shows its function value over the x_1-x_2 plane. Its gradient is calculated as $\nabla f(x_1, x_2)=2x_1-2x_2$. We can easily see that $(0,0)$ is the stationary point of the function in Eq. (3.13) since $\nabla f(0,0)=0$. However it is clear that $(0,0)$ is not a minimum point as shown in Fig. 3.4.

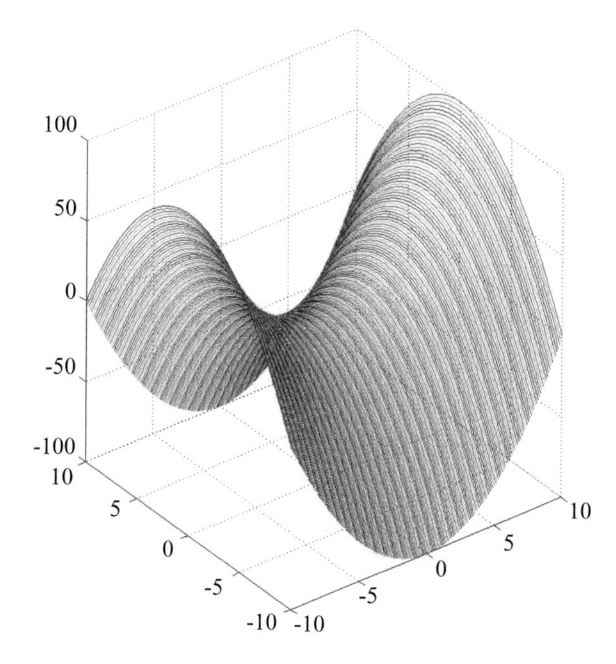

Fig. 3.4 An example of a saddle point, $f(x_1, x_2) = x_1^2 - x_2^2$

Definition 3.10: Level Sets The level set of a function $f(x) : R^N \rightarrow R$ at level c is the set of points

$$S = \{x|f(x) = c\}. \tag{3.14}$$

For a two-variable function, such as $f(x) : R^N \rightarrow R$, the level set is generally a curve. For a three-variable function, the level set is considered as a surface. For higher-dimensional functions, the level set is a hypersurface.

The gradient vector at x_0, $\nabla f(x_0)$ is orthogonal or normal to the level set corresponding to x_0. $\nabla f(x_0)$ is the direction of maximum rate of increase of $f(x)$ at x_0.

3.3 Conditions for the Minimum Point

According to Definitions 3.1 and 3.2, we can have the following necessary conditions for a local minimum. Although a local minimum usually means a weak local minimum, the theorem given below holds for both strong and weak local minima.

Theorem 3.3: First-Order Necessary Condition *Let* f(x) *have a continuous first derivative, that is,* $f(x) \in C^1$*. If* x^* *is a local minimum of* f(x)*, then*

$$\nabla f(x^*) = 0. \tag{3.15}$$

Proof For any $n \times 1$ unit vector d and a scalar ε, we can write the Taylor series expansion of $f(x)$ about x^*, as

$$\nabla f(x^* + \varepsilon d) = f(x^*) + \varepsilon d^{\mathrm{T}} \nabla f(x^*) + O(\varepsilon^2). \tag{3.16}$$

For ε is sufficiently small, it is clear that $f(x^*) \leq f(x^* + \varepsilon d)$, which implies

$$d^{\mathrm{T}} \nabla f(x^*) = 0, \quad \text{for all } d. \tag{3.17}$$

This in turn implies Eq. (3.14).

Example 3.2: First-Order Necessary Condition for Local Minimum Consider a continuous and convex function, $f(x_1, x_2) = x_1^2 + x_2^2$, as shown in Fig. 3.3. We can intuitively see that $x^* = [0, 0]^{\mathrm{T}}$ is the global minimum of the function. We can also have that $\nabla f(0,0) = 2x_1 + 2x_2|_{(0,0)} = 0$. This serves as a simple example of Theorem 3.3.

Theorem 3.4: Second-Order Necessary Condition *Let f(x) have a continuous second derivative, that is,* $f(x) \in C^2$. *If* x^* *is a weak local minimum of f(x), then* $\nabla f(x^*) = 0$, *and thus* $H(x^*)$ *is positive semidefinite, that is for all* $d \in R^N$,

$$d^{\mathrm{T}} H(x^*) d \geq 0. \tag{3.18}$$

Proof Since $\nabla f(x^*) = 0$, the Taylor series expansion of $f(x)$ about x^* reduces to

$$f(x^* + \varepsilon d) = f(x^*) + \frac{1}{2} \varepsilon^2 d^{\mathrm{T}} H(x^*) d + O(\varepsilon^3). \tag{3.19}$$

If there exists a vector, say $\hat{d}$, for which

$$\hat{d}^{\mathrm{T}} H(x^*) \hat{d} < 0, \tag{3.20}$$

we have that, for sufficiently small ε,

$$f(x^*) > f(x^* + \varepsilon \hat{d}), \tag{3.21}$$

which contradicts the assumption that x^* is a weak local minimum of $f(x)$. Therefore Eq. (3.18) must be satisfied. The necessary condition for a strong local minimum also holds if $H(x^*)$ is positive definite.

Example 3.3: Second-Order Necessary Condition for Local Minimum Consider the same function discussed in Example 3.2. We have already seen that $x^* = [0,0]^{\mathrm{T}}$ is the global minimum of the function and that $\nabla f(0,0) = 0$. The Hessian of $f(x)$ is calculated as

$$H(x) = \begin{bmatrix} 2 & 0 \\ 0 & 2 \end{bmatrix}, \tag{3.22}$$

which is clearly positive semidefinite. This is an example of the second-order necessary condition for a local minimum.

Theorem 3.5: Second-Order Sufficient Condition *Let f(x) have continuous second derivative, that is,* $f(x) \in C^2$. *If* $\nabla f(x^*) = 0$ *and* $H(x^*)$ *is positive definite, then* x^* *is a strong local minimum of* f(x).

Proof Since $f(x) \in C^2$, $H(x^*)$ is symmetric. Using the second assumption that $H(x^*)$ is positive definite and Rayleigh's inequality, it follows that if $d \neq 0$, then $0 < \lambda \|d\|^2 \leq d^{\mathrm{T}} H(x^*) d$, where λ represents the minimum eigenvalue of $H(x^*)$. From Taylor's theorem and the first assumption that $\nabla f(x^*) = 0$,

$$f(x^* + d) - f(x^*) = \frac{1}{2} d^{\mathrm{T}} H(x^*) d + o\left(\|d\|^2\right) \geq \frac{1}{2} \lambda_{\min} \|d\|^2 + o\left(\|d\|^2\right). \tag{3.23}$$

It follows that for all d, such that $\|d\|$ is sufficiently small, we have

$$f(x^* + d) > f(x^*), \tag{3.24}$$

and the proof is completed.

Corollary 3.1: Necessary and Sufficient Condition for a Local Minimum *Suppose that* $f(x) : R^N \rightarrow R$ *and* $f(x) \in C^2$. *The necessary and sufficient conditions for* x^* *to be a local minimum of* f(x) *is that* $\nabla f(x^*) = 0$ *and* $H(x^*)$ *is positive definite.*

Proof This corollary can be proved by combining Theorems 3.4 and 3.5. □

3.4 One-Dimensional Search

In this section we discuss the search methods that minimize a single-variable, unimodal function.[2] The minimization of a single-variable, unimodal function is called the one-dimensional search. Usually, our general goal is to minimize a

---

[2]A function is called unimodal if it has a single minimum point. A function is called multimodal if it has multiple local minima.

multivariable, multimodal function. However, the one-dimensional search (line search) plays an important role in solving such optimization problems. A straightforward method of minimizing a multivariable function is to find the minimum point along the direction of one variable, keeping the rest of the variables constant and repeating the same procedure for each variable. We should keep in mind that this is an approximated method for finding the minimum point.

3.4.1 Golden Section Search

Suppose a continuous, single-variable function $f(x)$ is defined on the interval $[a_0, b_0]$. If we assume that $f(x)$ is unimodal and the minimum point lies in the given interval, then we can find the minimum point by successively reducing the interval.

To narrow the interval, we need to evaluate the function at least at two points.[3] For a systematic approach, we choose two symmetric intermediate points, such as

$$a_1 - a_0 = b_0 - b_1 = \rho(b_0 - a_0), \text{ where } \rho < \frac{1}{2}, \text{ or equivalently } b_1 > a_1. \tag{3.25}$$

In the case when $f(a_1) < f(b_1)$, the minimum is in the interval $[a_0, b_1]$. Otherwise, the minimum is in the interval $[a_1, b_0]$. The first case is illustrated in Fig. 3.5.

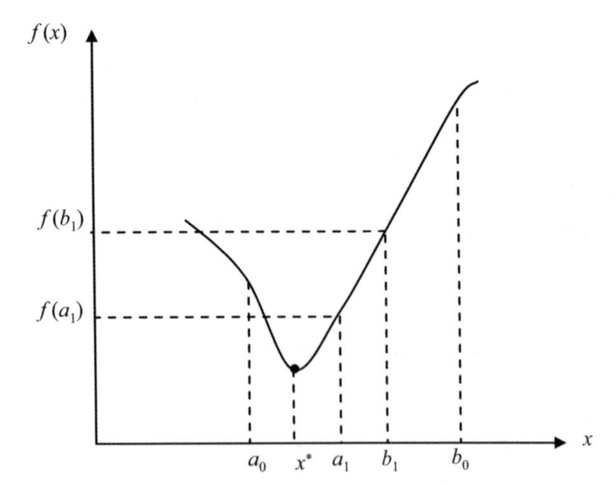

Fig. 3.5 A single-variable, unimodal function whose minimum is located in the interval $[a_0, b_1]$

[3]The bisection method, which is the similar approach for finding the solution of a nonlinear equation, requires only one function evaluation to reduce the interval by half.

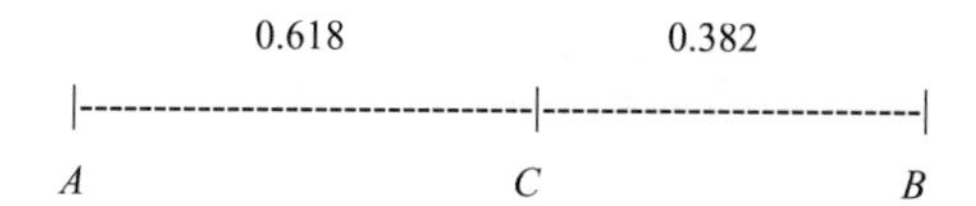

Fig. 3.6 A line segment divided by the golden section

To reduce the updated interval $[a_0, b_1]$ in the same manner, in concept, we need two new function evaluations at intermediate points, a_2 and b_2. Actually, the interval can be reduced to only one new function evaluation because we can utilize a_1 as one of the new intermediate points, for example, by making $b_2 = a_1$.

The following theorem gives the proper value of ρ, making possible the interval reduction using only one function evaluation.

Theorem 3.6: Golden Section *Consider the minimization problem of a single-variable, unimodal function in the given interval* $[a_0, b_0]$*. To reduce the search interval with one new function evaluation, the reduction ratio should be approximately equal to 0.618.*

Proof At first, the interval is reduced by comparing the function values at two intermediate points, which are defined as in Eq. (3.24). If $f(a_1) < f(b_1)$, the next reduction will be carried out by comparing the function values at a_2 and $b_2 = a_1$. Then the following relationship is satisfied:

$$(1-\rho)\rho = b_1 - a_1 = 1 - 2\rho, \quad \text{or } \rho^2 - 3\rho + 1 = 0. \tag{3.26}$$

Solving Eq. (3.26) and using the root for $\rho < \frac{1}{2}$, we have that $\rho \approx 0.382$. Therefore, the reduction ratio should be equal to $1 - \rho \approx 0.618$. □

Example 3.4: Golden Section Ratio Consider a line segment *AB* as shown in Fig. 3.6. If we divide it by two with the golden section ratio, we have that $AB : AC = AC : CB = 1 : 0.618$.

3.4.2 Newton's Method

For the purpose of minimization, derivative information is extremely important because the first and the second derivatives of the function can determine the change and the rate of change of the function value, respectively. Using this point of view, the major shortcoming of the golden section search method is that it does not utilize derivative information.

Newton's method has been devised for minimization by utilizing the first and the second derivatives of the function. Given a function of a single variable $f(x)$, we assume that at each measurement point x^k, we can calculate the function value, the first derivative, and the second derivative. We denote this by $f(x^k), f'(x^k)$, and $f''(x^k)$, respectively. Based on the idea that any valley of the function containing the minimum point at the bottom looks similar to a quadratic function, we can fit a

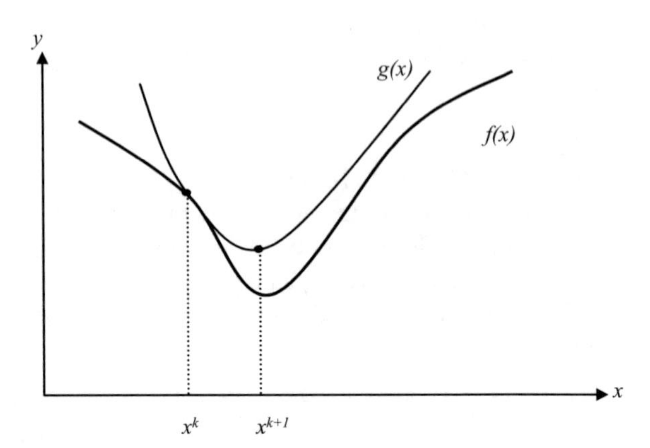

Fig. 3.7 Fitting a quadratic function at the measurement point x^k

quadratic function $g(x)$ through x^k that matches its first and second derivatives with that of the function $f(x)$, as shown in Fig. 3.7.

The constraint on the *fitting quadratic function* is that the function value, the first derivative, and the second derivative of the function should be the same as those of the given function $f(x)$ at x^k. This quadratic function then has the form

$$g(x) = f\left(x^k\right) + f'\left(x^k\right) + \left(x - x^k\right) + \frac{1}{2}f''\left(x^k\right)\left(x - x^k\right)^2. \tag{3.27}$$

Note that $g\left(x^k\right) = f\left(x^k\right)$, $g'\left(x^k\right) = f'\left(x^k\right)$, and $g''\left(x^k\right) = f''\left(x^k\right)$. We can easily minimize the quadratic function $g(x)$, instead of $f(x)$. The first-order necessary condition for a minimum of $g(x)$ is that $g'(x) = 0$, which yields

$$f'\left(x^k\right) + f''\left(x^k\right)\left(x - x^k\right) = 0. \tag{3.28}$$

Let the point that satisfies Eq. (3.25) be the next measurement point, x^{k+1}. We then utilize the iteration step of Newton's method as:

$$x^{k+1} = x^k - \frac{f'\left(x^k\right)}{f''\left(x^k\right)}. \tag{3.29}$$

The Newton's search algorithm is summarized in the following:

Algorithm 3.1: Newton's Method for the One-Dimensional Search

1. Set an initial guess as x^0 and $k \leftarrow 0$.
2. Evaluate $f'(x^k)$ and $f''(x^k)$.
3. Update the measurement point by $x^{k+1} \leftarrow x^k - \frac{f'\left(x^k\right)}{f''\left(x^k\right)}$.

4. If $|x^{k+1} - x^k|$ is smaller than a prespecified error limit, then stop the algorithm, and the current measurement point will be the estimate of the solution. Otherwise, $k \leftarrow k + 1$, and go to step 2.

Although Newton's method is very efficient in minimizing a single-variable function, it may not be used for the problem where the second derivative is not available. In that case, we can use an approximation for the second derivative. We call the approximated Newton's method the secant method, which will be described in the next subsection.

3.4.3 *Secant Method*

We may approximate the second derivative using the first derivative as

$$f''(x^k) = \frac{f'(x^k) - f'(x^{k-1})}{x^k - x^{k-1}}. \tag{3.30}$$

The algorithm for the secant method is summarized as follows:

Algorithm 3.2: Secant Method for the One-Dimensional Search

1. Set an initial guess as x^0 and $k \leftarrow 0$.
2. Evaluate $f'(x^k)$ and $f''(x^k)$.
3. Update the measurement point by $x^{k+1} \leftarrow \frac{f'(x^k)x^{k-1} - f'(x^{k-1})x^k}{f'(x^k) - f'(x^{k-1})}$.
4. If $|x^{k+1} - x^k|$ is smaller than a prespecified error limit, then stop the algorithm, and the current measurement point will be the estimate of the solution. Otherwise, $k \leftarrow k + 1$, and go to step 2.

In addition to the one-dimensional search methods described above, many variations can be found in [chong96, press92].

3.4.4 *Line Search*

Iterative algorithms for minimizing a function $f(x) : R^N \rightarrow R$ are generally of the form

$$x^{k+1} = x^k + \alpha_k s_k, \tag{3.31}$$

where x^k represents the current estimate of the solution, s_k represents the search direction or step vector, and α_k is chosen to minimize the function of α such as

$$\phi_k(\alpha) = f\left(x^k + \alpha s_k\right). \tag{3.32}$$

The one-dimensional search for the minimum of the function given in Eq. (3.31) is called the *line search.* In other words, the line search can be defined as the location of the minimum point on the line, which is given as the search direction.

References

[chong96] E.K.P. Chong, S.H. Zak, *An Introduction to Optimization* (Wiley, New York, 1996)
[jain89] A.K. Jain, *Fundamentals of Digital Image Processing* (Prentice-Hall, Upper Saddle River, 1989)
[murray72] W. Murray (ed.), *Numerical Methods for Unconstrained Optimization* (Academic Press, London, 1972)
[press92] W.H. Press, S.A. Teukolsky, W.T. Vetterling, B.R. Flannery, *Numerical Recipes in C*, 2nd edn. (Cambridge University Press, New York, 1992)

Additional References and Further Readings

[abd83] N.N. Abdelmalek, Restoration of images with missing high-frequency components using quadratic programming. Appl. Optics **22**(14), 2182–2188 (1983)
[abd80] N.N. Abdelmalek, T. Kasvand, Digital image restoration using quadratic programming. Appl. Optics **19**(19), 3407–3415 (1980)
[col90] T. Coleman, Y. Li, *Large Scale Numerical Optimization* (SIAM Books, Philadelphia, 1990)
[den83] J.E. Dennis, R.B. Schnabel, *Numerical Methods for Unconstrained Optimization and Nonlinear Equation* (Prentice Hall, Englewood Cliffs, 1983)
[du95] D.-Z. Du, P.M. Pardalos, *Minimax and Application* (Kluwer, Dordrecht, 1995)
[du94] D.-Z. Du, J. Sun, *Advances in Optimization and Approximation* (Kluwer, Dordrecht, 1994)
[duf67] R.J. Duffin, E. Peterson, C. Zener, *Geometric Programming* (Wiley, New York, 1967)
[fia] A.V. Fiacco, G.P. McCormick, Nonlinear Programming: *Sequential Unconstrained Minimization Techniques,* 1990, SIAM
[fle87] R. Fletcher, *Practical Methods of Optimization* (Wiley, New York, 1987)
[fl92] C.A. Floudas, P.M. Pardalos, *Recent Advances in Global Optimization* (Princeton University Press, Princeton, 1992)
[gill72] P.E. Gill, W. Murray, M.H. Wright, *Practical Optimization* (Academic Press, London, 1981)
[him72] D.M. Himmelblau, *Applied Nonlinear Programming* (McGraw-Hill, New York, 1972)
[hock81] W. Hock, K. Schittkowski, *Test Examples for Nonlinear Programming Code* (Springer, Berlin, 1981)
[hor95] R. Horst, P.M. Pardalos, *Handbook of Global Optimization* (Kluwer, Dordrecht, 1995)
[hor95] R. Horst, P.M. Pardalos, N.V. Thoai, *Introduction to Global Optimization* (Kluwer, Dordrecht, 1995)
[hor93] R. Horst, H. Tuy, *Global Optimization* (Springer, Berlin, 1993)
[lau94] H.T. Lau, *A Numerical Library in C for Scientists and Engineers* (CRC Press, Boca Raton, 1994)

[lue84] D.G. Luenberger, *Introduction to Linear and Nonlinear Programming* (Addison Wesley, Reading, 1984)

[torn89] A. Torn, A. Zilinskas, *Global Optimization* (Springer, Berlin, 1989)

[wis78] D.A. Wismer, R. Chattergy, *Introduction to Nonlinear Optimization* (North-Holland, New York, 1978)

[chong96] Edwin K.P. Chong, Stanislaw H. Zak, *An Introduction to Optimization* (Wiley, New York, 1996). (Already in Reference section)

[fl64] R. Fletcher, C.M. Reeves, Function minimization by conjugate gradients. Comp. J. **7**, 149–154 (1964)

[fl91] C.A. Floudas, P.M. Pardalos (eds.), *Recent Advances in Global Optimization*. Princeton Series in Computer Science (Princeton University Press, New Jersey, 1991)

[gill91] P.E. Gill, W. Murray, M.H. Wright, *Numerical Linear Algebra and Optimization*, vol 1 (Addison-Wesley, Redwood City, California, 1991)

[more84] J.J. Moré, D.C. Sorensen, Newton's method, in *Studies in Numerical Analysis*, ed. by G.H. Golub (Mathematical Association of America, Washington, D.C, 1984), pp. 29–82

[den89] J.E. Dennis, R.B. Schnabel, A view of unconstrained optimization, in *Optimization*, ed. by G.L. Nemhauser, A.H.G. Rinnooy Kan, M.J. Todd (North-Holland, Amsterdam, 1989), pp. 1–72

[col93] T.F. Coleman, Large scale numerical optimization: introduction and overview, in *Encyclopedia of Computer Science and Technology*, ed. by J. Williams, A. Kent (Marcel Dekker, New York, 1993), pp. 167–195

Chapter 4
Unconstrained Optimization

Abstract Most practical optimization problems arise with constraints on the solutions. Nevertheless, unconstrained optimization techniques serve as a major tool in finding solutions for both unconstrained and constrained optimization problems. In this chapter we present techniques for solving the unconstrained optimization problem. In Chap. 5 we will see how the unconstrained solution methods aid in finding the solutions to constrained optimization problems.

In order to find the solution of a given unconstrained optimization problem, a general procedure includes an initial solution estimate (guess) followed by iterative updates of the solution directed toward minimizing the objective function. If we consider the solution of an N-dimensional objective function as an N-vector, iterative updates can be performed by adding a vector to the current solution vector. We call the vector augmented to the current solution vector the *step vector* or simply the *step*.

Various unconstrained optimization methods can be classified by the method of determining step vectors. We place direct search methods, which are the simplest, in group one. In this group, the step vectors are randomly determined. The second group is known as derivative-based methods. In this case, the step vectors are determined based on the derivative of the objective function. Gradient methods and Newton's methods fall into this category.

In practice, more sophisticated methods, such as conjugate gradient and quasi-Newton methods, have shown comparable performance with greatly reduced computational cost and memory space. A brief discussion of the conjugate gradient algorithm will also be included in this chapter.

4.1 Direct Search Methods

If an objective function is very simple, for example, a second-order polynomial with a single variable, we can easily find the minimum or maximum at the point where the second-order derivative becomes zero. However, almost all numerical optimization problems have very complicated objective functions with a large

M.A. Abidi et al., *Optimization Techniques in Computer Vision*, Advances in Computer Vision and Pattern Recognition, DOI 10.1007/978-3-319-46364-3_4

number of variables. This means that the objective function may not be unimodal,[1] its derivative does not exist, or it is impossible or extremely difficult to compute its derivatives. For such complicated objective functions, there is no simple way to estimate the solution.

A straightforward approach for solving a minimization problem is to guess an arbitrary initial estimate of the solution and keep replacing it by a new estimate that gives the smaller functional value. Direct search methods are different from other methods by their strategies for producing the series of improved approximations. More specifically, direct search methods use the comparison of the objective function only, while others may use derivatives or their approximations [murray72].

For example, the Metropolis algorithm is a stochastic optimization algorithm which is based on the direct search strategy [metropolis53]. At each step of the Metropolis algorithm, a new estimate is generated at random. If the new solution decreases the objective function, it replaces the old one. If the new solution increases the objective function, it may or may not replace the old one, depending to the prespecified probability. If the probability is always equal to unity, the algorithm becomes equivalent to a deterministic direct search method. But the probability that it is less than unity keeps the solution from staying at a local minimum.

Some direct search methods are extremely efficient in practice, particularly in applications where the objective function is non-differentiable, has discontinuous first derivatives, or is subject to random error. Most of them, however, have been developed in a heuristic manner. Swann gives a good review of direct search methods in Chap. 2 of [murray72]. Here we will summarize several direct search methods.

4.1.1 Random Search Methods

Suppose the desired solution for an N-dimensional objective function is known to be within a finite area defined by upper and lower bounds on each independent variables, such that

$$l_i \leq x_i{}^* \leq u_i, \quad \text{for } i = 1, \ldots, N. \tag{4.1}$$

An illustration of the random search method for a two-dimensional objective function is shown in Fig. 4.1. Among four randomly generated points, $x^{(3)}$ gives the smallest function value. Although this method is extremely simple to implement, we need an extremely large number of function evaluations at random points.

---

[1] A function is called unimodal if it has a single minimum point. A function is called multimodal if it has multiple local minima.

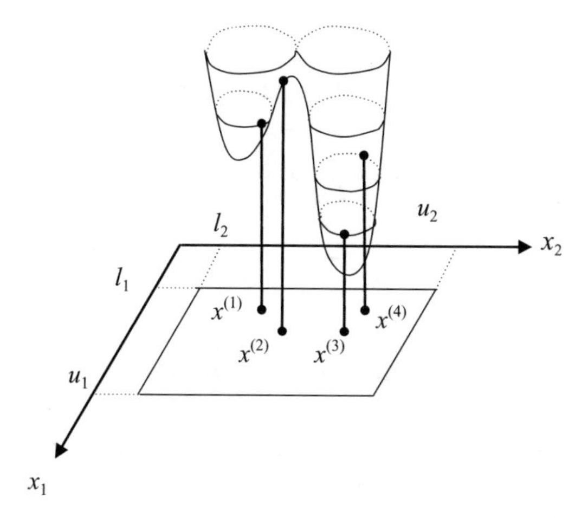

Fig. 4.1 A set of randomly chosen points for optimization by the random search method

In order to reduce the number of function evaluations, variations of this method might include (1) selection of reduced area of search and (2) the use of a grid over the area of interest and an evaluation of the function at each node of the grid.

The principal disadvantage of random search-based methods is that they do not utilize updated information of each new trial point. They depend on the predetermined strategy for a search, and because of this, they are inherently less efficient than other sophisticated methods that are described in later sections.

If we know the objective function is unimodal, a more systematic method for efficiently locating the minimum is the *generalized Fibonacci search*. In a one-dimensional case, suppose the minimum point lies in the given interval, $a \leq x^* \leq b$. Function values are evaluated at $x = c$ and $x = d$, where $a < c < d < b$. If $f(c) < f(d)$, we can conclude that the minimum point lies in the new, reduced interval, $a \leq x^* \leq d$. The procedure is repeated until the interval has been sufficiently reduced. In each successively reduced interval, we need to evaluate the function at one point only because the function value at the other point has already been evaluated in the previous interval.

The Fibonacci search method can be extended to deal with multivariable problems by combining multiple one-variable steps. Although the Fibonacci search method can considerably reduce the number of function evaluations compared to the random search method, its application is practically limited to small-dimensioned problems with well-shaped minimum.[2]

In order to solve a multivariable optimization problem, we can search for the minimum point along the direction of one variable while keeping the rest of them constant, and the same procedure will be repeated for each variable. We call this approach the *alternating variable method*. The multivariable Fibonacci search method discussed above can be used for this purpose. The alternating variable

[2]The function is said to have a well-shaped minimum if it has a sharp, narrow valley in the neighborhood of the minimum point.

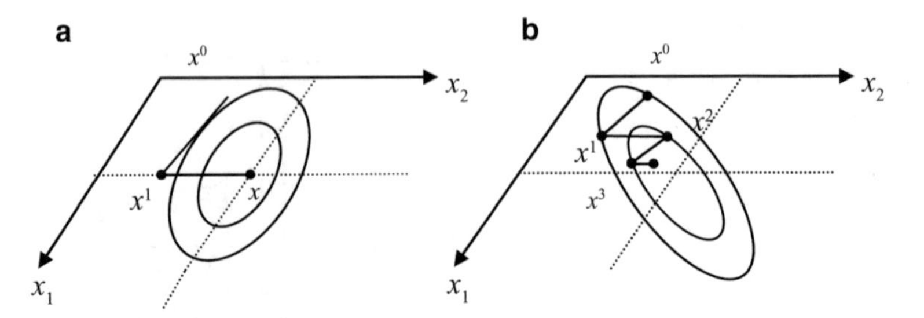

Fig. 4.2 Alternating variable method for two-variable functions of elliptic with the axes: (**a**) parallel to the coordinate axes and (**b**) skew with the coordinate axes

method is particularly efficient for functions whose contours are hyperspherical or elliptical along the axis parallel to the coordinate directions because the search process is made parallel to each coordinate direction in turn. On the other hand, if the function contours are skew with the coordinate axes, convergence of the alternating variable method is very slow. This is true because the minimum point in one coordinate direction changes each time with the minimum point in other coordinate directions. Figure 4.2 illustrates the performance of the alternating variable method for two different function contours. In this case, we assume that the proper line search method is performed at each directional search.

4.1.2 *Pattern Search Contours*

Although it is efficient to move the solution along the local principal axes of the contours of the function, the most simple and regular way is to move along the coordinate axes. We define the *exploratory move* as moving in each coordinate direction and the *pattern move* as any combination of previous exploratory moves for improved search directions. The pattern search method combines exploratory moves and pattern moves to take advantage of both moves.

In an exploratory move, a single step is taken along one coordinate direction and is considered successful if the function value has not been increased. The same procedure repeats for the remaining coordinate directions. Each estimated point obtained by the exploratory move is termed a base point, denoted by x^1, x^2, x^3, and so on.

Following one complete set of exploratory moves, one pattern move is used in order to find an improved search direction that cannot be obtained by any single exploratory move. The pattern search method for minimizing a function $f(x) : R^N \rightarrow R$ is summarized in Algorithm 4.1.

Algorithm 4.1: Pattern Search

1. Set an initial guess as x^0 and $k \leftarrow 0$.
2. Make $x^{k+1} = x^k$.

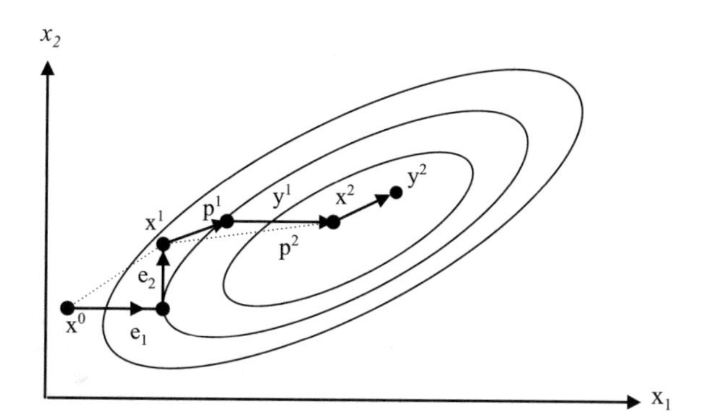

Fig. 4.3 Illustration of the pattern search algorithm for a two-dimensional function

3. Initiate one complete set of exploratory moves as follows: For $i = 1$ to N, for $x^{\text{temp}} = x^{k+1} + \alpha e_i$.
   If $f(x^{\text{temp}}) \leq f(x^{k+1})$, then $x^{k+1} = x^{\text{temp}}$.
   End for loop
4. If $\|x^{k+1} - x^k\|$ is smaller than a prespecified error limit, then stop the algorithm. This means x^{k+1} becomes the estimated solution. Otherwise, $k \leftarrow k+1$, and we proceed to the next step.
5. Make $p^k = x^k - x^{k-1}$ the point after the kth pattern move.
6. Set $y^k = x^k + p^k$, $x^k \leftarrow y^k$, and return to step 2.

The N-dimensional vector e_i represents the ith unit vector, and α represents a scalar value determining the length of the move. Usually, $\alpha = 1$ is used.

The performance of the pattern search algorithm on a function of two-variables is illustrated in Fig. 4.3. The initial estimate x^0 has been chosen arbitrarily. x^1 represents the point after the first turn of exploratory moves, and y^1 represents the point after the first pattern move. In the second turn of exploratory moves, the first move by e_1 is successful, but the second move by e_2 fails because the function value increases. x^2 represents the point after the second turn of exploratory moves.

4.1.3 Geometrical Search

The most successful direct search method is the simplex method. Although the name originated from the simplex method of linear programming, it should not be confused with the linear programming method.

A regular simplex is defined as a set of $N + 1$ points in R^N. If, for example, $N = 2$, the corresponding simplex is a set of three vertices of an equilateral triangle. If $N = 3$, the simplex is a set of four vertices of a tetrahedron, as shown in Fig. 4.4.

The simplest simplex method is now described for the case of $N = 2$. Given a two-variable objective function and an initial estimate, the simplex method begins by determining two other vertices, which form a regular simplex on R^2 together

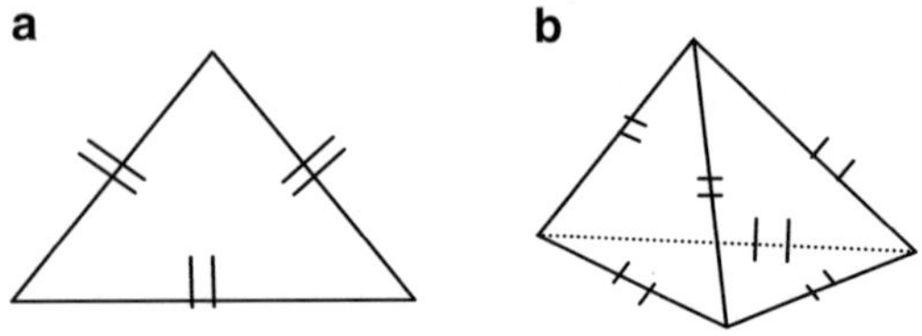

Fig. 4.4 Regular simplexes: (a) $N = 2$ and (b) $N = 3$

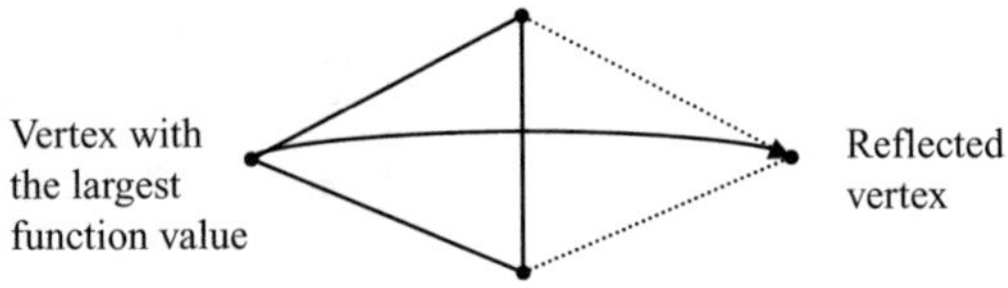

Fig. 4.5 Reflection of a simplex in R^2

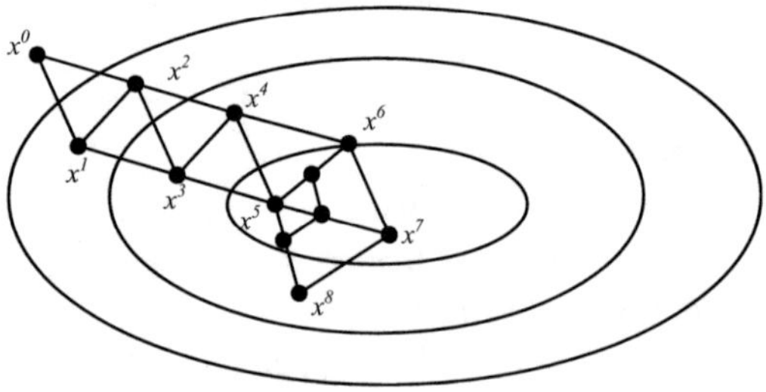

Fig. 4.6 The simplex method in two variables

with the initial estimate. On the first iteration of the simplex method, the function values of all three vertices are calculated, and the vertex with the largest function value is determined. This vertex is then replaced by its mirror image in the centroid of the remaining two vertices. We call this process the *reflection*, as shown in Fig. 4.5.

After the first iteration, only the function value of the new vertex is calculated and the procedure is repeated. The performance of the simplex algorithm for a two-variable function is illustrated in Fig. 4.6. The initial estimate is denoted by x^0. x^1 and x^2 are determined to form a simplex together with x^0. If we assume that $f(x^0)$ is larger than $f(x^1)$ and $f(x^2)$, vertex x^0 is reflected to x^3. The same procedure is repeated until vertex x^8 is obtained. If we further apply this iteration to simplex $\{x^5, x^7, x^8\}$, vertex x^8 is reflected to vertex x^6 again. Without an additional rule, the simplex method oscillates between vertices x^6 and x^8. Spendly et al. suggested that if a certain vertex has been in the current simplex for more than a fixed number of M iterations, the simplex should be contracted by replacing the other vertices with new ones that are halfway along the edge to the vertex. The reduced simplexes are shown in the original-size simplexes $\{x^5, x^7, x^8\}$ and $\{x^5, x^6, x^7\}$ [fletcher80]. They also deduced that the maximum expected age of any vertex could be approximately

$$M = 1.65N + 0.05N^2. \tag{4.2}$$

Given an initial estimate x^0, the simplex algorithm for minimizing an N-variable function is summarized in the following.

Algorithm 4.2: The Simplex Method

1. $x_0^0 \leftarrow x^0$, and $k \leftarrow 0$.
2. For $i = 1$ to N

$$x_i^0 = x_0^0 + \delta_i, \quad \text{where}$$

$$\delta_i = \begin{bmatrix} \delta_1 & \delta_1 & \dots & \delta_1 & \delta_2 & \delta_1 & \dots & \delta_1 \end{bmatrix}^{\mathrm{T}},$$
$$\quad \text{1st} \quad \text{2nd} \qquad\qquad i\text{-th} \qquad\qquad N\text{-th}$$

$$\delta_1 = \frac{(N+1)^{1/2} + N - 1}{N\sqrt{2}}, \text{ and } \delta_1 = \frac{(N+1)^{1/2} - 1}{N\sqrt{2}}.$$

3. Evaluate $N + 1$ function values at $\{x_0^0, \dots, x_N^0\}$, and determine the point, say x_j^0, which has the largest function value.
4. $x_j^{k+1} = \frac{2}{N}\left(x_0^k + x_1^k + \dots + x_{j-1}^k + x_{j+1}^k + \dots + x_N^k\right) - x_j^k$.
   For $i = 0$ to N (except $i = j$), $x_i^{k+1} = x_i^k$.
5. If the distance between any two vertices is smaller than a prespecified value, stop the algorithms, and x_j^{k+1} is the estimated solution.
6. Given N old function values $\left\{x_0^{k+1}, \dots, x_{j-1}^{k+1}, x_{j+1}^{k+1}, \dots, x_N^{k+1}\right\}$ and the newly evaluated $f\left(x_j^{k+1}\right)$, determine the point that has the largest function value, and, without loss of generality, consider it as x_j^{k+1}.
7. If x_j^{k+1} has been in the current simplex more than M times, shrink the simplex by half. Otherwise $k \leftarrow k + 1$, and go to step 4.

4.2 Derivative-Based Methods

The minimum point of an unconstrained optimization problem is determined by successively improving the estimated solution. To improve the solution point, we need to search for an appropriate step vector which moves the current estimated point so that the function value is reduced. In the direct search method, the improved estimated point is achieved based only on function values.

It is, however, more efficient to utilize the derivative of the function, in addition to the function value, because the first-order derivative (gradient) indicates the direction of change in the function value at the given point.

In the following two subsections, first-order derivative-based methods, referred to as the *gradient descent method* and the *steepest descent method*, will be described. Gradient-based methods give more efficient search directions than direct search methods because they use both function values and derivatives.

To refine (improve) the search direction, we can also utilize the second-order derivative. The corresponding minimization method is referred to as Newton's method and is described at the end of this section.

4.2.1 Gradient Descent Methods

In direct search methods, an improved solution is obtained by selecting a new point in an ad hoc way and comparing the function value with that of the current solution. Because the temporary solution may or may not decrease the function value, selection processes of direct search methods are often inefficient. In order to make the selection process more efficient, we need to utilize additional information as well as the function value.

If an objective function $f(x) : R^N \to R$ is differentiable with respect to all N variables at the current solution point x, we can calculate its gradient $\nabla f(x)$. The following theorem states that the gradient is the direction of maximum increase in the function value.

Theorem 4.1: Direction of Gradient *The function f(x) increases more in the direction of* $\nabla f(x)$ *than in any other direction.*

Proof Consider an arbitrary unit vector d, such as $\|d\| = 1$. We know that, by the Cauchy-Schwarz inequality, the inner product of an arbitrary vector with d cannot be greater than the norm of the vector. In addition, the maximum of the inner product occurs when two vectors are in the same direction. By applying these results to the rate of change of $f(x)$ at x in the direction d, we have

$$\nabla f(x)^{\mathrm{T}} d \leq \|\nabla f(x)\|, \tag{4.3}$$

where the equality holds when $d = \alpha \nabla f(x)$, for $\alpha = 1/\|\nabla f(x)\|$. Thus, the direction of $\nabla f(x)$ is the direction of the maximum rate of increase of $f(x)$ at x.

Based on Theorem 4.1, we can say the gradient vector $\nabla f(x)$ is the direction of maximum rate of increase of $f(x)$ at x. The negative gradient vector $-\nabla f(x)$ is the direction of maximum rate of decrease of $f(x)$ at x. The following theorem summarizes the functional decrease in the gradient direction.

Theorem 4.2: Functional Decrease in the Gradient Direction *Let* x^0 *be an initial point. Then a new point,* $x^0 - \alpha \nabla f(x^0)$*, always decreases the function value.*

Proof We obtain from Taylor's formula

$$f\left(x^0 - \alpha \nabla f\left(x^0\right)\right) = f\left(x^0\right) - \alpha \left\|\nabla f\left(x^0\right)\right\|^2 + O\left(\alpha^2\right). \tag{4.4}$$

For a nonzero $\nabla f(x^0)$ and a sufficiently small $\alpha > 0$, we have

$$f\left(x^0 - \alpha \nabla f\left(x^0\right)\right) < f\left(x^0\right). \tag{4.5}$$

□

In minimizing a function, it is reasonable to improve the estimated point along the gradient direction, based on Theorem 4.2. We refer to the following as a *gradient descent algorithm*.

Algorithm 4.3: Gradient Descent Algorithm

1. Set an initial guess as x^0 and $k \leftarrow 0$.
2. $x^{k+1} = x^k - \alpha_k \nabla f\left(x^k\right)$.
3. If $\left\|x^{k+1} - x^k\right\|$ is smaller than a prespecified error limit, terminate the algorithm with the converged solution x^{k+1}. Otherwise, $k \leftarrow k + 1$, and go to step 2.

The following theorem provides sufficient conditions for the gradient descent algorithm to converge to a local minimum of $f(x)$.

Theorem 4.3: Convergence of the Gradient Descent Algorithm *If* $\{x^k\}$ *converges to a point, denoted by* x^**, and if* $\nabla f(x)$ *is continuous in the neighborhood of* x^**, then* $f(\mathrm{x})$ *has a local minimum at* $x = x^*$.

Proof Assume that $f(x^*)$ is not a local minimum and then we have that $\nabla f(x^*) \neq 0$. There exists $\mu > 0$, such that

$$\nabla f(x)^{\mathrm{T}} \ \nabla f(y) \geq \mu, \quad for \ x, y \in S, \tag{4.6}$$

where S represents some neighborhood of x^*. There also exists $a > 0$, such that

$$x^k, x^{k+1} \in S, \quad \text{where} \tag{4.7}$$

$$x^{k+1} = x^k - \alpha \nabla f\left(x^k\right). \tag{4.8}$$

Applying the first-order mean value theorem to Eq. (4.8) yields

$$f\left(x^{k+1}\right) = f\left(x^k\right) - \alpha \nabla f\left(x^k\right) \nabla f(y), \quad \text{where } y = x^k - \theta\alpha\nabla f\left(x^k\right), \quad 0 \leq \theta \leq 1. \tag{4.9}$$

Equation (4.9) can be rewritten as

$$f\left(x^{k+1}\right) - f\left(x^k\right) = -\alpha\mu, \tag{4.10}$$

which means that $f(x^*)$ approaches $-\infty$ as the iteration continues. This statement contradicts the fact that $\nabla f(x)$ is continuous in the neighborhood of x^*. The original assumption that $f(x^*)$ is not a local minimum is therefore false, and the theorem is proved.

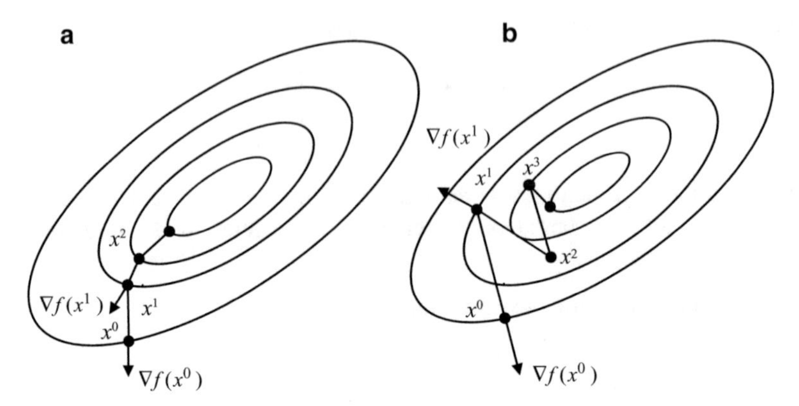

Fig. 4.7 Gradient descent algorithms with (**a**) small step size and (**b**) large step size

Variations of the gradient algorithm stem from how one determines the α_k at each iteration. If α_k is small, a great number of iterations are required for convergence, and the corresponding algorithm becomes very time-consuming. On the other hand, if α_k is large, the set of estimated solution points results in a zigzag pattern, which means convergence with oscillation, at best, or divergence will occur, at worst. The choice of the step size α_k is a compromise between accuracy and efficiency. The gradient descent algorithms, with two different step sizes, are illustrated in Fig. 4.7.

4.2.2 *Steepest Descent Method*

The steepest descent method is a special gradient descent method for which the step size α_k is chosen to achieve the maximum amount of decrease of the function at each iteration step. To achieve the maximum decrease along the given gradient direction, we perform the line search

$$\alpha_k = \arg\min_{\alpha \geq 0} f\left(x^k - \alpha \nabla f\left(x^k\right)\right). \tag{4.11}$$

The steepest descent method is summarized by the following algorithm.

Algorithm 4.4: Steepest Descent Algorithm

1. Set an initial guess as x^0 and $k \leftarrow 0$.
2. Evaluate $\nabla f\left(x^k\right)$, and determine the optimal step size by the line search as

$$\alpha_k = \arg\min_{\alpha \geq 0} f\left(x^k - \alpha \nabla f\left(x^k\right)\right).$$

3. $x^{k+1} = x^k - \alpha_k \nabla f\left(x^k\right)$.

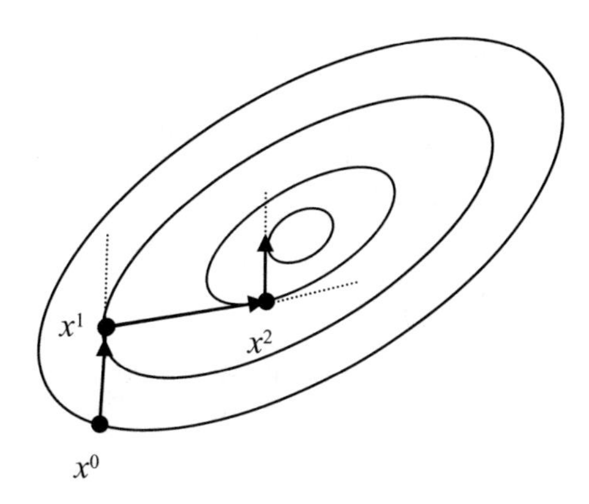

Fig. 4.8 Illustration of the steepest descent algorithm

4. If $\left\|x^{k+1} - x^k\right\|$ is smaller than a prespecified error limit, terminate the algorithm with the converged solution x^{k+1}. Otherwise, $k \leftarrow k + 1$, and go to step 2.

Figure 4.8 illustrates a typical sequence resulting from the steepest descent algorithm.

We consider convergence analysis by observing two ellipsoidal functions given in the following example.

Example 4.1: Convergence of Two Different Functions Consider two quadratic functions

$$f_1(x_1, x_2) = x_1^2 + x_2^2 \tag{4.12}$$

and

$$f_2(x_1, x_2) = \frac{1}{5}x_1^2 + x_2^2. \tag{4.13}$$

Because $f_1(x)$ has circular level sets, as shown in Fig. 4.9a, any initial point can reach the minimum point at once by the gradient vector with the line search. On the other hand, the steepest descent algorithm for minimizing $f_2(x)$ results in repeating orthogonal search directions, as shown in Fig. 4.9b. This is caused by a skew of the elliptic level set. In this case, as the solution approaches the minimum point, the step size becomes smaller as a result of the line search. This causes slow convergence of the steepest descent algorithm.

We now consider the steepest descent method for a quadratic function. Given a linear equation of the form

$$Ax = b, \tag{4.14}$$

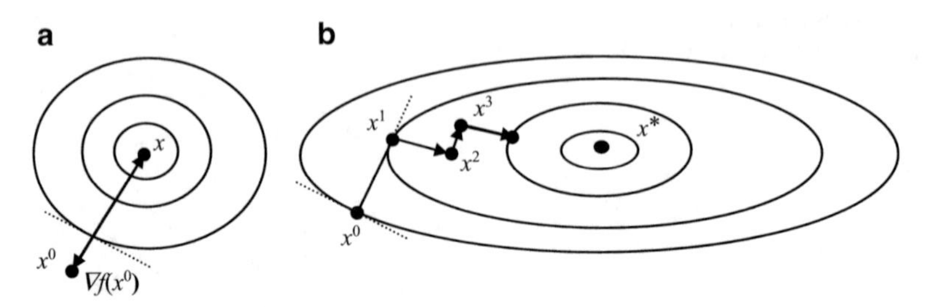

Fig. 4.9 Performance of the steepest descent method for two different quadratic functions: (**a**) $f_1(x_1, x_2) = {x_1}^2 + {x_2}^2$ and (**b**) $f_2(x_1, x_2) = (1/5){x_1}^2 + {x_2}^2$

where $A \in R^{N \times N}$ is a symmetric positive definite matrix and $x, y \in R^N$. The solution of Eq. (4.14) can be achieved by minimizing the quadratic function of the form

$$f(x) = \frac{1}{2} x^{\mathrm{T}} A x - b^{\mathrm{T}} x. \tag{4.15}$$

To verify that the solution of Eq. (4.15) is equal to that of Eq. (4.14), we set the gradient of Eq. (4.14) to zero, that is,

$$\nabla f(x) = Ax - b = 0, \tag{4.16}$$

yielding Eq. (4.14). This equivalence can be generalized to a nonsymmetric linear equation with $B \neq B^{\mathrm{T}}$ because, for any B, $A = B + B^{\mathrm{T}}$ is symmetric and $x^{\mathrm{T}} B x = \frac{1}{2} x^{\mathrm{T}} A x$.

To find the optimal α_k in step 2 of Algorithm 4.4, we need to minimize the following one-dimensional function with respect to α.

$$\phi_k(a) = f\left(x^k - \alpha \nabla f\left(x^k\right)\right). \tag{4.17}$$

Setting the derivative of Eq. (4.17) to zero, we have

$$\frac{d}{d\alpha} \phi_k(a) = \left(x^k - \alpha \nabla f\left(x^k\right)\right)^{\mathrm{T}} A\left(-\nabla f\left(x^k\right)\right) - y^{\mathrm{T}}\left(-\nabla f\left(x^k\right)\right) = 0. \tag{4.18}$$

Since $x^{k^{\mathrm{T}}} A - y^{\mathrm{T}} = \nabla f\left(x^k\right)^{\mathrm{T}}$, the optimal α_k is determined as

$$\alpha_k = \frac{\nabla f\left(x^k\right)^{\mathrm{T}} \nabla f\left(x^k\right)}{\nabla f\left(x^k\right)^{\mathrm{T}} A \nabla f\left(x^k\right)}. \tag{4.19}$$

The steepest descent algorithm for a quadratic function can then be summarized in the following steps.

Algorithm 4.5: Steepest Descent Algorithm for a Quadratic Function

1. Set an initial guess as x^0 and $k \leftarrow 0$.
2. $x^{k+1} = x^k - \left(\frac{\nabla f(x^k)^{\mathrm{T}} \nabla f(x^k)}{\nabla f(x^k)^{\mathrm{T}} A \nabla f(x^k)}\right) \nabla f(x^k)$.
3. If $\left\| x^{k+1} - x^k \right\|$ is smaller than a prespecified error limit, terminate the algorithm with the converged solution x^{k+1}. Otherwise, $k \leftarrow k+1$, and go to step 2.

The steepest descent algorithm can be used for image restoration as shown in the following example.

Example 4.2: Iterative Image Restoration As described in Sect. 2.1, most cases of image degradation can be represented by the linear equation,

$$y = Dx + \eta, \tag{4.20}$$

where y, x, and η, respectively, represent the observed, the original, and noise images. D represents the degradation matrix.

Because restoration of the original image x by using Eq. (4.20) is an ill-posed problem, we may instead minimize the following functional:

$$f(x) = \frac{1}{2} x^{\mathrm{T}} T x - b^{\mathrm{T}} x, \tag{4.21}$$

where $T = D^{\mathrm{T}} D + \lambda C^{\mathrm{T}} C, b = D^{\mathrm{T}} y$, C represents a high-pass filter, and λ represents the regularization parameter which controls the fidelity to the original image and smoothness of the restored solution.

According to Algorithm 4.5, we use $D^{\mathrm{T}} y$ as the initial guess for x^0 and perform iterations of the form

$$x^{k+1} = x^k - \left(\frac{r_k^{\mathrm{T}} r_k}{r_k^{\mathrm{T}} T r_k}\right) r^k, \quad \text{where } r_k = \nabla f(x^k) = b - Tx^k. \tag{4.22}$$

Figure 4.10 shows the simulation results of iterative image restoration. Figure 4.10a represents the 256×256 digital image with 8 bit gray scale. Figure 4.10b is the simulated degraded image by using a 7×7 uniform blur with 40 dB of additive noise. Figure 4.10c, d is iterative image restoration results after 10 and 100 iterations [moon96].

4.3 Newton's Method

Newton's method for minimizing an N-variable function, $f(x) : R^N \rightarrow R$, utilizes a straightforward extension of the single-variable Newton's method described in Sect. 3.4. We can state that the simplest function having a strong minimum is a

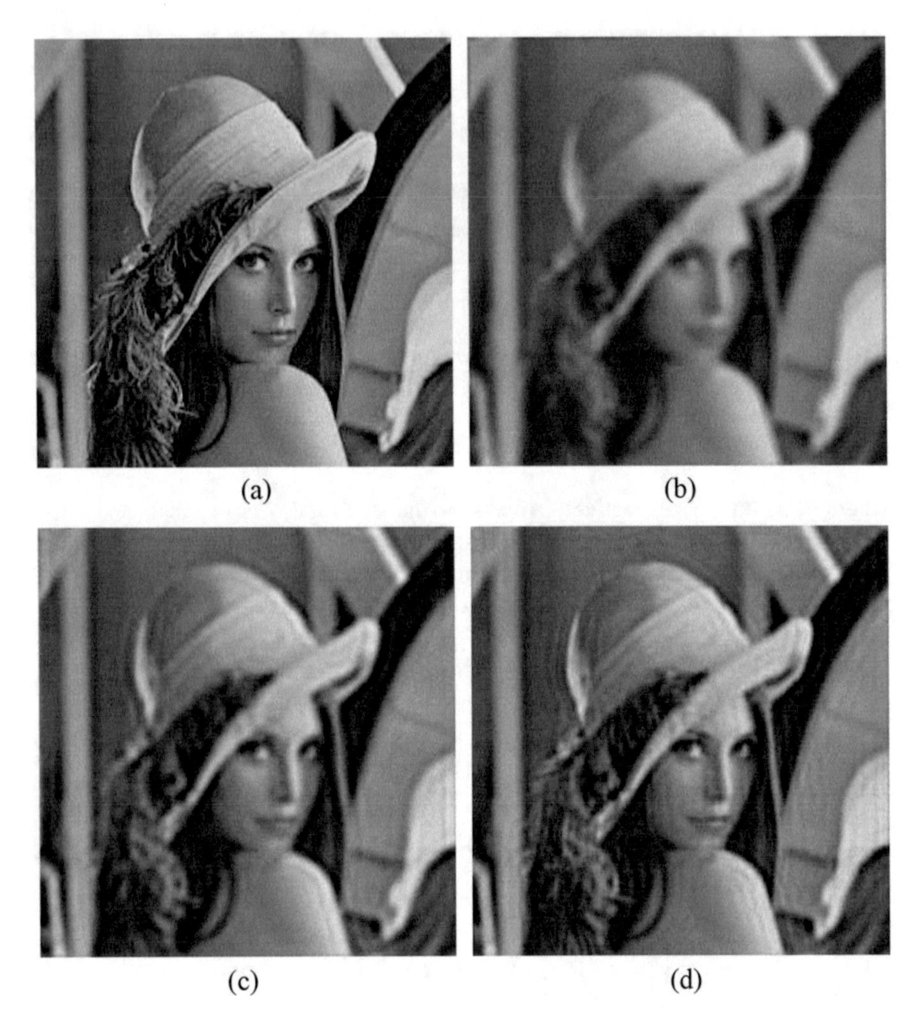

Fig. 4.10 Simulation results of image restoration (in the raster scanning order): (**a**) the original 256×256, 8 bit gray-level image, (**b**) the simulated degraded image by a 7×7 uniform blur and 40 dB additive noise, (**c**) the restored image after 10 iterations, and (**d**) the restored image after 100 iterations

quadratic function with a positive definite Hessian matrix. The basic idea of Newton's method is that any valley of the function containing the minimum point at the bottom looks similar to a quadratic function and that the minimum point of the quadratic function replaces the current estimation.

More rigorously, given an estimated solution point x_k, we assume that the function value $f(x^k)$, the gradient vector $\nabla f(x^k)$, and the Hessian matrix $H(x^k)$ can be evaluated. We can then find a quadratic function fitting those three quantities at x^k, such as

$$g(x) = f(x^k) + \nabla f(x^k)(x - x^k) + \frac{1}{2}(x - x^k)^{\mathrm{T}} H(x^k)(x - x^k). \tag{4.23}$$

Note that the objective function $f(x)$ and the quadratic function $g(x)$ have the same function value, the gradient vector, and the Hessian matrix at x^k.

For $g(x)$ to have the minimum at $x = x^m$, the first-order necessary condition should be satisfied as

$$\nabla g(x)|_{x=x^m} = \nabla f(x^k) + H(x^k)(x^m - x^k) = 0, \tag{4.24}$$

which yields

$$x^m = x^k - H(x^k)^{-1} \nabla f(x^k). \tag{4.25}$$

Because x^m is the minimum point of the quadratic function $g(x)$, it should replace the current estimate for the minimum of the given function $f(x)$. This minimization algorithm is termed *Newton's algorithm* and can be summarized as follows.

Algorithm 4.6: Newton's Algorithm

1. Set an initial guess as x^0 and $k \leftarrow 0$.
2. Solve $H(x^k)d^k = -\nabla f(x^k)$ for d^k.
3. $x^{k+1} = x^k + d^k$.
4. If $\|x^{k+1} - x^k\|$ is smaller than a prespecified error limit, terminate the algorithm with the converged solution x^{k+1}. Otherwise, $k \leftarrow k + 1$, and go to step 2.

Since step 2 requires the solution of an $N \times N$ system of linear equations, an efficient method for solving systems of linear equations is necessary for running Newton's algorithm.

Example 4.3 We use the Newton's method to find the minimizer of

$$f(x_1, x_2, x_3) = (x_1 - 4)^2 + (x_2 - 3)^2 + 4(x_3 + 5)^2. \tag{4.26}$$

Use the starting point $x^0 = [0, 0, 0]^{\mathrm{T}}$. Note that $f(x^0) = 125$. We have that

$$\nabla f(x) = [2x_1 - 8, 2x_2 - 6, 8x_3 + 40]^{\mathrm{T}} \quad \text{and} \tag{4.27}$$

$$H(x) = \begin{bmatrix} 2 & 0 & 0 \\ 0 & 2 & 0 \\ 0 & 0 & 8 \end{bmatrix}. \tag{4.28}$$

To perform the first iteration, we calculate $\nabla f(x^0) = [-8, -6, 40]^{\mathrm{T}}$ and $H(x^0) = H(x)$. By solving

$$\begin{bmatrix} 2 & 0 & 0 \\ 0 & 2 & 0 \\ 0 & 0 & 8 \end{bmatrix} d^0 = -\nabla f\left(x^0\right) = \begin{bmatrix} 8 \\ 6 \\ -40 \end{bmatrix}, \tag{4.29}$$

we obtain $d^0 = [4, 3, -5]^{\mathrm{T}}$ and $x^1 = x^0 + d^0 = [4, 3, -5]^{\mathrm{T}}$. Note that $f(x^1) = 0$, which is the minimum of the given function.

Because the minimizer of the function in Eq. (4.26) coincides with that of the fitting quadratic function of Newton's algorithm, only one iteration makes the solution converge to the minimum point. Also, the first Newton's direction vector d^0 was easily calculated because the Hessian matrix is diagonal. If, in general, the Hessian matrix is nondiagonal, we need to solve Eq. (4.29) using an efficient linear equation solving method.

Newton's algorithm is guaranteed to converge only when the Hessian matrix is positive definite. The single-variable counterpart of this statement is shown in Fig. 4.11.

Note that the second-order derivative of a single-variable function $f''(x)$ is equivalent to the Hessian of a multiple-variable function $H(x)$. Based on this observation, every update by Newton's direction does not guarantee a decrease in the function value. Moreover, even if the Hessian is positive definite, Newton's method may not be the best method for finding the minimum. This may occur if the starting point x^0 is far away from the solution.

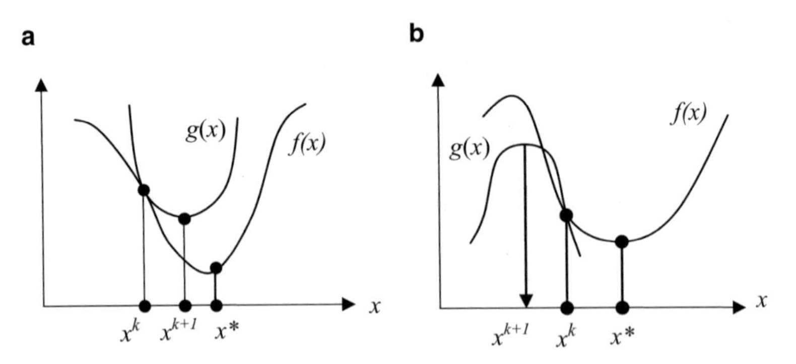

Fig. 4.11 Single-variable Newton's algorithms with (**a**) $f''(x^k) > 0$ and (b) $f''(x^k) < 0$

Possible remedies of this problem include (1) scaling each Newton direction by the result of a line search and (2) modifying the given objective function so that its Hessian is always positive definite. The first method guarantees the descent direction at each iteration with computational overhead for the line search. On the other hand, the second method results in a slightly different minimum point at the cost of the function modification. One example of modifying the objective function is regularization, which will be discussed in Part III of this book.

4.4 Conjugate Direction Methods

The basic idea of the conjugate direction methods is that the minimum point of an n-variable quadratic function can be found by searching in only n different directions. For example, the minimum point of a two-variable function, $f(x_1,x_2) = x_1^2 + x_2^2$, can be found by searching two directions, $[1,0]^{\mathrm{T}}$ and $[0,1]^{\mathrm{T}}$, respectively. We observe that the two directions coincide with the principal axes of the function's level curves and are mutually orthogonal.

Such search directions, for a general n-variable quadratic function, should satisfy *conjugacy*, as described in the following subsection. The minimization methods based on the conjugate search directions are called the *conjugate direction methods*.

The conjugate direction methods have the following properties: (1) they can find the minimum of an n-variable quadratic function in n steps, (2) the searching directions are determined without evaluating the Hessian matrix, and (3) neither matrix inversion nor storage of an $n \times n$ matrix is required.

Definition of conjugacy and the relating property is given in the following subsection.

4.4.1 Conjugate Directions

Consider the quadratic function

$$f(x) = \frac{1}{2}x^{\mathrm{T}}Ax - b^{\mathrm{T}}x, \tag{4.30}$$

where A is symmetric and positive definite. Suppose we search for a minimum of a two-variable function whose level curves are concentric ellipses as shown in Fig. 4.12.

Given an arbitrary initial estimate x^0 and a search direction d_0, the minimum along the given direction occurs at point x^1. If we can find the optimal direction at x^1, we can find the minimum point in two iteration steps. We say that such direction is conjugate to d_0.

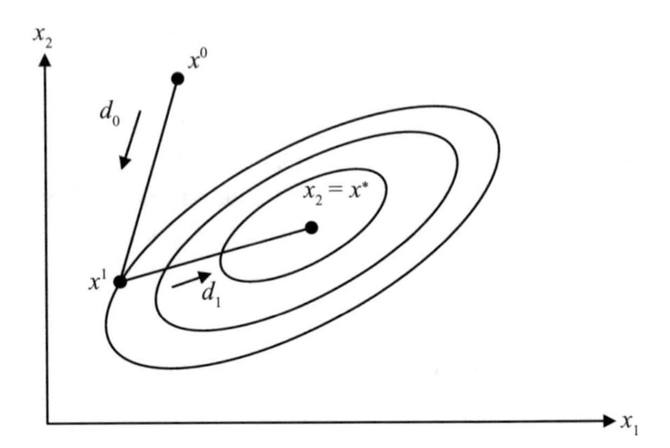

Fig. 4.12 A set of conjugate directions on the level sets of a quadratic function

The idea of conjugate directions can be extended to n-variable functions in the following definition.

Definition 4.1: Conjugate Directions Let A be a real symmetric $N \times N$ matrix. The directions $d_0, d_1, \ldots, d_m$ are A-conjugate or mutually conjugate with respect to A if, for all $i \neq j$,

$$d_i^{\mathrm{T}} A d_j = 0. \tag{4.31}$$

For a set of mutually conjugate vectors $\{d_j\}$ with respect to A, suppose that

$$\sum_j \lambda_j d_j = 0, \tag{4.32}$$

for some scalars λ_j. Multiplying both sides of Eq. (4.32) by $d_k^{\mathrm{T}} A$ yields

$$d_k^{\mathrm{T}} A \sum_j \lambda_j d_j = \lambda_k d_k^{\mathrm{T}} A d_k = 0, \tag{4.33}$$

which implies $\lambda_k = 0$. Therefore, mutually conjugate vectors are linearly independent.

The following theorem describes an important property of conjugate directions.

Theorem 4.4: Conjugate Directions *Let $\{d_1, \ldots, d_M\}$ be a set of mutually conjugate directions. Let $N > M$. Then the global minimum of the N-variable, quadratic function of Eq. (4.30), in the subspace R^M, containing the point x^0 and the directions d_M, can be found by searching only once along each direction.*

Proof Let x^0 be the initial estimate of the minimum point. The global minimum will then occur at the point

$$x^* = \left(\left(\left(x^0 + \lambda_0 d_0\right) + \lambda_1 d_1\right) + \cdots + \lambda_m d_m\right) = x^0 + \sum_{i=0}^{m} \lambda_i d_i, \tag{4.34}$$

where λ_i are chosen so as to minimize

$$f(x^*) = \frac{1}{2}\sum_{i=1}^{m} \lambda_i^2 d_i^{\mathrm{T}} A d_i + \sum_{i=1}^{m} \lambda_i^2 d_i^{\mathrm{T}} \left(Ax^0 - b\right) + f\left(x^0\right). \tag{4.35}$$

There is no term in $\lambda_i \lambda_j \, (i \neq j)$ on the right-hand side because the directions d_i are mutually conjugate. Hence λ_k is independent of all the other $\lambda_i \, (i \neq k)$ and minimizes

$$\frac{1}{2}\lambda_k^2 d_k^{\mathrm{T}} A d_k + \lambda_k d_i^{\mathrm{T}} \left(Ax^0 - b\right) \tag{4.36}$$

along the direction d_r.

In other words, the global minimum of $f(x)$ in the M-dimensional subspace can be found by searching only once along each direction.

Given an arbitrary set of linearly independent vectors $\{p_0, \ldots, p_{N-1}\}$, a systematic way to find a set of N-conjugate vectors is devised based on the Gram-Schmidt orthogonalization procedure as follows:

Algorithm 4.7: Gram-Schmidt-Based Algorithm for Generating Conjugate Vector 7

1. $d_0 \leftarrow p_0$, $k \leftarrow 0$.
2. For $k = 0$ to $N - 2$ do

$$d_{k+1} = p_{k+1} - \sum_{i=0}^{k} \frac{p_{k+1}^{\mathrm{T}} A d_i}{d_i^{\mathrm{T}} A d_i} d_i.$$

The vectors $d_0, \ldots, d_{N-1}$ generated by Algorithm 4.7 have been proven to be conjugate with respect to A in [luenberger84].

4.4.2 *Conjugate Gradient Algorithm for a Quadratic Function*

Although conjugate direction methods are guaranteed to converge after N iterations, where N represents the number of variables of a quadratic function, it is not, in many large-scale problems, practical to perform all N iterations. For example, if we remove noise on a 256×256 digital image by minimizing the high-frequency components, the conjugate direction method requires 65,536 iterations to converge. Furthermore, generating such a large number of conjugate directions by Algorithm 4.7 is almost impossible in the sense of memory space and computation time.

On the other hand, the *conjugate gradient algorithm*, which will be described in this subsection, computes each conjugate direction as the algorithm progresses without using the prespecified set of conjugate directions. More specifically, at each stage of the algorithm, the direction is calculated as a linear combination of the previous direction and the current gradient so that all the directions are conjugate with respect to the Hessian of the quadratic function.

Consider the quadratic function

$$f(x) = \frac{1}{2}x^{\mathrm{T}}Ax - b^{\mathrm{T}}x, \tag{4.37}$$

where A is symmetric and positive definite. Given an initial point x^0, the steepest descent direction is used as the first search direction as

$$x^1 = x^0 + \alpha_0 d_0, \tag{4.38}$$

where

$$d_0 = -\nabla f\left(x^0\right) \tag{4.39}$$

and α_0 is determined by the optimal line search in Eqs. (4.18) and (4.19) as

$$\alpha_0 = -\frac{\nabla f(x^0)d_0}{d_0^T A d_0}. \tag{4.40}$$

In the next stage, we choose d_1 as a linear combination of $\nabla f(x^1)$ and d_0 in such a way that d_1 is conjugate to d_0 with respect to A. In general, at the $k+1$th step, we choose

$$d_{k+1} = -\nabla f\left(x^{k+1}\right) + \beta_k d_k, \quad \text{for } k = 0, 1, \ldots. \tag{4.41}$$

Directions $d_1, d_2, \ldots, d_{k+1}$ generated by Eq. (4.41) are A-conjugate if β_k are chosen as

$$\beta_k = \frac{\nabla f\left(x^{k+1}\right)^{\mathrm{T}} A d_k}{d_k^{\mathrm{T}} A d_k}. \tag{4.42}$$

The proof of this statement is rather cumbersome and omitted from this text. Interested readers can find a proof in [chong96].

The conjugate gradient algorithm is summarized below.

Algorithm 4.8: Conjugate Gradient Algorithm

1. Set an initial guess x^0, $k \leftarrow 0$, and $d_0 = -\nabla f(x^0)$.
2. $\alpha_k = -\frac{\nabla f\left(x^k\right) d_k}{d_k^{\mathrm{T}} A d_k}$.
3. $x^{k+1} = x^k + \alpha_k d_k$.

4. If $\nabla f(x^{k+1}) = 0$, stop the algorithm. Otherwise, proceed to the next step.
5. $\beta_k = \dfrac{\nabla f(x^{k+1})^{\mathrm{T}} A d_k}{d_k^{\mathrm{T}} A d_k}$.
6. $d_{k+1} = -\nabla f(x^{k+1}) + \beta_k d_k$.
7. $k \leftarrow k + 1$, go to step 2.

4.4.2.1 Conjugate Gradient Algorithm for Nonquadratic Problems

The conjugate gradient algorithm can minimize a nonquadratic problem with some modifications. If we assume that near the solution the valley of a nonquadratic function is similar to that of an appropriate quadratic function, the conjugate gradient algorithm may converge in the same way to minimize the nonquadratic function.

Let $f(x)$ be a nonquadratic function to be minimized and $g(x)$ be the fitting quadratic function at $x = x^k$. Now $g(x)$ should have the same function value, gradient vector, and Hessian matrix as those of $f(x)$ at $x = x^k$. This yields the Taylor series expansion of $f(x)$ about x^k as

$$g(x) = f(x^k) + (x - x^k)^{\mathrm{T}} \nabla f(x^k) + \frac{1}{2}(x - x^k)^{\mathrm{T}} A(x^k)(x - x^k). \tag{4.43}$$

Observing Algorithm 4.8, we can find that the Hessian appears only in computing α_k and β_k. For a quadratic function, the Hessian is a constant matrix. On the other hand, for a nonquadratic function, we must evaluate the Hessian at each iteration, which is very computationally expensive.

To avoid evaluating the Hessian at each iteration, computing α_k is replaced by a numerical line search procedure. Computing β_k is replaced by one of the following approximations.

Recall that

$$\beta_k = \frac{\nabla f(x^{k+1})^{\mathrm{T}} A d_k}{d_k^{\mathrm{T}} A d_k}. \tag{4.44}$$

In the quadratic case, we have that

$$A d_k = \frac{\nabla f(x^{k+1}) - \nabla f(x^k)}{\alpha_k}. \tag{4.45}$$

Substituting Eq. (4.45) into Eq. (4.44) yields the Hestenes-Stiefel formula as

$$\beta_k = \frac{\nabla f(x^{k+1})^{\mathrm{T}} \{\nabla f(x^{k+1}) - \nabla f(x^k)\}}{d_k^{\mathrm{T}} \{\nabla f(x^{k+1}) - \nabla f(x^k)\}}. \tag{4.46}$$

Using the property that $d_k^{\mathrm{T}} \nabla f(x^{k+1}) = 0$, we can approximate the Hestenes-Stiefel formula as

$$\beta_k = \frac{\nabla f(x^{k+1})^{\mathrm{T}} \{\nabla f(x^{k+1}) - \nabla f(x^k)\}}{\nabla f(x^k)^{\mathrm{T}} \nabla f(x^k)}, \tag{4.47}$$

which is called the Polak-Ribiere formula.

Using the property that $\nabla f(x^{k+1})^{\mathrm{T}} \nabla f(x^k) = 0$, we can further approximate the Polak-Ribiere formula as

$$\beta_k = \frac{\nabla f(x^{k+1})^{\mathrm{T}} \nabla f(x^{k+1})}{\nabla f(x^k)^{\mathrm{T}} \nabla f(x^k)}. \tag{4.48}$$

For nonquadratic problems, search directions are not guaranteed to be A-conjugate, and the algorithm will not converge in N steps. To solve this problem, we replace the direction vector by the steepest gradient vector after every few iterations. This modified algorithm is termed the *conjugate gradient algorithm with restarts*.

References

[chong96] E.K.P. Chong, S.H. Zak, *An Introduction to Optimization* (Wiley, New York, 1996)

[fletcher80] R. Fletcher, *Practical Methods of Optimization*. Unconstrained Optimization, vol 1 (Wiley, Chichester, 1980)

[luenberger84] D.B. Luenberger, *Linear and Nonlinear Programming*, 2nd edn. (Addison-Wesley, Reading, 1984)

[metropolis53] N. Metropolis, A. Rosenbluth, M. Rosenbluth, H. Teller, E. Teller, Equation of state calculations by fast computing machines. J. Chem. Phys. **21**, 1087–1092 (1953)

[moon96] J.I. Moon, J.K. Paik, Fast iterative image restoration algorithm. J. Electr. Eng. Inf. Sci. **1** (2), 67–75 (1996)

[murray72] W. Murray (ed.), *Numerical Methods for Unconstrained Optimization* (Academic Press, New York, 1972)

Additional References and Further Readings

[ahu93] R.K. Ahuya, T.L. Magnanti, J.B. Orlin, *Network Flows: Theory, Algorithms, and Applications* (Prentice Hall, Englewood Cliffs, NJ, 1993)

[den83] J.E. Dennis, R.B. Schnabel, *Numerical Methods for Unconstrained Optimization and Nonlinear Equations* (Prentice Hall, Englewood Cliffs, NJ, 1983)

[den89] J.E. Dennis, R.B. Schnabel, A view of unconstrained optimization, in *Optimization*, ed. by G.L. Nemhauser, A.H.G. Rinnooy Kan, M.J. Todd (North-Holland, Amsterdam, 1989), pp. 1–72

[gill91] P.E. Gill, W. Murray, M.H. Wright, *Numerical Linear Algebra and Optimization* (Addison-Wesley, Reading, MA, 1991)

[gol89] G.H. Golub, C.F. Van Loan, *Matrix Computations*, 2nd edn. (The Johns Hopkins University Press, Baltimore, MD, 1989)

[more84] J.J. Moré, D.C. Sorensen, Newton's method, in *Studies in Numerical Analysis*, ed. by G.H. Golub (Mathematical Association of America, Washington, DC, 1984), pp. 29–82

[mur81] B.A. Murtagh, *Advanced Linear Programming: Computation and Practice* (McGraw-Hill, New York, 1981)

[nem88] G.L. Nemhauser, L.A. Wolsey, *Integer and Combinatorial Optimization* (Wiley, New York, 1988)

[noc92] J. Nocedal, Theory of algorithms for unconstrained optimization, in *Acta Numerica 1992* (Cambridge University Press, Cambridge, 1992), pp. 199–242

[bel99] A. Belegundu, T.R. Chandrapatla, *Optimization Concepts and Applications in Engineering* (Prentice Hall, Englewood Cliffs, 1999)

[lue84] D.G. Luenberger, *Introduction to Linear and Nonlinear Programming* (Addison Wesley, Reading, 1984)

[rek83] G.V. Reklaitis, A. Ravindran, K.M. Ragsdell, *Engineering Optimization: Methods and Applications* (Wiley, Hoboken, 1983)

[bei79] C.S. Beightler, D.T. Phillips, D.J. Wilde, *Foundations of Optimization*, 2nd edn. (Prentice-Hall, Englewood Cliffs, NJ, 1979)

[bev70] G.S. Beveridge, R.S. Schechter, *Optimization: Theory and Practice* (McGraw-Hill Book Company, New York, 1970)

[rek83] G.V. Reklaitis, A. Ravindran, K.M. Ragsdell, *Engineering Optimization: Methods and Applications* (Wiley, New York, 1983)

[per88] A.L. Peressini, F.E. Sullivan, J.J. Uhl Jr., *The Mathematics of Nonlinear Programming* (Springer, New York, 1988)

[baz93] M.S. Bazaraa, H.D. Sherali, C.M. Shetty, *Nonlinear Programming: Theory and Algorithms*, 2nd edn. (Wiley, New York, 1993)

[man94] O.L. Mangasarian, *Nonlinear Programming*. Classics in Applied Mathematics series, vol 10 (SIAM, Philadelphia, 1994)

[bor2000] J.M. Borwein, A.S. Lewis, *Convex Analysis and Nonlinear Optimization* (Springer, Berlin, 2000)

[hir93] J.-B. Hiriart-Urruty, C. Lemarechal, *Convex Analysis and Minimization Algorithms* (Springer, New York, 1993)

[rock96] R.T. Rockafellar, *Convex Analysis* (Princeton University Press, Princeton, 1996)

[hor95] R.P. Horst, M. Pardalos, N.V. Thoai, *Introduction to Global Optimization* (Kluwer Academic Publishers, Dordrecht, 1995)

[hor96] R. Horst, H. Tuy, *Global Optimization: Deterministic Approaches*, 3rd edn. (Springer, Heidelberg, 1996)

[ber95] D.P. Bertsekas, *Nonlinear Programming* (Athena Scientific, Boston, MA, 1995)

[32] J.R. Birge, F. Louveaux, *Introduction to Stochastic Programming* (Springer, New York, NY, 1997)

[bor2000] J.M. Borwein, A.S. Lewis, *Convex Analysis and Nonlinear Optimization: Theory and Examples* (Springer, New York, 2000)

[chv83] V. Chvátal, *Linear Programming* (W. H Freeman and Company, New York, 1983)

[cook98] W.J. Cook, W.H. Cunningham, W.R. Pulleyblank, A. Schrijver, *Combinatorial Optimization* (Wiley, New York, 1998)

[dan63] G.B. Dantzig, *Linear Programming and Extensions* (Princeton University Press, Princeton, NJ, 1963)

[edw73] C.H. Edwards, *Advanced Calculus of Several Variables* (Academic Press, New York, 1973)
[fang93] S.-C. Fang, S. Puthenpura, *Linear Optimization and Extensions* (Prentice-Hall, Englewood Cliffs, NJ, 1993)
[fia68] A.V. Fiacco, G.P. McCormick, *Nonlinear Programming: Sequential Unconstrained Minimization Techniques* (Wiley, New York, 1968)
[noc99] J. Nocedal, S.J. Wright, *Numerical Optimization* (Springer, New York, 1999)
[roc97] R.T. Rockafellar, R.J.-B. Wets, *Variational Analysis* (Springer, New York, 1997)
[strang88] G. Strang, *Linear Algebra and Its Applications*, 3rd edn. (Harcourt Brace Jovanovich, San Diego, 1988)
[ber97] D. Bertsimas, J. Tsitsiklis, *Introduction to Linear Optimization* (Athena Scientific, Belmont, 1997)
[gill91] P.E. Gill, W. Murray, M.H. Wright, *Numerical Linear Algebra and Optimization*, vol 1 (Addison-Wesley, Redwood City, CA, 1991)
[wright91] M.H. Wright, Optimization and large-scale computation, in *Very Large-Scale Computation in the 21st Century*, ed. by J.P. Mesirov (Society for Industrial and Applied Mathematics, Philadelphia, 1991), pp. 250–272
[wright96] M.H. Wright, Direct search methods: once scorned now respectable, in *Numerical Analysis 1995 (Proceedings of the 1995 Dundee Biennial Conference in Numerical Analysis)*, ed. by D.F. Griffiths, G.A. Watson (Longman, Addison Wesley, 1996), pp. 191–208
[wright98] M.H. Wright, Ill-conditioning and computational error in primal-dual interior methods for nonlinear programming. SIAM J. Optim. **9**, 84–111 (1998)
[kel99] C.T. Kelley, *Iterative Methods for Optimization* (SIAM, Philadelphia, 1999)
[hest80] M.R. Hestness, *Conjugate Direction Methods in Optimization* (Springer, Berlin, 1980)

Chapter 5
Constrained Optimization

Abstract In most practical image processing and computer vision problems involving optimization, the solutions do not span the entire range of real values. This resulting bounding property is often imposed upon the solution process in the form of constraints placed upon system parameters. Optimization problems with constraints are generally referred to as constrained optimization problems.

The basic idea of solving most constrained optimization problems is to determine the solution that satisfies the stated performance index, subject to the given constraints. In this chapter, proper methods of finding the solutions to constrained optimization problems, according to the characteristics of various types of constraints, are discussed.

If the solution and its local perturbation do not violate the given constraints, we consider the problem as being unconstrained. Therefore, the important issue of constrained optimization is analyzing the local behavior of the solution on or near the boundary of constraints.

This chapter includes three sections. The first section describes the constrained optimization problem and defines the terminology commonly used with this problem, both in the literature and in this chapter. The second and the third sections describe optimality conditions and the corresponding optimization methods with linear and nonlinear constraints, respectively.

5.1 Introduction

A general constrained optimization problem can be formulated as

$$\min_{x \in R^N} f(x), \quad \text{subject to } x \in C, \tag{5.1}$$

where C represents the constraint set which is described by, for example, the solution of a set of linear equations, linear inequalities, nonlinear equations, nonlinear inequalities, or their mixture.

M.A. Abidi et al., *Optimization Techniques in Computer Vision*, Advances in Computer Vision and Pattern Recognition, DOI 10.1007/978-3-319-46364-3_5

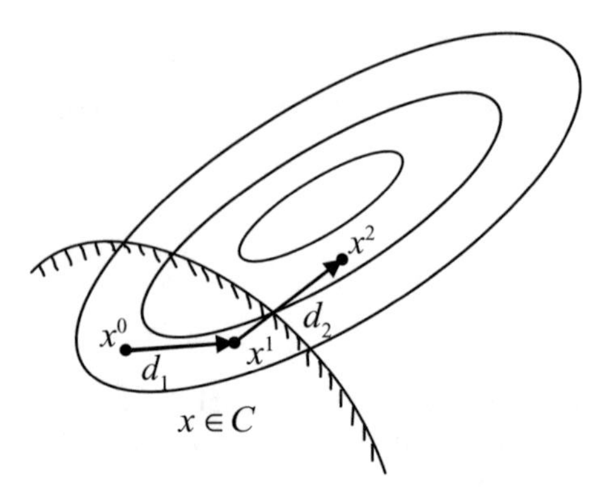

Fig. 5.1 Feasible points and feasible directions

Any solution outside the constraint set is meaningless. Therefore, each estimated solution, including the initial guess, should be in the constraint set. Definitions of feasible points and feasible direction are now presented.

Definition 5.1: Feasible Points Any point $\hat{x}$ that is in the constraint set C or, equivalently, that satisfies all the constraints of a constrained optimization problem is said to be feasible.

Definition 5.2: Feasible Direction Any vector $p = \bar{x} - \hat{x}$ is said to be a feasible direction if both $\bar{x}$ and $\hat{x}$ are feasible points.

An illustration of a constraint set, feasible points, and a feasible direction is shown in Fig. 5.1. The set of ellipses represents the level set of a function to be minimized. Points x^0 and x^1 are feasible points, while x^2 is not. Vector d_1 is a feasible direction, while d_2 is not.

Constrained optimization problems can be classified according to the types of constraints. Typical classifications include linear equality, linear inequality, nonlinear equality, and nonlinear inequality constraints.

5.2 Constrained Optimization with Linear Constraints

A linear function of the variable is the simplest type of constraint. Because of its simplicity, solutions of many optimization problems are constrained by the linear equations or inequalities. We first consider the single linear equality constraint.

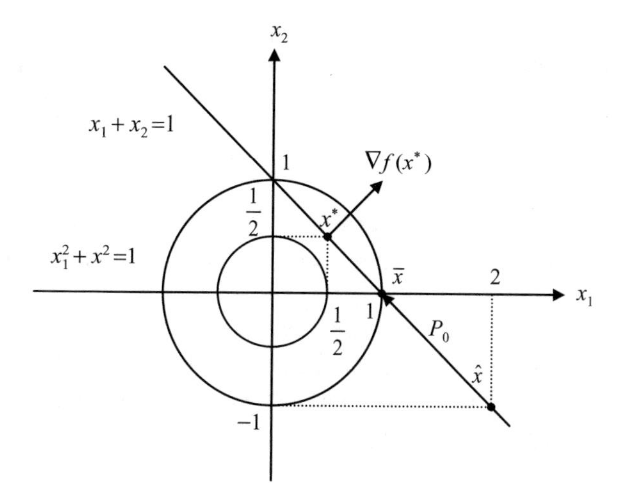

Fig. 5.2 An example of a constrained optimization problem with a single linear equality constraint

5.2.1 *Single Linear Equality Constraint*

An optimization problem with a single linear equation constraint can be formulated as

$$\min_{x \in R^N} f(x), \quad \text{subject to} \quad a^{\mathrm{T}}x = b, \tag{5.2}$$

where a represents an N-dimensional vector.

Figure 5.2 shows an example of the constrained optimization problem with a single linear equality constraint,

$$\text{minimize } f(x_1, x_2) = x_1^2 + x_2^2, \quad \text{subject to} \quad x_1 + x_2 = 1. \tag{5.3}$$

Let $\hat{x}$ be a feasible point of the linear constraint described by the linear equation. We have that

$$a^{\mathrm{T}}\hat{x} = b. \tag{5.4}$$

Suppose that p is a feasible direction. We then see that

$$a^{\mathrm{T}}p = a^{\mathrm{T}}(\bar{x} - \hat{x}) = b - b = 0, \tag{5.5}$$

where $\bar{x}$ represents another feasible point. In Fig. 5.2, since $a = [1, 1]^{\mathrm{T}}$ and $p_0 = \bar{x} - \hat{x} = [-1, 1]^{\mathrm{T}}$, we see that $a^{\mathrm{T}}p_0 = 0$.

By applying the mean value theorem to the Taylor series expansion of $f(x^* + \varepsilon p)$ at x^*, we have

$$f(x^* + \varepsilon p) = f(x^*) + \varepsilon p^{\mathrm{T}}\nabla f(x^*) + \frac{1}{2}\varepsilon^2 p^{\mathrm{T}}H(x^* + \varepsilon\theta p)p, \quad 0 \le \theta \le 1, \tag{5.6}$$

where ε is taken as a positive scalar and $H(x)$ represents the Hessian matrix of $f(x)$.

To satisfy the first-order optimality condition, we need to show that x^* is a stationary point. Suppose the second-order term in Eq. (5.6) is sufficiently small, then the first-order term should vanish for every p, such that

$$p^{\mathrm{T}} \nabla f\left(x^*\right) = 0. \tag{5.7}$$

The scalar quantity $p^{\mathrm{T}} \nabla f\left(x^*\right)$ is termed the projected gradient of $f(x)$ at x^*. Any point at which the projected gradient vanishes is termed a constrained stationary point.

In Fig. 5.2, we can intuitively see that the point $x^* = [1/2, 1/2]^{\mathrm{T}}$ is the solution of the constrained minimization problem in Eq. (5.3). Since $\nabla f(x) = [2x_1, 2x_2]^{\mathrm{T}}$, $p^{\mathrm{T}} \nabla f\left(x^*\right) = 0$ for any feasible direction, $p = \beta p_0$, $\beta \neq 0$.

Equation (5.7) implies that

$$\nabla f\left(x^*\right) = \lambda^* a, \tag{5.8}$$

which can be proven by multiplying the transpose of the feasible direction vector p^{T} on the left of both sides in Eq. (5.8). As a result, Eq. (5.8) is a sufficient condition for x^* to be a constrained stationary point. The scalar value λ^* that satisfies Eq. (5.8) is called the Lagrange multiplier.

If p is a feasible direction, Eq. (5.6) can be reduced to

$$f\left(x^* + \varepsilon p\right) = f\left(x^*\right) + \frac{1}{2}\varepsilon^2 p^{\mathrm{T}} H\left(x^* + \varepsilon\theta p\right)p. \tag{5.9}$$

From this result, the second-order necessary condition for x^* to be a local minimum is that the matrix $p^{\mathrm{T}}H(x^*)p$, termed the projected Hessian, must be positive semidefinite. Note that the Hessian matrix itself, $H(x^*)$, is not required to be positive semidefinite.

From the result of Eq. (5.8), we can formulate an unconstrained optimization problem whose solution is the same as that of the constrained optimization problem given in Eq. (5.2) as

$$\min_{x \in R^N} L(x), \quad \text{where } L(x) = f(x) - \lambda a^{\mathrm{T}} x. \tag{5.10}$$

$L(x)$ is called the Lagrangian function. We can use any kind of unconstrained optimization methods for solving the problem in Eq. (5.10). The constrained optimization problem, stated in Eq. (5.3), gives the corresponding Lagrangian function

$$L(x) = {x_1}^2 + {x_2}^2 - \lambda(x_1 + x_2). \tag{5.11}$$

If $\lambda = 1$, the minimum of the Lagrangian function occurs at $[1/2, 1/2]^{\mathrm{T}}$, which is the same as the solution of Eq. (5.3).

Example 5.1 Consider the two-variable problem minimizing the function $f(x) = -x_1{}^2 + x_2{}^2$, subject to $x_2 = 1$. We have $\nabla f(x) = [-2x_1, 2x_2]^{\mathrm{T}}$, $a = [0, 1]^{\mathrm{T}}$, and $H(x) = \begin{bmatrix} -2 & 0 \\ 0 & 2 \end{bmatrix}$. We know that $\hat{x} = [0,\ 1]^{\mathrm{T}}$ is a feasible point and that $\lambda = 2$ satisfies Eq. (5.8). Hence $\hat{x} = [0,\ 1]^{\mathrm{T}}$ is a constrained stationary point.

Consider another feasible point, $\bar{x} = [\delta,\ 1]^{\mathrm{T}}$, for any nonzero δ. The corresponding feasible direction is $p = \bar{x} - \hat{x} = [\delta,\ 0]^{\mathrm{T}}$. The projected Hessian is computed as $p^{\mathrm{T}} G(\hat{x})p = -2\delta^2 < 0$. Since the projected Hessian is negative, $\hat{x}$ is not a local minimum.

Example 5.2 Consider the two-variable constrained minimization problem

$$f(x) = x_1{}^2 + x_2{}^2, \quad \text{subject to } x_2 = 1. \tag{5.12}$$

We have $\nabla f(x) = [2x_1, 2x_2]^{\mathrm{T}}$, $a = [0,\ 1]^{\mathrm{T}}$, and $H(x) = \begin{bmatrix} 2 & 0 \\ 0 & 2 \end{bmatrix}$. We know that $\hat{x} = [0,\ 1]^{\mathrm{T}}$ is a feasible point and that $\lambda = 2$ satisfies Eq. (5.8). Hence $\hat{x} = [0,\ 1]^{\mathrm{T}}$ is a constrained stationary point.

Consider another feasible point, $\bar{x} = [\delta, 1]^{\mathrm{T}}$, for any nonzero δ. The corresponding feasible direction is $p = \bar{x} - \hat{x} = [\delta,\ 0]^{\mathrm{T}}$. The projected Hessian is computed as $p^{\mathrm{T}} G(\hat{x})p = 2\delta^2 > 0$, for nonzero δ. Since the projected Hessian is nonnegative, $\hat{x}$ is a local minimum.

According to Eq. (5.10), the corresponding unconstrained version of the optimization problem can be written as

$$\min L(x), \quad \text{where } L(x) = x_1{}^2 + x_2{}^2 - 2x_2. \tag{5.13}$$

We can easily find that the solution of the unconstrained minimization problem in Eq. (5.13) is equal to $[0,\ 1]^{\mathrm{T}}$, which is equal to the solution of the constrained minimization problem in Eq. (5.12).

5.2.2 *Multiple Linear Equality Constraints*

An optimization problem having multiple, say M, linear equality constraints can be formulated as

$$\min_{x \in R^N} f(x), \quad \text{subject to } Ax = b, \tag{5.14}$$

where A represents an $M \times N$ matrix and b an M-dimensional vector. Let both $\bar{x}$ and $\hat{x}$ be feasible points. Then $p = \bar{x} - \hat{x}$ is the corresponding feasible direction, and we see that

$$Ap = A(\overline{x} - \hat{x}) = b - b = 0. \tag{5.15}$$

Note that the vector p is orthogonal to the rows of A. Any movement from a feasible point along the corresponding feasible direction does not violate the constraints since $A(\hat{x} + \alpha p) = A\hat{x} = b$. Let Z be a matrix whose columns form a basis for the subspace of vectors that satisfies $Ap = 0$. If A has k independent rows, or equivalently is of rank k, the dimension of Z should be $N \times (N - k)$. Columns of Z span the null space to a subspace spanned by k independent rows of A. Now, every feasible direction can be written as a linear combination of the columns of Z, as

$$p = Zp_z. \tag{5.16}$$

To determine the optimality of a given feasible point x^*, we apply the mean value theorem to the Taylor expansion of $f(x^* + \varepsilon Zp_z)$ about x^* and write

$$f(x^* + \varepsilon Zp_z) = f(x^*) + \varepsilon p_z^{\mathrm{T}} Z^{\mathrm{T}} \nabla f(x^*) + \frac{1}{2}\varepsilon^2 p_z^{\mathrm{T}} Z^{\mathrm{T}} H(x^* + \varepsilon\theta Zp_z) Zp_z, \tag{5.17}$$

where θ satisfies $0 \le \theta \le 1$ and ε is taken as a positive scalar. A necessary condition for x^* to be a local minimum of the constrained optimization problem in Eq. (5.14) is that the first-order derivative term in Eq. (5.17) must vanish for every p_z, which implies that

$$Z^{\mathrm{T}} \nabla f(x^*) = 0. \tag{5.18}$$

The $N - k$ vector $Z^{\mathrm{T}} \nabla f(x^*)$ is termed the projected gradient[1] of $f(x)$ at x^*, and any point at which the projected gradient vanishes is termed a constrained stationary point.

According to the definition of the matrix Z, Eq. (5.18) implies that $\nabla f(x^*)$ must be a linear combination of the rows of A, which can be described as

$$\nabla f(x^*) = \sum_{i=1}^{M} \lambda_i^* a_i = A^{\mathrm{T}} \lambda^*, \tag{5.19}$$

where $\lambda^* = [\lambda_1^*, \ldots, \lambda_M^*]^{\mathrm{T}}$ is termed the vector of Lagrange multipliers and a_i represents the ith row of the matrix A. If the rows of A are linearly independent, the Lagrange multipliers are unique.

[1] We note that in the single linear equation constraint case, the projected gradient is a scalar quantity, while in the multiple linear equations constraint case, it is an $M \times 1$ vector, where M represents the number of constraints.

According to Eq. (5.18), Eq. (5.17) can be reduced to

$$f(x^* + \varepsilon Z p_z) = f(x^*) + \frac{1}{2}\varepsilon^2 p_z^{\mathrm{T}} Z^{\mathrm{T}} H(x^* + \varepsilon\theta Z p_z) Z p_z. \tag{5.20}$$

From this result, the second-order necessary condition for x^* to be a local minimum is that the matrix $Z^{\mathrm{T}}H(x^*)Z$ termed the projected Hessian matrix must be positive semidefinite. Note that the Hessian matrix itself, $H(x^*)$, is not required to be positive semidefinite.

5.2.3 *Optimization Methods for Linear Equality Constraints*

As a summary of the previous section, a first-order necessary condition for the feasible point x^* to be a local minimum of the optimization problem, with linear equality constraints given in Eq. (5.14), is that the projected gradient at x^* is equal to zero, such as

$$Z^{\mathrm{T}} \nabla f(x^*) = 0. \tag{5.21}$$

A second-order necessary condition is that the projected Hessian $Z^{\mathrm{T}}H(x^*)Z$ is positive semidefinite.

For a problem with N variables and M linearly independent equality constraints, the dimension of optimization becomes $N - M$. Figure 5.3 shows an example of an optimization problem with $N = 3$ and $M = 2$. The function value varies throughout the three-dimensional space. One will note that each of the two constraints is illustrated by using corresponding planes. The solution must lie on the line generated by the intersection of two planes, which means optimization is performed on the line that is the intersection of two planes defined by the constraints.

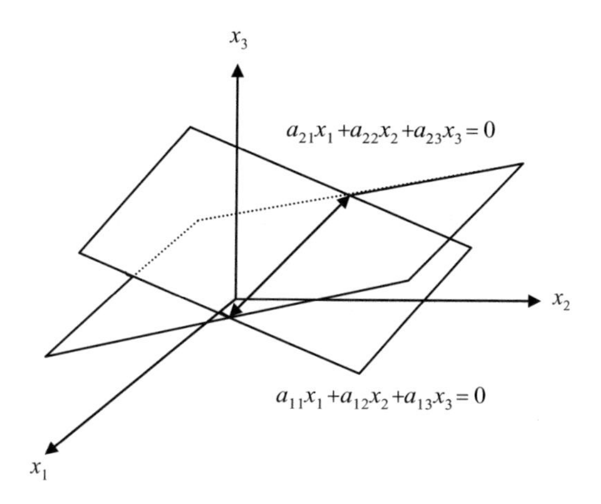

Fig. 5.3 A three-variable optimization problem with two linearly independent equality constraints

More rigorously, any N-dimensional vector can be decomposed into the range-space portion and the null space portion of matrix A, such as

$$x = Rx_r + Zx_z, \tag{5.22}$$

where R represents the $N \times M$ matrix whose columns form a basis for the range space of A^T and Z represents the $N \times (N - M)$ matrix whose columns form a basis for the null space of A^T.

For a feasible point x^*, it holds that

$$Ax^* = A\left(Rx_r^* + Zx_z^*\right) = ARx_r^* = b, \tag{5.23}$$

since $AZ = 0$. By definition, AR is nonsingular, and x_r^* is then uniquely determined. In other words, there is only $N - M$-dimensional freedom in searching for the minimum point.

The above discussion states that an N-dimensional optimization problem, with M linearly independent equality constraints, can be viewed as an $N - M$ variable unconstrained optimization problem.

The typical algorithm for an optimization problem with linear equality constraints is summarized in the following:

Algorithm 5.1: Constrained Optimization Algorithm with Linear Equality Constraints

1. Set an initial feasible point as x^0 and $k \leftarrow 0$.
2. Compute a feasible search direction as $p_k = Zp_z$, where Z is the $N \times (N - M)$ vector described in Eq. (5.23).
3. Compute a step length α_k such that $f\left(x^k + \alpha_k p_k\right) < f\left(x^k\right)$.
4. $x^{k+1} = x^k + \alpha_k p_k$.
5. If $\left|\left|x^{k+1} - x^k\right|\right|$ is smaller than a prespecified error limit, terminate the algorithm with the converged solution x^{k+1}. Otherwise, $k \leftarrow k + 1$, and go to step 2.

5.2.4 *Linear Inequality Constraints*

Consider the optimization problem with linear inequality constraints

$$\min_{x \in R^N} f(x), \quad \text{subject to} \quad a_i^T x \geq b_i, \quad \text{for} \quad i = 1, \ldots, M. \tag{5.24}$$

The M constraints in Eq. (5.24) can be expressed as

$$Ax \geq b, \tag{5.25}$$

where the $M \times N$ matrix A has a_i^{T} as the ith row and $b = [b_1, \ldots, b_M]^{\mathrm{T}}$. To determine the optimality condition for the optimization problem as stated in Eq. (5.24), we need to define the following terminology.

Definition 5.3: Active and Inactive Feasible Points Let $\hat{x}$ be a feasible point of the constraints in Eq. (5.24). $\hat{x}$ is said to be an active feasible point with respect to the ith constraint if $a_i^{\mathrm{T}}\hat{x} = b_i$, and the corresponding equation is called the active constraint. A feasible point is said to be an inactive feasible point with respect to the ith constraint if $a_i^{\mathrm{T}}\hat{x} > b_i$, and the corresponding inequality is called the inactive constraint.

Consider, for example, the simple optimization problem with linear inequality constraints,

$$\text{minimize } f(x_1, x_2) = {x_1}^2 + {x_2}^2, \quad \text{subject to } x_1 - x_2 \geq 1 \text{ and } x_2 \geq 0. \tag{5.26}$$

The level curves and constraints are illustrated in Fig. 5.4, where x_{a} represents an active feasible point with respect to the first constraint, x_i represents an inactive feasible point with respect to both the first and the second constraints, and x_{v} violates the constraints.

As shown in Fig. 5.4, the feasible perturbation from the active feasible point, x_{a}, is restricted along the line $x_1 - x_2 = 1$ or downward from the line. All p_1, p_2, and p_3 are feasible perturbations from x_{a}. On the other hand, the feasible perturbation from the inactive feasible point, x_{i}, can take any direction if the length is sufficiently small.

Note that the active feasible point plays an important role in determining the optimality condition of constrained optimization problems. This is true because the optimality condition for the inactive feasible point is the same to that of an unconstrained problem. Two different perturbations from an active feasible point are defined in the following:

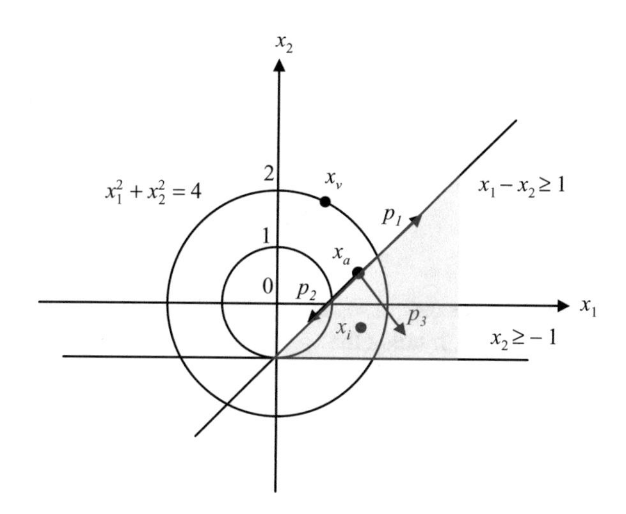

Fig. 5.4 Active and inactive feasible points

Definition 5.4: Binding and Nonbinding Perturbations Let x_a be an active feasible point with respect to the ith constraint, such as $a_i^T x_a = 0$. A vector p is called a binding perturbation from x_a, with respect to the ith constraint, if

$$a_i^T p = 0. \tag{5.27}$$

The vector p is called a nonbinding perturbation from x_a, with respect to the ith constraint, if

$$a_i^T p > 0. \tag{5.28}$$

To have the first-order optimality condition, we need to determine whether the given feasible point, say x^*, is stationary. If x^* is an inactive feasible point, the optimality condition should be the same as that of an unconstrained optimization problem. On the other hand, if x^* is an active feasible point with respect to M constraints, we have that $Ax^* = b$, and we need special consideration for its optimality condition. There are two conditions for the active feasible point x^* to be a local minimum. The conditions are as follows: (1) the function is stationary for any binding perturbation from x^*, and (2) any nonbinding perturbation does not decrease the function value.

Note that every binding perturbation p, satisfying $Ap = 0$, can be written as a linear combination of the columns of Z, yielding $p = Zp_z$. For mathematical analysis of the first case, we apply the mean value theorem to the Taylor expansion of $f(x^* + \varepsilon Z p_z)$ about x^*, yielding

$$f(x^* + \varepsilon Z p_z) = f(x^*) + \varepsilon p_z^T Z^T \nabla f(x^*) + \frac{1}{2}\varepsilon^2 p_z^T Z^T H(x^* + \varepsilon\theta Z p_z) Z p_z, \tag{5.29}$$

where θ satisfies $0 \le \theta \le 1$ and ε is taken as a positive scalar. If the first-order term is not zero for any p_z, then x^* cannot be a local minimum. Thus the optimality condition for x^* to be a local minimum is that

$$Z^T \nabla f(x^*) = 0 \tag{5.30}$$

or

$$\nabla f(x^*) = A^T \lambda^* = \sum_{i=1}^{M} \lambda_i^* a_i. \tag{5.31}$$

For the second case, any nonbinding perturbation p satisfying $a_i^T p \ge 0$, for $i = 1, \ldots, M$, it holds that

$$\nabla f(x^*)^T p \ge 0. \tag{5.32}$$

From Eq. (5.31), we see that the gradient of the function $f(x)$ is a linear combination of a_i. Inequality (Eq. 5.32) is then rewritten as

$$\nabla f\left(x^*\right)^{\mathrm{T}} p = \lambda_1{}^* a_1^{\mathrm{T}} p + \cdots + \lambda_M{}^* a_M^{\mathrm{T}} p \geq 0. \tag{5.33}$$

Since $a_i^{\mathrm{T}} p \geq 0$, for $i = 1, \ldots, M$, the condition in Eq. (5.32) is satisfied only if $\lambda_i{}^* \geq 0$, for all i. In other words, if there are any negative Lagrange multipliers, x^* will not be optimal.

5.2.5 *Optimization Methods for Linear Inequality Constraints*

Consider the optimization problem with linear inequality constraints, described in Eqs. (5.24) and (5.25). If we know that the initial feasible point is active with respect to all M constraints, the solution of the first iteration is the same as that obtained by the algorithm for a linear equality problem given in Algorithm 5.1. Furthermore, if we assume that all iterates are active with respect to all M constraints, the solution of the linear inequality problem is equivalent to that of the linear equality problem.

However, it is not usual for the given feasible point to be active with respect to all the constraints, and information about the set of active constraints is not available. Hence, the optimization algorithm for linear inequality problems predicts the set of active constraints. Since it is not guaranteed that the predicted set has correct active constraints, an additional procedure is needed which tests whether the current prediction is correct. If the predicted set is proved to be wrong, a new prediction is made and tested.

The abovementioned algorithm is called the active set method, or the working set method, since the solution is searched on the set of active constraints.

5.3 Constrained Optimization with Nonlinear Constraints

To understand the methods for constrained optimization problems with nonlinear constraints, we will discuss the optimality condition for the feasible point and present typical optimization methods.

The first subsection deals with a simple single constraint problem for the purpose of intuitively understanding the nature of the solution process. The second and the third subsections then generalize the optimality conditions and present the method of optimization.

5.3.1 Single Nonlinear Equality Constraint

A constrained optimization problem with a single nonlinear equality constraint can be formulated as

$$\min_{x \in R^N} f(x), \ \text{subject to} \ c(x) = 0. \tag{5.34}$$

In case of linear constraints, we have seen that all feasible perturbations can be defined in terms of a linear subspace. However, in the case of nonlinear constraints, such as $c(x^*) = 0$, there is, in general, no feasible direction p such that $c(x^* + \alpha p) = 0$ for any small $|\alpha|$. Therefore, in order to retain feasibility with respect to the nonlinear function, $c(x)$, it is necessary to move along a feasible arc instead of a linear feasible direction.

For $N = 2$ in the optimization problem of Eq. (5.34), a simple example for $f(x_1, x_2)$ is illustrated as shown in Fig. 5.5. The corresponding level set is shown on the x_1x_2-plane, and the single constraint, denoted by $c(x_1, x_2) = 0$, is represented by the curve on the plane.

The curve, defined by $c(x_1, x_2) = 0$, can also be represented by a parameterized curve $\{x(t)\}$, such as

$$x(t) = [x_1(t), x_2(t)]^{\mathrm{T}}, \quad \text{for} \ t \in (a, b). \tag{5.35}$$

Define the two-dimensional vector

$$\dot{x}(t) = \frac{dx}{dt}(t) = [\dot{x}_1(t), \dot{x}_2(t)]^{\mathrm{T}}, \tag{5.36}$$

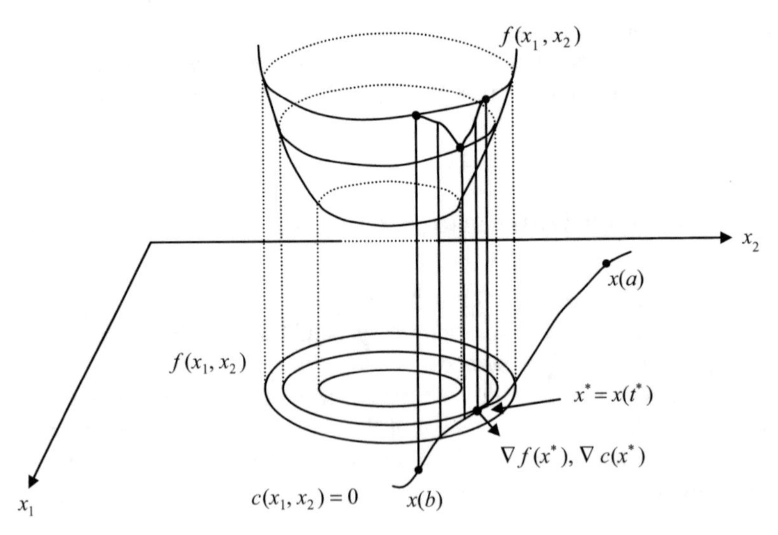

Fig. 5.5 Level sets of a two-variable function $f(x_1, x_2)$ and a single nonlinear constraint $c(x_1, x_2) = 0$

which can be thought of as the velocity of a point traversing the curve, defined by $c(x_1, x_2) = 0$, with position $x(t)$ at time t. The vector $\dot{x}(t)$ points in the direction of the instantaneous motion of $x(t)$. Therefore, the vector $\dot{x}(t^*)$ is tangent to the curve at x^*.

Let us choose a point $x^* = x(t^*)$, for $t^* \in (a, b)$. We can then show that $\nabla c(x^*)$ is orthogonal to $\dot{x}(t^*)$. Since $c(x)$ has the constant value of zero, its derivative with respect to t should be zero. By applying the chain rule we get

$$\frac{d}{dt} c(x(t)) = \nabla c(x(t))^{\mathrm{T}} \dot{x}(t) = 0, \tag{5.37}$$

which proves $\nabla c(x^*)$ is orthogonal to $\dot{x}(t^*)$.

We can also show that $\nabla f(x^*)$ is orthogonal to $\dot{x}(t^*)$ if x^* is a minimizer of $f(x) : R^2 \to R$. Since $f(x)$ can be considered as a function of t, the first-order necessary condition requires that its derivative with respect to t should be zero at $t = t^*$. Applying the chain rule yields

$$\frac{d}{dt} f(x(t^*)) = \nabla f(x(t^*))^{\mathrm{T}} \dot{x}(t^*) = 0, \tag{5.38}$$

which proves that $\nabla f(x^*)$ is orthogonal to $\dot{x}(t^*)$.

From Eqs. (5.37) and (5.38), we see that $\nabla f(x^*)$ and $\nabla c(x^*)$ are parallel if x^* is a minimizer of the function $f(x)$. That is, there exists a scalar λ^* such that

$$\nabla f(x^*) = \lambda^* \nabla c(x^*), \tag{5.39}$$

where λ^* is called the Lagrange multiplier.

5.3.2 *Multiple Nonlinear Equality Constraints*

A constrained optimization problem, with multiple nonlinear equality constraints, can be formulated as

$$\min_{x \in R^N} f(x), \quad \text{subject to } c_i(x) = 0, \quad i = 1, \ldots, M. \tag{5.40}$$

Before discussing the optimality condition for the problem in Eq. (5.40), we need to first define the related terminology.

The $N \times N$ Hessian matrix of $c_i(x)$ is denoted by $G_i(x)$, and the $M \times N$ Jacobian matrix of $c_i(x)$ is denoted by

$$J(x) = \begin{bmatrix} \nabla c_1(x)^{\mathrm{T}} \\ \nabla c_2(x)^{\mathrm{T}} \\ \vdots \\ \nabla c_M(x)^{\mathrm{T}} \end{bmatrix}. \tag{5.41}$$

Definition 5.5: Regular Point A feasible point that satisfies $c_i(x) = 0$, $i = 1, \ldots, M$, is a regular point of the constraints if the vectors of $\nabla c_i(x^*)$ are linearly independent.

Each constraint forms a surface on the N-dimensional space. A mathematical definition of the surface is given below.

Definition 5.6: A Surface The set of nonlinear equality constraints describes a surface

$$S = \{x \in R^N | c_i(x) = 0, \quad i = 1, \ldots, M\}. \tag{5.42}$$

If the Jacobian matrix $J(x)$ is of full rank, the dimension of the surface S is $N - M$. Figure 5.6 shows examples of two different sets of constraints and the corresponding surfaces. Figure 5.6a shows a two-dimensional surface, where $N = 3$ and $M = 1$. On the other hand, Fig. 5.6b shows a one-dimensional surface, where $N = 3$ and $M = 2$.

If the points in S are regular and the number of constraints is equal to $N - 1$, the intersection of the constraint line with the surface forms a curve that is characterized by a single parameter as defined below.

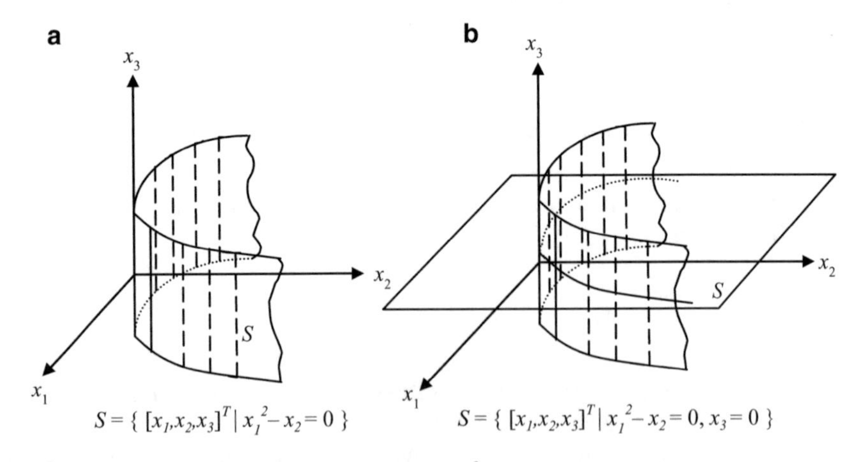

Fig. 5.6 Illustration of two different surfaces in R^3: (**a**) a two-dimensional surface and (**b**) a one-dimensional surface

Definition 5.7: A Curve A set of points characterized by $t \in (a, b)$ is said to be a curve, such as

$$C = \{x(t) \in S | t \in (a, b)\}. \quad (5.43)$$

We can think of a curve as the path traversed by a point x with respect to time t. Therefore, $x(t)$ represents the position of the point at time t.

Definition 5.8: Tangent Space and Tangent Plane The tangent space at a point on the surface of Eq. (5.42) is a subspace of R^N, which satisfies

$$\begin{aligned} T(x^*) &= \{y \in R^N | \nabla c_i(x^*)^{\mathrm{T}} y = 0, \quad i = 1, \ldots, M\} \\ &= \{y \in R^N | J(x^*)^{\mathrm{T}} y = 0\}. \end{aligned} \quad (5.44)$$

The tangent plane at a point x^* is defined as

$$\mathrm{TP}(x^*) = T(x^*) + x^*. \quad (5.45)$$

Note that the tangent space passes through the origin, while the tangent plane passes through the point x^*. Since $\nabla c_i(x^*)$ represents the direction of increase of the function $c_i(x)$ at x^*, any vector on the tangent plane TP(x^*) is a tangential vector of the surface S. The dimension of the tangent plane at x^* is equal to $N - M$ if x^* is a regular point.

For example, the tangent plane for the surface shown in Fig. 5.6a is the two-dimensional plane $x_2 = 0$, and the tangent plane for the surface shown in Fig. 5.6b is the one-dimensional line, the x_1-axis.

Consider a directed curve $x(t)$ in R^N parameterized by a single variable t as a feasible curve. Let $x^* = x(t^*)$ and p denote the tangent to the curve at x^*. For this curve to be feasible, the rate of change, or derivative, of $c_i(x)$ along the curve must be zero at x^*, which can be described as

$$\frac{d}{dt} c_i(x(t))|_{t=t^*} = \nabla c_i(x(t^*))^{\mathrm{T}} p = \nabla c_i(x^*)^{\mathrm{T}} p = 0. \quad (5.46)$$

This means that the tangent to a feasible curve of the nonlinear constraint must be orthogonal to its gradient, yielding

$$J(x^*) p = 0. \quad (5.47)$$

As a linear counterpart, Eq. (5.15) is a necessary and sufficient condition for p to be a feasible direction. Note, however, that Eq. (5.47) is a necessary condition for p to be tangent to a feasible curve. In order for Eq. (5.47) to completely characterize p, the point x^* should be a regular point.

To have the first-order optimality condition, $f(x)$ must be stationary at x^* along any feasible curve. The sufficient condition is that

$$\nabla f\left(x^*\right)^{\mathrm{T}} p = 0, \tag{5.48}$$

which implies that the projected gradient at x^* should be zero, that is,

$$Z\left(x^*\right)^{\mathrm{T}} \nabla f\left(x^*\right) = 0, \tag{5.49}$$

where $Z(x^*)$ represents a matrix whose columns form a basis for the set of vectors orthogonal to the rows of $J(x^*)$. The linear counterpart of this condition has been described in Eq. (5.16).

The condition of Eq. (5.49) is satisfied if $\nabla f\left(x^*\right)$ is a linear combination of the rows of $J(x^*)$, expressed as

$$\nabla f\left(x^*\right) = J\left(x^*\right)^{\mathrm{T}} \lambda^*, \tag{5.50}$$

where λ^* represents an M-vector of Lagrange multipliers.

Define the Lagrangian function

$$L(x, \lambda) = f(x) - \sum_{i=1}^{M} \lambda_i c_i(x) = f(x) - \lambda^{\mathrm{T}} c(x), \tag{5.51}$$

where $\lambda = [\lambda_1 \cdots \lambda_M]^{\mathrm{T}}$ and $c(x) = [c_1(x) \cdots c_M(x)]^{\mathrm{T}}$. We see that a stationary point, say x^*, satisfies the condition in Eq. (5.50) when $\lambda = \lambda^*$. This satisfies the first-order necessary condition for x^* to be a feasible stationary point of the constrained optimization problem given in Eq. (5.40).

The second-order necessary condition will be satisfied only if $f(x)$ has nonnegative curvature at x^* along any feasible arc, which results from

$$p^{\mathrm{T}} W(x, \lambda) p \geq 0, \tag{5.52}$$

where

$$W(x, \lambda) = H(x) - \sum_{i=1}^{M} \lambda_i G_i(x). \tag{5.53}$$

The condition of Eq. (5.52) is equivalent to the matrix

$$Z\left(x^*\right)^{\mathrm{T}} W\left(x^*, \lambda\right) Z\left(x^*\right) \tag{5.54}$$

being positive semidefinite.

5.3.3 Nonlinear Inequality Constraints

In this subsection we consider the following problem:

$$\min_{x \in R^N} f(x), \quad \text{subject to} \quad c_i(x) \geq 0, \quad i = 1, \ldots, M. \tag{5.55}$$

Let x^* be a feasible point of the problem in Eq. (5.55). As in the linear case, an active feasible point plays an important role in determining the optimality condition of the problem, since an inactive feasible point will remain strictly satisfied within a sufficiently small neighborhood.

The conditions for the active feasible point x^* to be a local minimum of $f(x)$ can be summarized as (1) the function $f(x)$ is stationary for any binding perturbation form x^*, and (2) any nonbinding perturbation does not decrease the function value.

In order to satisfy the above conditions, the gradient of $f(x)$ should be a linear combination of the Jacobian matrix of $c_i(x)$, such as

$$\nabla f(x^*) = J(x^*)\lambda^*, \tag{5.56}$$

where $\lambda_i^* > 0$, for $i = 1, \ldots, M$.

The second-order necessary condition is that $Z(x^*)^{\mathrm{T}} W(x^*, \lambda^*) Z(x^*)$ is positive semidefinite, where matrix $W(x, \lambda)$ has been defined in Eq. (5.53).

References and Further Readings

[ber82] D.P. Bertsekas, *Constrained Optimization and Lagrange Multiplier Methods* (Academic, New York, 1982)

[gill89] P.E. Gill, W. Murray, M.A. Saunders, M.H. Wright, Constrained nonlinear programming, in *Optimization*, ed. by G.L. Nemhauser, A.H.G. Rinnooy Kan, M.J. Todd (North-Holland, Amsterdam, 1989), pp. 171–210

[gold93] J. Goldsmith, A.H. Barr, Applying constrained optimization to computer graphics. SMPTE J. **102**(10), 910–912 (1993)

[par87] P.M. Pardalos, J.B. Rosen, *Constrained Global Optimization: Algorithms and Applications* (Springer, New York, 1987)

[flou90] C.A. Floudas, P.M. Pardalos, *A Collection of Test Problems for Constrained Global Optimization Algorithms*. Lecture Notes in Computer Science (Springer, Berlin, 1990)

[gill82] P.E. Gill, W. Murray, M.A. Saunders, M.H. Wright, Linearly constrained optimization, in *Nonlinear Optimization 1981*, ed. by M.J.D. Powell (Academic Press, London, 1982), pp. 123–139

[gill88] P.E. Gill, W. Murray, M.A. Saunders, M.H. Wright, Recent developments in constrained optimization. J. Comput. Appl. Math. **22**, 257–270 (1988)

[gill89a] P.E. Gill, W. Murray, M.A. Saunders, M.H. Wright, Constrained nonlinear programming, in *Optimization*, ed. by G.L. Nemhauser, A.H.G. Rinnooy Kan, M.J. Todd (Elsevier, Amsterdam, 1989), pp. 171–210

[gill89b] P.E. Gill, W. Murray, M.A. Saunders, M.H. Wright, A practical anti-cycling procedure for linearly constrained optimization. Math. Program. **45**, 437–474 (1989)

[wright92] M.H. Wright, Interior methods for constrained optimization, in *Acta Numerica*, vol 1 (Cambridge University Press, Cambridge, 1992), pp. 341–407

[wright98] M.H. Wright, The interior-point revolution in constrained optimization, in *High-Performance Algorithms and Software in Nonlinear Optimization*, ed. by R. DeLeone, A. Murli, P.M. Pardalos, G. Toraldo (Kluwer Academic Publishers, Dordrecht, 1998), pp. 359–381

[conn89] A.R. Conn, N.I.M. Gould, P.L. Toint, Large-scale optimization. Math. Program. B **45**, 3 (1989)

[conn90] A.R. Conn, N.I.M. Gould, P.L. Toint, Large-scale optimization—applications. Math. Program. B **48**, 1 (1990)

[smo96] Gert Smolka, Problem solving with constraints and programming. *ACM Comput. Surv.* **28** (4es) (December 1996). Electronic Section

[brent73] R.P. Brent, *Algorithms for Minimization Without Derivatives* (Prentice Hall, Englewood Cliffs, NJ, 1973)

[chong96] E. Chong, S. Zak, *An Introduction to Optimization* (Wiley, New York, 1996)

[fang93] S.C. Fang, S. Puthenpura, *Linear Optimization and Extensions* (Prentice Hall, Englewood Cliffs, NJ, 1993)

[far87] R.W. Farebrother, *Linear Least Squares Computation* (Marcel Dekker, New York, 1987)

[golub65] G.H. Golub, Numerical methods for solving linear least squares problems. Numer. Math. **7**, 206–216 (1965)

[strang86] G. Strang, *Introduction to Applied Mathematics* (Wellesley-Cambridge, Wellesley, MA, 1986)

[strang88] G. Strang, *Linear Algebra and Its Applications*, 3rd edn. (W. B. Saunders, New York, 1988)

Part III

Chapter 6
Frequency-Domain Implementation of Regularization

Regularization methods play an important role in solving linear equations of the form

$$y = Hx, \tag{6.1}$$

with prior knowledge about the solution. The corresponding regularization results in minimization of

$$f(x) = \|y - Hx\|^2 + \lambda\|Cx\|^2. \tag{6.2}$$

In image processing and computer vision applications, the linear equation in Eq. (6.1) is used to estimate the original complete data x, given the incomplete data y, and the relationship between x and y, which is represented by H. When Eq. (6.1) represents the two-dimensional image degradation model, for example, matrix H represents an operation that degrades the original high-quality image x and results in the degraded image y.

Although a quadratic function is the simplest type of objective function, practical implementation of its minimization is still complicated and time-consuming, as its size becomes larger. In this context, if both matrices H and C in Eq. (6.2) have circulant structure, minimization of the quadratic function can be dramatically simplified by using frequency-domain implementation.

6.1 Space-Invariance and Circulant Matrices

The linear equation in Eq. (6.1) can be considered as the relationship between the original complete data and the observed incomplete data. We first present the one-dimensional time-invariant systems. The two-dimensional case is the

M.A. Abidi et al., *Optimization Techniques in Computer Vision*, Advances in Computer Vision and Pattern Recognition, DOI 10.1007/978-3-319-46364-3_6

Kronecker product-based extension of the one-dimensional case and is presented in the following subsection.

6.1.1 One-Dimensional Time-Invariant Systems

Let $x(n)$ be a vector whose elements represent a one-dimensional discrete-time signal. If the signal is the input to a time-invariant system, with impulse response denoted by $h(n)$, the corresponding output is determined by

$$y(n) = h(n)^* x(n), \tag{6.3}$$

where * represents one-dimensional linear convolution [oppenheim83]. If the output is the average of three consecutive input samples, the corresponding impulse response of the system should be

$$h(n) = \frac{1}{3}\{\delta(n+1) + \delta(n) + \delta(n-1)\}, \tag{6.4}$$

where $\delta(\cdot)$ represents the Kronecker delta function. The circular convolution equation representation of Eq. (6.3) is equivalent to the following vector-matrix expression,[1]

$$y = Hx, \tag{6.5}$$

where

$$H = \begin{bmatrix} 1 & 1 & 0 & \cdots & 1 \\ 1 & 1 & 1 & \cdots & 0 \\ 0 & 1 & 1 & \cdots & 0 \\ \vdots & \vdots & \vdots & \ddots & \vdots \\ 1 & 0 & 0 & \cdots & 1 \end{bmatrix}. \tag{6.6}$$

As shown in the above example, any one-dimensional linear time-invariant system can be expressed with a corresponding circulant matrix. By using the fundamental property that any circulant matrix can be diagonalized by the DFT, the matrix-vector multiplication in Eq. (6.5) can be simplified as a set of element-by-element multiplications. Furthermore, the diagonalization property of a

[1] In order to determine boundary samples of output signal, we assumed circularly symmetric or periodic input with period N.

circulant matrix makes it possible to provide a closed-form solution of the minimization given in Eq. (6.2).

Let F_1 be the $N \times N$ DFT matrix that transforms a one-dimensional signal of length N into the same number of corresponding discrete Fourier transform (DFT) coefficients. For more detail about the DFT, see Appendix B. According to the orthogonal property of the DFT matrix, we note that

$$F_1^{-1} = F_1^{*\mathrm{T}} = F_1^*, \tag{6.7}$$

where F_1^* is the conjugate of F_1.

Pre-multiplying the DFT matrix expression of Eq. (6.5) by F_1 gives

$$F_1 y = F_1 H x = \left(F_1 H F_1^*\right)\left(F_1 x\right). \tag{6.8}$$

Now, let $\widetilde{y} = F_1 y$ and $\widetilde{x} = F_1 x$ be vectors formed by the DFTs of y and x, respectively. Also, let $\widetilde{H} = F_1 H F_1^*$. Then, $\widetilde{H}$ becomes a diagonal matrix whose main diagonal elements are equal to the DFT of the first column of the circulant matrix H, such as

$$\mathrm{diag}\left\{\widetilde{H}\right\} = F_1(He_1), \tag{6.9}$$

where $e_1 = [1, 0, \ldots, 0]^{\mathrm{T}}$. Then we can rewrite Eq. (6.8) as

$$\widetilde{y} = \widetilde{H}\widetilde{x}. \tag{6.10}$$

In Eq. (6.10), matrix-vector multiplication can be performed by simple element-by-element multiplication because $\widetilde{H}$ is a diagonal matrix. Finally, y can be obtained by simply multiplying the inverse of the DFT matrix F_1^{-1} or, equivalently, the conjugate of the DFT matrix $F_1{*}$ by $\widetilde{y}$, such that

$$F_1^*\widetilde{y} = F_1^* F_1 y = y. \tag{6.11}$$

6.1.2 Two-Dimensional Space-Invariant Systems

Let $x(m, n)$, $y(m, n)$, and $h(m, n)$ be the two-dimensional input, output, and impulse response of a space-invariant system, respectively. In this case, the output is determined by the two-dimensional convolution

$$y(m, n) = h(m, n)^{**}x(m, n). \tag{6.12}$$

We can express the two-dimensional convolution equation in matrix-vector multiplication form as

$$y = Hx, \tag{6.13}$$

where x and y represent row-ordered vectors for the input and the output, with x being the input and y the output. H represents the doubly block circulant matrix for the space-invariant system.

If we define the two-dimensional DFT matrix as the Kronecker product of two one-dimensional DFT matrices, such as

$$F = F_1 \otimes F_1, \tag{6.14}$$

then it is well known that the doubly block circulant matrix H in Eq. (6.13) can be diagonalized using the matrix F, giving

$$FHF^* = \widetilde{H}, \tag{6.15}$$

where $\widetilde{H}$ represents a diagonal matrix whose diagonal elements are equal to the row-ordered vector of the two-dimensional DFT of $h(m, n)$.

By pre-multiplying the two-dimensional DFT matrix of Eq. (6.13) by F on both sides, we have

$$\widetilde{y} = \widetilde{H}\widetilde{x}, \tag{6.16}$$

where $\widetilde{y} = Fy$ and $\widetilde{x} = Fx$ represent the row-ordered vectors formed by the two-dimensional DFTs of $y(m, n)$ and $x(m, n)$, respectively.

6.2 Frequency-Domain Implementation of the Constrained Least Square Filter

In the previous section, we reviewed the circulant structure of one-dimensional time-invariant and two-dimensional space-invariant systems. The circulant structure transforms the matrix-vector multiplication, which represents the corresponding convolution operation given in Eq. (6.13), into multiplication of the diagonalized matrix and the transformed vector as shown in Eq. (6.16).

The diagonalized matrix-vector multiplication significantly simplifies the complicated signal restoration filtering operations, such as the constrained least squares (CLS) and the Wiener filters. We first present the CLS filter, which restores the original signal from the observed signal with prior knowledge about the solution in the form of constraints. The Wiener filter will be presented in the subsequent section.

6.2.1 Derivation of the Constrained Least Square Filter

There exists a closed-form solution that minimizes the regularized objective function in Eq. (6.2), if both the degradation matrix H and the stabilizing matrix C are doubly block circulant. The doubly block circulant structure results from space-invariant degradation and stabilizing operations.

In order to derive the closed-form solution in the frequency domain, we can rewrite Eq. (6.2) as

$$f(x) = y^{\mathrm{T}}y - y^{\mathrm{T}}Hx - x^{\mathrm{T}}H^{\mathrm{T}}y + x^{\mathrm{T}}H^{\mathrm{T}}Hx + \lambda x^{\mathrm{T}}C^{\mathrm{T}}Cx. \tag{6.17}$$

By setting

$$\frac{\partial}{\partial x}f(x) = 0, \tag{6.18}$$

we have

$$Tx = b, \tag{6.19}$$

where

$$T = H^{\mathrm{T}}H + \lambda C^{\mathrm{T}}C \quad \text{and} \quad b = H^{T}y. \tag{6.20}$$

The solution of Eq. (6.19) can be given as

$$x = T^{-1}b, \tag{6.21}$$

on condition that matrix T is nonsingular. Even if T is nonsingular, it is not easy to compute its inverse because the size of the matrix is usually very large.

By using the property that the two-dimensional DFT matrix can diagonalize any doubly block circulant matrix, we can easily compute the solution of Eq. (6.19) in the frequency domain.

By multiplying the two-dimensional DFT matrix of Eq. (6.21) on the left-hand side by F, using Eq. (6.20) and the orthogonal property of the DFT matrix, we have

$$\begin{aligned} Fx &= F\left(H^{\mathrm{T}}H + \lambda C^{\mathrm{T}}C\right)^{-1}\left(F^{*}F\right)H^{\mathrm{T}}y \\ &= \left\{F\left(H^{\mathrm{T}}H + \lambda C^{\mathrm{T}}C\right)F^{*}\right\}^{-1}\left(FH^{\mathrm{T}}F^{*}\right)(Fy), \end{aligned} \tag{6.22}$$

which can be rewritten as

$$\widetilde{x} = \left(\widetilde{H}^{\mathrm{T}}\widetilde{H} + \lambda\widetilde{C}^{\mathrm{T}}\widetilde{C}\right)^{-1}\widetilde{H}^{\mathrm{T}}\widetilde{y}, \tag{6.23}$$

where $\widetilde{C}$ and $\widetilde{C}^{\mathrm{T}}$ represent the diagonal matrices whose diagonal elements are equal to the row-ordered vector of the two-dimensional DFT of $c(m,n)$ and its conjugate, respectively. Because all matrices in Eq. (6.23) have been diagonalized, the two-dimensional DFT of x can be computed using scalar multiplications and divisions, such as, for $k, \ l = 0, \ 1, \ \ldots, \ N-1,$

$$X(k,l) = \mathrm{DFT}\{x(m,n)\} = \frac{\mathrm{DFT}\{h(m,n)\}^{*}}{|\mathrm{DFT}\{h(m,n)\}|^{2} + \lambda|\mathrm{DFT}\{c(m,n)\}|^{2}}\mathrm{DFT}\{y(m,n)\}, \tag{6.24}$$

where $\mathrm{DFT}\{\cdot\}$ represents the (k,l)th coefficient of the two-dimensional DFT.

Finally, the solution for x is obtained from the inverse transform as

$$F^{*}\widetilde{x} = F^{*}Fx = x. \tag{6.25}$$

In Eq. (6.24), if the regularization parameter λ is equal to zero, the solution is equivalent to that achieved by the inverse filter. On the other hand, a nonzero λ can control the amount of smoothness in the solution. The frequency-domain implementation of regularization in Eq. (6.24) is also called the constrained least square filter.

6.2.2 Image Restoration Using the CLS Filter

A simple, linear, space-invariant, image restoration problem was described in Chap. 3. Here, we consider the same problem but use the method given above for designing a CLS filter for restoring the degraded image.

Suppose we are given a degraded image that has been obtained by defocusing and injecting additive noise. Let $x(m,n)$, $0 \le m,n \le 255$, represent pixel intensity values of the 256×256 digital image. The corresponding model of image degradation and restoration is shown in Fig. 6.1.

In this diagram, $h(m,n)$ represents defocusing, $\eta(m,n)$ additive noise, and $g(m,n)$ the impulse response of the constrained least square filter.

If we assume that the same defocusing occurs for each pixel of the image, the corresponding blur can be modeled by convolving the input image with a two-dimensional, isotropic low-pass filter. Although the low-pass filter usually

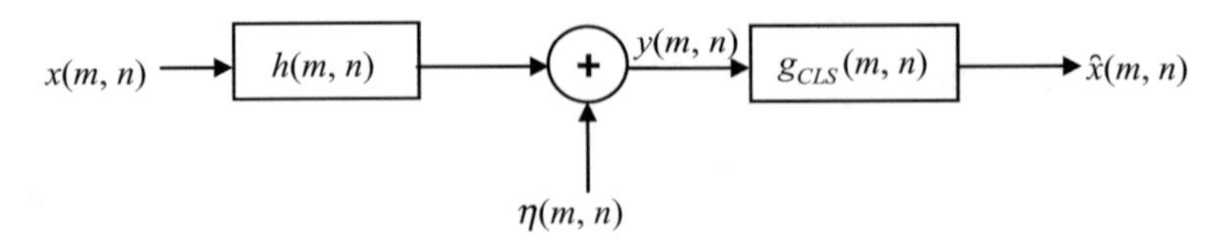

Fig. 6.1 Image degradation and restoration model

has iso-circular shape, we will use the 7×7 rectangular uniform averaging filter for simplicity. Since we are assuming that the blur occurs in a space-invariant fashion, we can simulate the blurring process in the two-dimensional DFT domain, as described in Appendix B. The two-dimensional, noncausal[2] point spread function (PSF)[3] of the averaging filter can be given as

$$\frac{1}{49}\left\{\begin{matrix} \delta(m+3,n+3) & \cdots & +\delta(m+3,n-3) \\ \vdots & \ddots & \vdots \\ +\delta(m-3,n+3) & \cdots & +\delta(m-3,n-3) \end{matrix}\right\}. \tag{6.26}$$

The simulated defocused image can be obtained from the convolution, $h(m,n) \otimes x(m,n)$. The $\otimes$ symbol represents the circular convolution. We assume that $x(m,n)$ is circularly periodic and $h_e(m,n)$ is the extended circularly periodic PSF with zero filling as shown in Fig. 6.2. The (k,l)th DFT coefficients of the averaging filtered imaged can then be obtained by multiplying the (k,l)th DFT coefficients of $h_e(m,n)$ and $x(m,n)$.

The degraded image $y(m,n)$ in Fig. 6.1 is obtained by adding the noise image $\eta(m,n)$ to the simulated defocused image. The simulation process for defocusing and adding noise is summarized as follows:

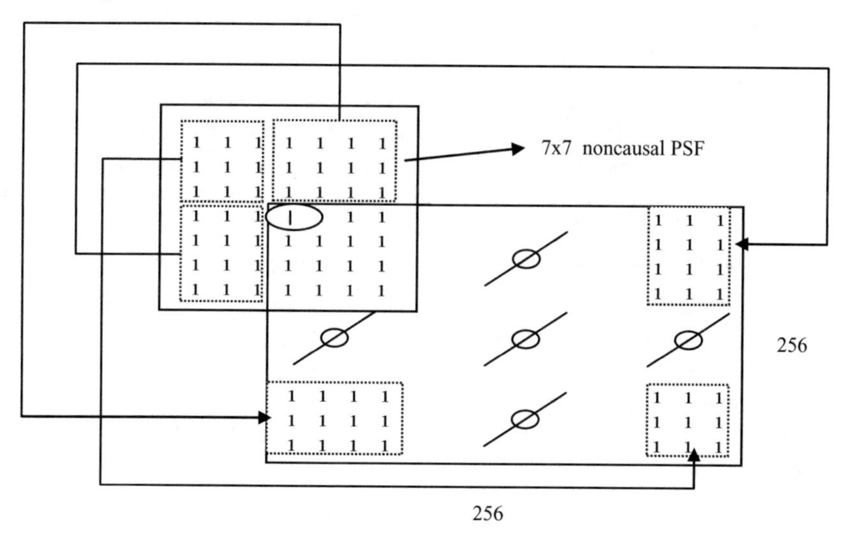

Fig. 6.2 Periodically extended PSF based on the 7×7 averaging filter equation in Eq. (6.26)

[2]Most one-dimensional filters have a causal impulse response because the future input is not available for convolution with the filter. In this case, the filtered output comes with a certain amount of delay. On the other hand, in two-dimensional image processing, noncausal filters are used in order to avoid a shifted output image, caused by the two-dimensional delay.

[3]The impulse response of a two-dimensional filter is called the point spread function if each coefficient has a nonnegative value.

(a) (b)

Fig. 6.3 (**a**) The original 256×256 Lena image and (**b**) the defocused and noisy image with 7×7 uniform blur and 40 dB additive noise

Algorithm 6.1: Simulation for Image Defocusing and Noise Addition Let the $N \times N$ original undistorted image $x(m,n)$, the PSF $h(m,n)$, for defocusing, and the noise image $\eta(m,n)$ be given.

1. Compute the DFT of the original image $X(k,l) = \text{DFT}\{x(m,n)\}$.
2. Make the periodically extended PSF based on $h(m,n)$ as shown in Fig. 6.2.
3. Compute the DFT of the extended PSF as $H(k,l) = \text{DFT}\{h_e(m,n)\}$.
4. For $k, l = 0, \ldots, N-1$, compute the DFT for $W(k,l)$ as $W(k,l) = H(k,l)X(k,l)$.
5. Compute the inverse DFT of $W(k,l)$ as $w(m,n) = \text{IDFT}\{W(k,l)\}$.
6. The defocused and noisy image is obtained as $y(m,n) = w(m,n) + \eta(m,n)$.

The process of using this algorithm is illustrated with the conventional Lena test file given in Fig. 6.3. The original scene, without any alteration, is shown in Fig. 6.3a. Figure 6.3b presents the image after defocusing with a 7×7 uniform blur and adding 40 dB of noise.

From Eq. (6.24) we can deduce that the frequency response of the CLS restoration filter should be

$$G_{\text{CLS}}(k,l) = \frac{H^*(k,l)}{|H(k,l)|^2 + \lambda |C(k,l)|^2}, \tag{6.27}$$

where $H(k,l)$ represents the (k,l)th DFT coefficient of the extended PSF $h(m,n)$, $H^*(k,l)$ the conjugate of $H(k,l)$, $C(k,l)$ the (k,l)th DFT coefficient of the extended high-pass filter $c(m,n)$, and λ the regularization parameter.

The CLS restoration algorithm can be summarized as:

Algorithm 6.2: Image Restoration Using the CLS Filter Let the defocused, noisy image $y(m,n)$, the high-pass filter $c(m,n)$, and the suitable regularization parameter λ be given.

1. Compute the DFT of $y(m,n)$ as $Y(k,l) = \text{DFT}\{y(m,n)\}$.
2. Compute the DFT of $h_e(m,n)$ as $H(k,l) = \text{DFT}\{h_e(m,n)\}$.
3. Make the periodically extended high-pass filter $c_e(m,n)$ as shown in Fig. 6.2.
4. Compute the DFT of $c_e(m,n)$ as $C(k,l) = \text{DFT}\{c_e(m,n)\}$.
5. Determine the frequency response of the CLS restoration filter $G_{CLS}(K,L)$, as in Eq. (6.27).
6. Multiply the results of step 1 and step 5.
7. Determine the restored image from $\hat{x}(m,n) = \text{IDFT}\{Y(k,l)G_{CLS}(k,l)\}$.

As shown in Fig. 6.4, the regularization parameter λ plays an important role in restoring the image using the CLS filter. If λ is too small, the resulting image does

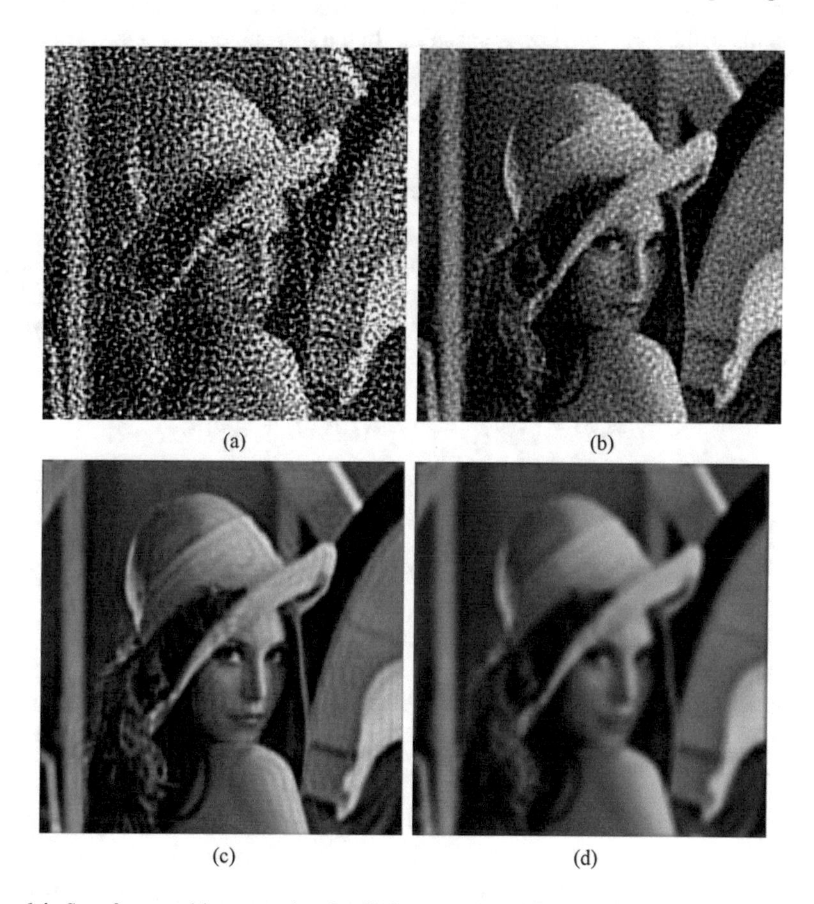

Fig. 6.4 Set of restored images using the CLS restoration filter and different regulation parameters, in the raster scanning order: **(a)** $\lambda = 0.001$, **(b)** $\lambda = 0.01$, **(c)** $\lambda = 0.1$, **(d)** $\lambda = 0.5$

not sufficiently reflect a priori smoothness constraints, and therefore it cannot avoid noise amplification. On the other hand, if λ is too large, the resulting image becomes too smooth, and it cannot keep detailed information of the original image. By experiments, we can find that Fig. 6.4c with $\lambda = 0.1$ shows the best restored image.

6.3 Wiener Filtering

The CLS filter has been derived under the assumption that we know important characteristics of the desired solution. Using appropriate constraints, the CLS filter provides satisfactory restoration results with a significantly reduced amount of computation. It is, however, not always possible to have appropriate constraints about the desired solution. The most popular and systematic way to obtain the characteristics of the solution is to estimate its power spectrum.

The Wiener filter can be said to be a stochastic optimization process with constraints in the form of a power spectrum. In the first subsection below, we derive the Wiener filter, and various practical spectrum estimation techniques follow.

6.3.1 Derivation of the Wiener Filter

Let $x(m,n)$ and $y(m, n)$ be arbitrary, zero mean, and random sequences with sizes $M_1 \times M_2$ and $N_1 \times N_2$, respectively. If we assume that $y(m,n)$ is the output of the deterministic linear system with impulse response $h(m,n)$ and additive noise $\eta(m,n)$, then we have

$$y(m,n) = \sum_i \sum_j h(m,n;i,j)x(i,j) + \eta(m,n). \tag{6.28}$$

This equation can be rewritten in the matrix-vector expression as

$$y = Hx + \eta, \tag{6.29}$$

where y and η represent $N_1N_2 \times 1$ vectors, x an $M_1M_2 \times 1$ vector, and H an $N_1N_2 \times M_1M_2$ block matrix.

Consider the linear estimate problem, for which the original undegraded image x is to be estimated, given the noisy degraded observation y. Intuitively, we may think that the following is the solution of the linear equation in Eq. (6.29)

$$x = H^{-1}(y - \eta). \tag{6.30}$$

Whether it is possible to compute H^{-1} or not, the estimate given in Eq. (6.30) is meaningless because y and η represent only one sample of the corresponding random sequence. In other words, in order to estimate a random sequence, it is reasonable to use statistical characteristics of given sequences as well as the information of the given sample sequence.

The most popular way to estimate a random sequence is to compute the best linear estimate of $\hat{x}$ from

$$\hat{x} = Gy, \tag{6.31}$$

which minimizes the average mean square error

$$\frac{1}{M_1 M_2} E\left\{ (x - \hat{x})^{\mathrm{T}} (x - \hat{x}) \right\}. \tag{6.32}$$

In Eq. (6.31), G represents an $M_1 M_2 \times N_1 N_2$ block matrix, and $E\{\cdot\}$, in Eq. (6.32), represents the averaging operation of the corresponding random sequence.

By the orthogonality property of the random sequences, the estimate $\hat{x}$ that minimizes Eq. (6.32) must satisfy

$$E\{ (x - \hat{x}) y^{\mathrm{T}} \} = 0. \tag{6.33}$$

Substituting Eq. (6.31) into Eq. (6.33) for $\hat{x}$ gives

$$R_{xy} - G R_{yy} = 0, \tag{6.34}$$

where R_{xy} and R_{yy}, respectively, represent the cross covariance matrix of x and y and the auto-covariance matrix of y, such that

$$R_{xy} = E\{ x y^{\mathrm{T}} \}, \text{ and } R_{yy} = E\{ y y^{\mathrm{T}} \}. \tag{6.35}$$

Using the relationships in Eqs. (6.29) and (6.35) and assuming that x is uncorrelated with η, we can obtain the linearly estimated matrix G from Eq. (6.34) as

$$G = R_{xy} {R_{yy}}^{-1} = R_{xx} H^{\mathrm{T}} \left(H R_{xx} H^{\mathrm{T}} + R_{\eta\eta} \right)^{-1}, \tag{6.36}$$

which is called the Wiener filter.

If the degradation operation occurs in a space-invariant manner, H is a doubly block Toeplitz. According to their definitions, both R_{xx} and $R_{\eta\eta}$ are also doubly block Toeplitz. If we assume a doubly block circulant approximation for each doubly block Toeplitz matrix and assume that $M_1 = M_2 = N_1 = N_2 = N$, Eq. (6.36) can be diagonalized by the two-dimensional DFT matrix as

$$D_G = D_{xx} {D_H}^{*} \left(D_H D_{xx} {D_H}^{*} + D_{\eta\eta} \right)^{-1}, \tag{6.37}$$

where $D_G = FGF^*$, $D_{xx} = FR_{xx}F^*$, $D_H = FHF^*$, and $D_{\eta\eta} = FR_{\eta\eta}F^*$. The (k,l)th diagonal element of D_G in Eq. (6.37) can be obtained as

$$\widetilde{G}(k,l) = \frac{S_{xx}(k,l)\widetilde{H}^*(k,l)}{\left|\widetilde{H}(k,l)\right|^2 S_{xx}(k,l) + S_{\eta\eta}(k,l)} = \frac{\widetilde{H}^*(k,l)}{\left|\widetilde{H}(k,l)\right|^2 + \dfrac{S_{\eta\eta}(k,l)}{S_{xx}(k,l)}}, \tag{6.38}$$

where $\widetilde{G}$ and $\widetilde{H}$, respectively, represent the two-dimensional DFTs of the impulse responses of the Wiener filter and the degradation system. S_{xx} and $S_{\eta\eta}$ represent spectral density functions of x and η, which are the two-dimensional DFTs of R_{xx} and $R_{\eta\eta}$, respectively.

The two-dimensional DFT of the estimate is obtained as

$$F\hat{x} = FGy = FGF^*Fy = D_G\widetilde{y}, \tag{6.39}$$

which requires only N^2 complex multiplications. Therefore, a random sequence can be estimated from the space-invariant degraded observation by using the frequency-domain Wiener filter.

6.3.2 *Practical Implementation of the Wiener Filter Using Spectrum Estimation*

The Wiener filter, or the minimum mean square error filter, is known to be optimum in the sense of minimizing the mean squared error between the original and estimated images. However, the optimality of the Wiener filter holds only when the power spectra of the original image and noise are given in addition to the PSF of the imaging system [na94]. This statement can be justified by investigating the filter transfer function given in Eq. (6.38).

6.3.2.1 Two-Dimensional Spectrum Estimation

Classical, one-dimensional power spectrum estimation methods have been applied to estimate two-dimensional power spectra [pitas93].

Let $x(m,n)$ represent a sample of a two-dimensional random field with size $N \times N$, then the power spectrum of $x(m,n)$ is defined as

$$S_{xx}(k,l) = \mathrm{S}_{xx}(z_1,z_2)\big|_{z_1=e^{mk},\,z_2=e^{nl}}, \tag{6.40}$$

where

$$S_{xx}(z_1, z_2) = \sum_m \sum_n R_{xx}(m, n) {z_1}^{-m} {z_2}^{-n}. \tag{6.41}$$

Since the autocorrelation function $R_{xx}(m, n)$ is determined under the assumption that the joint probability density function of the ensemble images is known, it is impossible to obtain the accurate autocorrelation function using only one sample of the random field. For this reason, there have been many methods introduced for estimating the power spectrum using a sample of the random image. Among them, we will introduce two popular methods in the following paragraphs.

First, the periodogram estimation, denoted by P_{xx}^{PER}, for the power spectrum of $x(m, n)$ is defined as

$$S_{xx}(k, l) \cong P_{xx}^{\text{PER}}(k, l) = \frac{1}{N^2} |X(k, l)|^2, \tag{6.42}$$

where $X(k, l)$ represents the DFT of $x(m, n)$ [welch67].

Second, the Blackman-Tukey estimation, denoted by P_{xx}^{BT}, for the power spectrum is defined as

$$S_{xx}(k, l) \cong P_{xx}^{\text{BT}}(k, l) = \sum_m \sum_n R_{xx}(m, n) w(m, n) e^{-j\frac{2\pi}{N}(mk+nl)}, \tag{6.43}$$

where

$$R_{xx} \cong \frac{1}{N^2} \sum_{\alpha=1}^{N-m} \sum_{\tau=1}^{N-l} x(\alpha, \tau) x(m + \alpha, n + \tau), \quad \text{for } 0 \le m \le \frac{N}{2}, \quad 0 \le n \le \frac{N}{2} \tag{6.44}$$

and the rest part of R_{xx} is filled by the two-dimensional periodic extension.

A power spectrum estimation is often evaluated by its frequency resolution and stability. The frequency resolution is a measure of the level of detail in an estimated spectrum. In other words, it represents the ability to distinguish closely spaced sinusoids. Stability, as used here, means the reciprocal of variance of the estimated power spectrum. According to the uncertainty principle, the product of resolution and stability is constant. Evaluation of various spectrum estimation methods, including the periodogram and Blackman-Tukey methods, has been summarized in [kay81].

6.3.2.2 Implementation of the Wiener Filter Based on Approximated Spectra

We can have various kinds of Wiener filters depending on the type of approximation of the original signal and noise spectra. In all cases, we assume that the noise is white Gaussian and its power spectrum is

$$S_{\eta\eta}(k,l) = \sigma_\eta^2, \quad \text{for all } k,l, \tag{6.45}$$

where σ_η^2 represents the variance of noise.

Wiener Filter Using a Constant Signal-to-Noise Power Ratio

The simplest method for implementing the Wiener filter is to assume that the power spectrum of the original image is constant over the entire frequency domain. Since both power spectra of the original image and of the noise are constant, we can have the constant signal-to-noise power ratio (SNPR) as [pratt78]

$$\frac{S_{xx}(k,l)}{S_{\eta\eta}(k,l)} \cong \frac{1}{\Gamma}. \tag{6.46}$$

From Eqs. (6.37) and (6.45), the transfer function of the approximated Wiener filter is given as

$$\widetilde{G}_A(k,l) = \frac{\widetilde{H}^*(k,l)}{\left|\widetilde{H}(k,l)\right|^2 + \Gamma}, \tag{6.47}$$

where $\widetilde{H}(k,l)$ represents the (k,l)th DFT coefficient of the periodically extended PSF[4] and $\widetilde{H}^*(k,l)$ the conjugate of $\widetilde{H}(k,l)$.

Although the approximated Wiener filter is not optimal, it gives acceptable restoration results when compared with other simple and efficient implementations.

Wiener Filter Using the Original Periodogram

In most practical image restoration problems, we do not have full information of the original image. However, when comparing performances of different restoration algorithms, it is often useful to assume that we know the original image. By using the original image, we can estimate its periodogram and implement an approximated Wiener filter as

$$\widetilde{G}_B(k,l) = \frac{\widetilde{H}^*(k,l)}{\left|\widetilde{H}(k,l)\right|^2 + \frac{\sigma_\eta^2}{P_{xx}^{\text{PER}}(k,l)}}. \tag{6.48}$$

[4]A sample procedure to make the periodically extended PSF is illustrated in Fig. 6.2.

Wiener Filter Using the Observed Periodogram

Although the approximated Wiener filter in Eq. (6.48) gives almost optimal restoration results, we cannot use this approximated filter in practical applications because the original image is inherently unavailable in the image restoration problem.

Without any information of the original image, the observed image becomes the best estimation of the original image. In [pitas93], the observed spectrum $S_{yy}(k,l)$ is therefore substituted for the original spectrum, which results in another approximated Wiener filter as

$$\widetilde{G}_C(k,l) = \frac{\widetilde{H}^*(k,l)}{\left|\widetilde{H}(k,l)\right|^2 + \dfrac{\sigma_\eta^2}{P_{yy}^{\mathrm{PER}}(k,l)}}. \tag{6.49}$$

In spite of its practical feasibility, the performance of the filter in Eq. (6.49) is poorer than filters given in Eqs. (6.47) and (6.48).

Wiener Filter Using the Modified Periodogram

The basic idea of the power spectrum estimation using the periodogram is that a stochastic quantity, power spectrum, is evaluated by using a deterministic quantity, a real sampled image. In a similar manner, we can develop relationships between the stochastic model and a given real sampled image as presented in the following development.

Based on the image degradation model shown in Fig. 6.1, taking the DFT of both sides of Eq. (6.28) yields

$$Y(k,l) = \widetilde{H}(k,l)X(k,l) + E(k,l). \tag{6.50}$$

$E(k,l)$ represents the DFT of noise $\eta(m,n)$. Since Eq. (6.50) holds for all discrete frequency (k,l) it can be rewritten as

$$Y = \widetilde{H}X + E. \tag{6.51}$$

By temporarily assuming that $\widetilde{H}(k,l) \neq 0$, $\forall(k,l)$, we have that

$$X = \frac{Y - E}{\widetilde{H}}. \tag{6.52}$$

By taking the squared magnitude on both sides, we have

$$|X|^2 = \frac{1}{\left|\widetilde{H}\right|^2}\left\{|Y|^2 + |E|^2 - 2\mathrm{Re}(EY)\right\}. \tag{6.53}$$

Since Eq. (6.53) has the form of the periodogram of $x(m, n)$, it can be substituted for $P_{xx}^{\mathrm{PER}}(k, l)$ in Eq. (6.48). The resulting Wiener filter is

$$\widetilde{G}_D = \frac{\widetilde{H}^*}{\left|\widetilde{H}\right|^2 + \dfrac{\sigma_\eta^2}{\dfrac{1}{\left|\widetilde{H}\right|^2}\left(|Y|^2 + |E|^2 - 2\mathrm{Re}\{EY\}\right)}}. \tag{6.54}$$

If we further assume that y and η are uncorrelated, Eq. (6.54) can then be simplified as

$$\widetilde{G}_D = \frac{\widetilde{H}^*}{\left|\widetilde{H}\right|^2\left(1 + \dfrac{\sigma_\eta^2}{|Y|^2 + \sigma_\eta^2}\right)}. \tag{6.55}$$

One critical disadvantage in using the filter in Eq. (6.55) is that it is defined only for $\widetilde{H}(k, l) \neq 0$. As a practical solution for that problem, we can use the following modified approximation:

$$\widetilde{G}_M(k, l) = \begin{cases} \widetilde{G}_D(k, l), & \widetilde{H}(k, l) \geq \varepsilon \\ \widetilde{G}_C(k, l), & \widetilde{H}(k, l) < \varepsilon \end{cases}, \tag{6.56}$$

where ε represents a practical lower bound of $\widetilde{H}(k, l)$, which is not considered zero.

6.3.2.3 Performance Evaluation

We shall use the 256×256 Lena image to compare the performance of various approximated Wiener filters. Based on the image degradation and restoration model shown in Fig. 6.1, the two-dimensional Gaussian function

$$h(m, n) = \frac{1}{2\pi\sigma^2}\exp\left(-\frac{m^2 + n^2}{2\sigma^2}\right) \tag{6.57}$$

is used as the PSF of the imaging system with $\sigma = 2.0$. Simulated noise is also added with SNR $= 40$ dB.

For objective comparison purposes, we use the quantity called improvement in signal-to-noise ratio (ISNR) defined as

$$\mathrm{ISNR} = 10\log\frac{\|x - y\|^2}{\|x - \hat{x}\|^2} \ (\mathrm{dB}), \tag{6.58}$$

where $\hat{x}$ represents the restored image. Four differently approximated Wiener filters, $\widetilde{G}_A$, $\widetilde{G}_B$, $\widetilde{G}_C$, and $\widetilde{G}_M$, are evaluated in the sense of ISNR as shown in Table 6.1.

Some experimental results using the modified Wiener restoration filter, $\widetilde{G}_M$, are summarized in the following figures (Figs. 6.5 and 6.6).

Table 6.1 Image restoration results in ISNR

Filter types	$\widetilde{G}_A$	$\widetilde{G}_B$	$\widetilde{G}_C$	$\widetilde{G}_M$
ISNR (dB)	9.25	9.81	6.21	8.44
Miscellaneous parameters	$\Gamma = 0.01$	–	–	$\varepsilon = 0.0006$

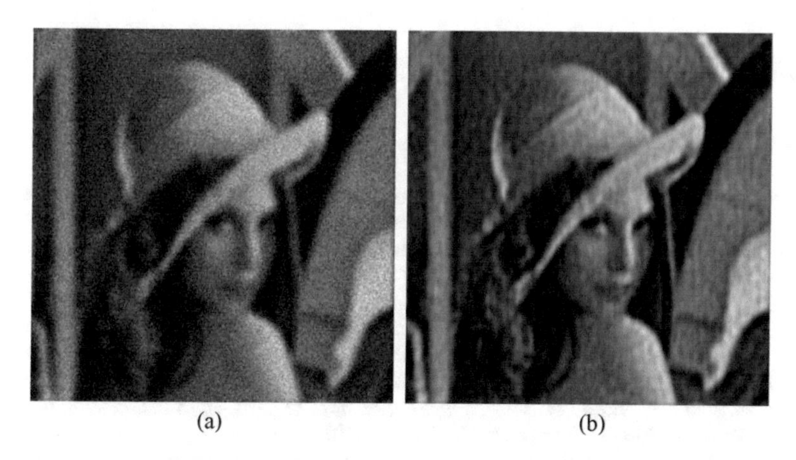

Fig. 6.5 (**a**) Defocused image by the two-dimensional Gaussian PSF in Eq. (6.56) with 10 dB noise, and (**b**) the restored image using the Wiener filter

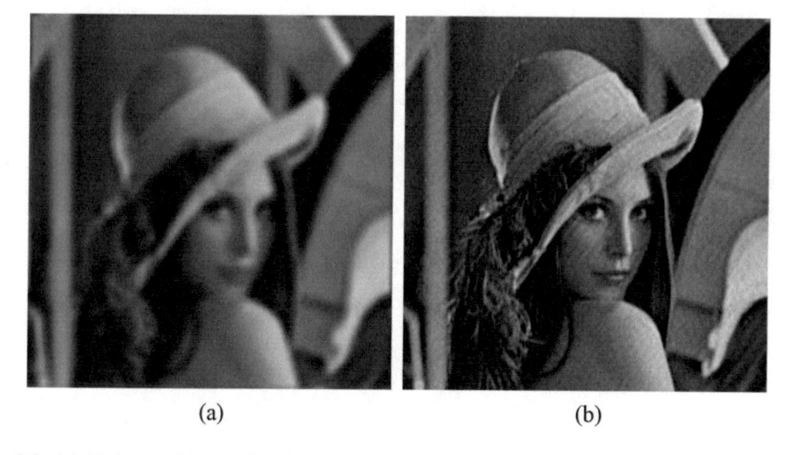

Fig. 6.6 (**a**) Defocused image by the two-dimensional Gaussian PSF in Eq. (6.56) with 40 dB noise and (**b**) the restored image using the Wiener filter

References

[kay81] S.M. Kay, S.L. Marple Jr., Spectrum analysis—a modern perspective. Proc. IEEE **69**(11), 1380–1419 (1981)
[na94] W. Na, J.K. Paik, Image restoration using spectrum estimation. Proc. 1994 Visual Commun. Image Process. **2308**(2), 1313–1321 (1994)
[oppenheim83] A.V. Oppenheim, A.S. Willsky, I.T. Young, *Signals and Systems* (Prentice-Hall, Englewood Cliffs, 1983)
[pitas93] I. Pitas, *Digital Image Processing Algorithms* (Prentice-Hall, Englewood Cliffs, 1993)
[pratt78] W.K. Pratt, *Digital Image Processing* (Wiley, New York, 1978)
[welch67] P.D. Welch, The use of fast fourier transform for the estimation of power spectra. IEEE Trans. Audio Electroacoust. **AU-15**(2), 70–73 (1967)

Additional References and Further Readings

[hill91] A.D. Hillery, R.T. Chin, Iterative Wiener filters for image restoration. IEEE Trans. Signal Process. **39**(8), 1892–1899 (1991)
[kang94] M.G. Kang, A.K. Katsaggelos, Frequency domain adaptive iterative image restoration and evaluation of the regularization parameter. Opt. Eng. **33**(10), 3222–3232 (1994)
[kim94] J. Kim, J.W. Woods, Image identification and restoration in the subband domain. IEEE Trans. Image Process. **3**, 312–314 (1994)
[press92] W.H. Press, B.P. Flannery, S.A. Teukolsky, W.T. Vetterling, Optimal (Wiener) filtering with the FFT. §13.3, in *Numerical Recipes in FORTRAN: The Art of Scientific Computing*, 2nd edn. (Cambridge University Press, Cambridge, England, 1992), pp. 539–542
[cast96] K.R. Castleman, *Digital Image Processing* (Prentice Hall, Englewood Cliffs, NJ, 1996)
[son99] M. Sonka, V. Hlavac, R. Boyle, *Image Processing, Analysis, and Machine Vision* (Brooks/Cole Publishing Company, Pacific Grove, 1999)
[lim91] H. Lim, K.-C. Tan, B.T.G. Tan, Edge errors in inverse and Wiener filter restorations of motion-blurred images and their windowing treatment. CVGIP: Graph. Model. Image Process. **53**(2), 186–195 (1991)
[bov00] Al Bovik, *Handbook of Image and Video Processing*, 2000. ISBN:0-12-119790-5

Chapter 7
Iterative Methods

Abstract In solving optimization problems, proper constraints play an important role in both making the problem well posed and in making the solution approach the desired and appropriate results. Once constraints are imposed on the solution of a problem, it becomes a constrained optimization problem. Since there are a variety of methods available for solving unconstrained optimization problems, rather than the constrained ones, we usually replace the constrained problem by an unconstrained counterpart by using regularization techniques.

In many image processing and computer vision applications, optimization problems are used for estimating the complete data. This assumes that the incomplete data, and the transformation that relates the complete and the incomplete data, are given. If the transformation occurs in a space-invariant manner, the optimization process can be performed in the discrete Fourier transform domain, as discussed in the previous chapter.

Although frequency domain implementation is extremely efficient for solving space-variant optimization problems, its application is limited because the space-invariance assumption does not hold for many problems. The use of iterative type methods for solving general optimization problems is widely accepted. They include (a) direct search methods, (b) derivative-based methods, (c) conjugate-gradient methods, and (d) quasi-Newton methods.

Iterative type methods in the above list have been developed for solving rather general-purpose numerical optimization problems. For this reason, they are not very efficient in solving more stringent optimization problems. This is especially true for image processing and computer vision applications, where special constraints, such as non-negativity and smoothness, are widely used.

In this chapter we introduce the regularized iterative method for solving optimization problems in image processing and computer vision applications. Particular emphasis is given to incorporation of constraints and discussion of convergence issues.

M.A. Abidi et al., *Optimization Techniques in Computer Vision*, Advances in Computer Vision and Pattern Recognition, DOI 10.1007/978-3-319-46364-3_7

7.1 Solution Space

Let the solution of an optimization problem be a N-dimensional vector. This vector may reside in N-dimensional subspace as

$$x \in S \subset R^N, \tag{7.1}$$

where x represents the solution and S the solution space.

Example 7.1 Solution space for 256×256 images.

Let x be an original image, y the degraded image, and H the matrix that transforms x into y, such as $y = Hx$. This relationship can be drawn in solution space, as shown in Fig. 7.1. A point in solution space represents a 256×256 image. Because the solution space is a subspace of R^N, where $N = 65,536$, a point on the space represents an N-dimensional vector.

If the original image x is degraded by matrix H, the resulting degraded image occurs on a different position, denoted by y. In Fig. 7.1, the arrow from x to y represents the direction of degradation by matrix H, not a vector.

Given a degraded image y and the degradation matrix H, the image restoration problem is to find the original image x. If we can find a matrix G which approximates the inverse of H, then we may estimate x by $x' = Gy$.

Note that the solution space drawn on a two-dimensional plane, as shown in Fig. 7.1, is not an exact form, but it gives an intuitive concept that each image can be represented by the corresponding vector and a vector moves to a different position by multiplication of a matrix.

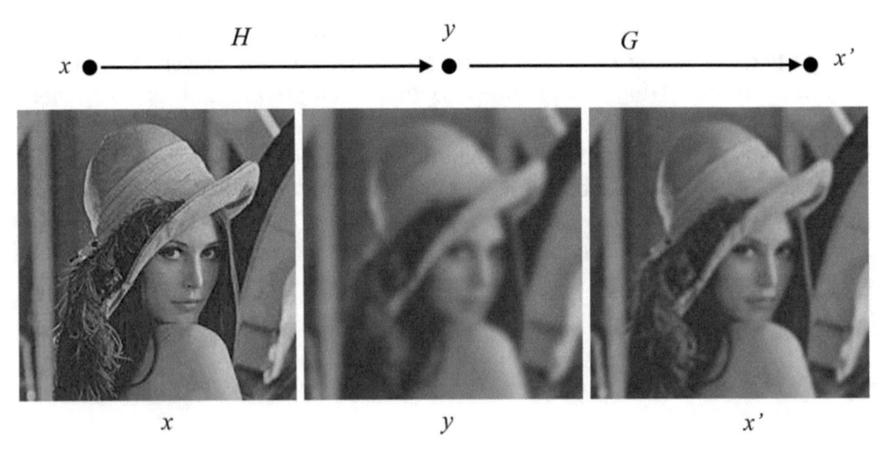

Fig. 7.1 Image degradation-restoration process on the $256^2 \times 1$-dimensional solution space

7.2 Formulation of Iterative Methods

In this section we consider the linear equation

$$y = Hx, \tag{7.2}$$

as applied to both well-posed and ill-posed cases.

7.2.1 Well-Posed Case

If H is nonsingular and well-posed, the solution of Eq. (7.2) can be estimated by $x' = Gy$, where $G = H^{-1}$. This process is shown in Fig. 7.1.

In many practical applications, however, it may be difficult, or even impossible, to determine and implement the inverse operator, especially when the dimensions of the matrix are large. In this case, iterative methods are particularly efficient in solving the linear equation. The formulation of gradient descent iterative methods is presented in the following paragraphs.

We know that the solution of the linear equation expressed in Eq. (7.2) minimizes the following function:

$$g(x) = \frac{1}{2}||y - Hx||^2. \tag{7.3}$$

The ellipsoid in Fig. 7.2 shows a contour on which the function $g(x)$ has the same value. As an estimate comes closer to the desired solution x^*, the function value decreases. The gradient of $g(x)$, when expressed as

$$\nabla g(x) = -H^{\mathrm{T}}(y - Hx), \tag{7.4}$$

represents the direction of increasing $g(x)$. On any point on the ellipsoid, the gradient vector points outward.

Based upon the characteristics of the quadratic function, we can formulate an iterative algorithm. In order to estimate the algorithmic solution, which minimizes

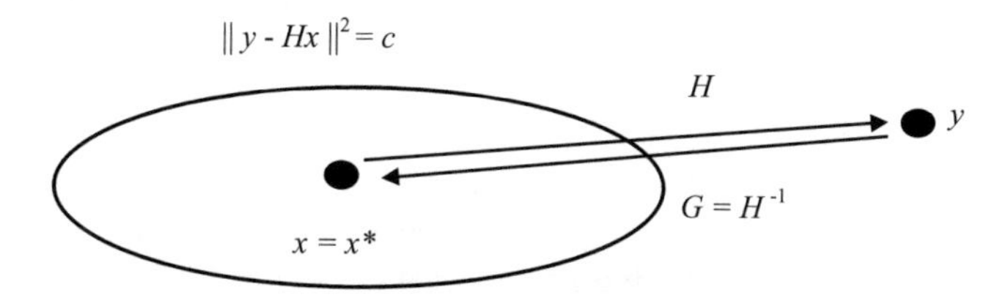

Fig. 7.2 Solution space description for well-posed image degradation and restoration process

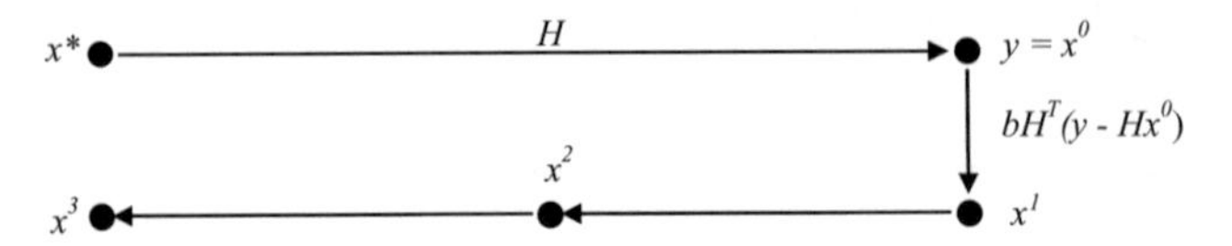

Fig. 7.3 A direct iterative procedure for minimizing $g(x) = ½ \|y - Hx\|^2$

the function in Eq. (7.1), we set the initial estimate as $x^0 = y$, as shown in Fig. 7.3. The gradient of $g(x)$ at the initial estimate is determined as $\nabla g(x^0) = -H^T(y - Hx^0)$. In order to update x^0 to a new estimate that decreases $g(x)$, we may add a portion of opposite direction to that of the gradient, such that

$$x^1 = x^0 + \beta H^T(y - Hx^0). \tag{7.5}$$

In the same manner, we can have the following iteration step:

$$x^{k+1} = x^k + \beta H^T(y - Hx^k), \quad for\ k = 0, 1, 2, \ldots \tag{7.6}$$

The solution of each iteration step, denoted by x^k, continues to decrease the function $g(x)$ as the iterations continue.

7.2.2 *Ill-Posed Case*

In order to solve Eq. (7.2), where the distortion operator H is ill posed, we must estimate the solution which minimizes

$$f(x) = \frac{1}{2}||y - Hx||^2 + \frac{\lambda}{2}||Cx||^2 \tag{7.7}$$

instead of $g(x)$ as expressed in Eq. (7.3). The gradient of $f(x)$ is determined from

$$\nabla f(x) = -\{H^T y - (H^T H + \lambda C^T C)x\}. \tag{7.8}$$

Let x^k be the estimated solution at the k-th iteration. The following iteration, x^{k+1}, can be determined by adding a portion of opposite direction to that of the gradient, such that

$$x^{k+1} = x^k + \beta(b - Tx^k), \tag{7.9}$$

where $b = H^T y$ and $T = H^T H + \lambda C^T C$. This represents the regularized version of the iterative algorithm for solving the linear equation in Eq. (7.2). The regularized iterative procedure is depicted in Fig. 7.4.

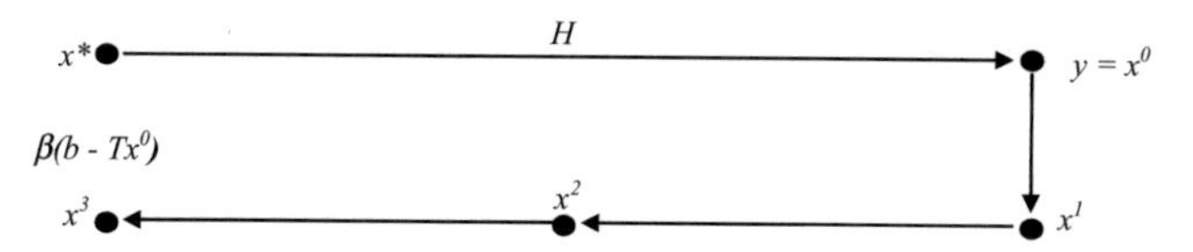

Fig. 7.4 A regularized iterative procedure for minimizing $f(x) = \frac{1}{2}\,\|y - Hx\|^2 + {}^{\lambda}/_{2}\,\|Cx\|^2$, where $b = H^T y$ and $T = H^T H + \lambda C^T C$

7.3 Convergence of Iterative Algorithms

Equation (7.9) can be expressed in transformation form by an operator in iterative form as

$$x^{k+1} = Fx^k, \tag{7.10}$$

where the operator F is defined by

$$Fx = x + \beta(b - Tx) = \beta b + Gx, \quad \text{where } G = I - \lambda T. \tag{7.11}$$

The solution x that satisfies

$$x = Fx \tag{7.12}$$

is called a fixed point of the transformation F.

Convergence of iterative algorithms can be analyzed by using the contraction mapping theorem as presented below.

Theorem 7.1: Contraction Mapping *Suppose that*

$$\|Fx^i - Fx^j\| \le r\|x^i - x^j\| \tag{7.13}$$

for x^i and x^j in some closed subspace of the space of signals. If $0 \le r < 1$, the operator F is said to be a contraction mapping, or simply a contraction, in that subspace. If $r = 1$, the operator is said to be non-expansive. If $r = 1$ and, in addition, Eq. (7.13) holds (with the equality only if $x^i = x^j$), then the operator F is strictly non-expansive.

The norm can be interpreted as the distance between two signals. Thus, we can say that contraction operators have the property that the distance between two signals tends to decrease as the operator transforms the signals.

If the operator F is a contraction in some subspace, then it has a unique fixed point x *in that subspace such that Eq. (7.12) holds. Furthermore, every sequence of successive approximations defined by Eq. (7.10) converges to x for any choice of the initial estimate x^0 in the subspace. In other words, $x^k \to x$ as $k \to \infty$.*

A further consequence of the contraction mapping theorem is

$$\|x - x^k\| \leq \frac{r^{k+1}}{1-r}\|x - x^0\| \tag{7.14}$$

for any x^0 in the subspace. That is, every sequence of iterations converges geometrically to the unique fixed point x in the sense that $\lim_{k\to\infty}\|x - x^k\| = 0$.

This is a very powerful theorem that not only guarantees convergence of the iteration in Eq. (7.10) but also guarantees the existence and uniqueness of the solution.

Based on the contraction mapping theorem, it can be proven that the iteration expressed in Eq. (7.10) converges if G is contractive, because

$$\|Fx^i - Fx^j\| = \|Gx^i - Gx^j\|. \tag{7.15}$$

7.4 Accelerating Convergence by Preconditioning

In order to analyze convergence of an iterative algorithm, the spectral radius of the iteration matrix is used. First, the iteration can be written as

$$x^+ = \beta b + (I - \beta T)x = \beta b + Gx, \tag{7.16}$$

where $G = I - \beta T$ is called the "iteration matrix." In analyzing the convergence of iterative algorithms, the spectral radius (which represents the maximum absolute eigenvalue of the iteration matrix) should be less than unity in order to guarantee convergence. When $\beta = 1$ in the iteration equation given in Eq. (7.16), the convergence accelerates as the maximum eigenvalue of T becomes closer to unity. In other words, as T becomes closer to the identity matrix, in some sense, the convergence will be accelerated. The preconditioned equivalent system for $Tx = b$ is the following:

$$\widetilde{T}\widetilde{x} = \widetilde{b}, \tag{7.17}$$

where

$$\widetilde{T} = PTP, \quad \widetilde{x} = P^{-1}x, \quad \text{and } \widetilde{b} = Pb. \tag{7.18}$$

In order to solve Eq. (7.17), the corresponding iteration is written as

$$\widetilde{x}^+ = \widetilde{x} + \beta\left(\widetilde{b} - \widetilde{T}\widetilde{x}\right). \tag{7.19}$$

By substituting Eq. (7.18) into Eq. (7.19), we have the preconditioned version of the iteration as

$$x^{+} = x + \beta\left(P^2 b - P^2 T x\right). \tag{7.20}$$

It is easily found that the ideal preconditioning matrix is

$$P \approx T^{-\frac{1}{2}}. \tag{7.21}$$

Therefore, we need $P^2 = T^{-1}$ pre-multiplied by b and T as shown in Eq. (7.20). If we approximate the block-Toeplitz matrix T to be block circulant, a 2D DFT can diagonalize T. Based on this approximation, we present an algorithm to obtain a preconditioning matrix P^2 as follows:

Algorithm 7.1: Preconditioning ($P^2 = T^{-1}$)

1. Place T's circulant coefficients symmetric to the DC axis in the frequency domain, and apply a 2D DFT.
2. Compute the reciprocal of the real DFT coefficients.
3. Compute the inverse 2D DFT of the reciprocal of the DFT coefficients obtained in step 2.
4. Truncate step 3's coefficients to a desired length with a raised cosine window.
5. Form the block circulant matrix $P \approx T^{-\frac{1}{2}}$ by using the IDFT values obtained in step 4.

In summary, in this chapter we have presented some of the fundamental theories for applying regularized iterative methods of optimization to image processing. As appropriate, algorithms for executing the iterative methods have also been presented.

Reference

[schafer81] R.W. Schafer, R.M. Mersereau, M.A. Richards, Constrained iterative restoration algorithms. Proc. IEEE **69**(4), 432–450 (1981)

Additional References and Further Readings

[hanke96] M. Hanke, J.G. Nagy, Restoration of atmospherically blurred images by symmetric indefinite conjugate gradient techniques. Inverse Prob. **12**, 157–173 (1996)

[hanke95] Martin Hanke, Conjugate gradient type methods for ill-posed problems, Pitman Research Notes in Mathematics Longman Scientific and Technical Longman House Harlow, Essex CM20 2JE UK, 1995

[reeves94] S.J. Reeves, Optimal space-varying regularization in iterative image restoration. IEEE Trans. Image Process. **3**, 319–324 (1994) [reeves93] S.J. Reeves, K.M. Perry, A stopping rule for iterative image restoration with constraints, in *Proceedings of the 1993 International Symposium on Circuits and Systems*, pp. 411–414, 1993

[bam93] R.H. Bamberger, S.L. Eddins, S.J. Reeves, An instructional image database package for image processing, in *Proceedings of the 1993 I.E. International Conference on Acoustics, Speech, and Signal Processing*, vol. I, pp. 16–19, 1993

[perry93] K.M. Perry, S.J. Reeves, A practical stopping rule for iterative signal restoration, in *Proceedings of the 1993 I.E. International Conference on Acoustics, Speech, and Signal Processing*, vol. III, pp. 440–443, 1993

[reeves92] S.J. Reeves, Optimal regularized image restoration with constraints, in *Proceedings of the 1992 I.E. International Conference on Acoustics, Speech, and Signal Processing*, pp. 301–304, 1992

[reeves92] S.J. Reeves, K.M. Perry, A practical stopping rule for iterative image restoration, in *SPIE—Image Processing Algorithms and Techniques III*, vol. 1657, pp. 192–200, 1992

[reeves91a] S.J. Reeves, Assessing the validity of constraint sets in image restoration, in *Proceedings of the 1991 I.E. International Conference on Acoustics, Speech, and Signal Processing*, pp. 2929–2932, 1991

[reeves91b] S.J. Reeves, R.M. Mersereau, Optimal regularization parameter estimation for image restoration, in *SPIE Image Processing Algorithms Techniques II*, vol. 1452, pp. 127–138, 1991

[reeves90a] S.J. Reeves, R.M. Mersereau, Optimal constraint parameter estimation for constrained image restoration, in *SPIE Visual Communications and Image Processing*, pp. 1372–1380, 1990

[reeves90b] S.J. Reeves, R.M. Mersereau, Identification of image blur parameters by the method of generalized cross-validation, in *Proceedings of the 1990 International Symposium on Circuits and Systems*, pp. 223–226, 1990

[kat89] A.K. Katsaggelos, Iterative image restoration algorithms. Opt. Eng., special issue on Visual Communications and Image Processing **28**(7), 735–748 (1989)

[pia97] M. Piana, M. Bertero, Projected Landweber method and preconditioning. Inverse Prob. **13**, 441–464 (1997)

[ber97] M. Bertero, D. Bindi, P. Boccacci, M. Cattaneo, C. Eva, V. Lanza, Application of the projected Landweber method to the estimation of the source time function in seismology. Inverse Prob. **13**, 465–486 (1997)

[lan51] L. Landweber, Am. J. Math. **73**, 615 (1951)

Chapter 8
Regularized Image Interpolation Based on Data Fusion

Abstract This chapter presents an adaptive regularized image interpolation algorithm, which is developed in a general framework of data fusion, to enlarge noisy-blurred, low-resolution (LR) image sequences. Initially, the assumption is made that each LR image frame is obtained by subsampling the corresponding original high-resolution (HR) image frame. Then the mathematical model of the subsampling process is obtained. Given a sequence of LR image frames and the mathematical model of subsampling, the general regularized image interpolation estimates HR image frames by minimizing the residual between the given LR image frame and the subsampled estimated solution with appropriate smoothness constraints.

The proposed algorithm adopts spatial adaptivity which can preserve the high-frequency components along the edge orientation in a restored HR image frame. This multiframe image interpolation algorithm is composed of two levels of data fusion. At the first level, an LR image is obtained and used as an input of the adaptive regularized image interpolation. At the second level, the spatially adaptive, fusion-based regularized interpolation is implemented by using steerable orientation analysis.

In order to apply the regularization approach to the interpolation procedure, an observation model of the LR video formation system is first presented. Based on the observation model, an interpolated image can be obtained, where the residual between the original HR and the interpolated images is minimized under a priori constraints. In addition, directional high-frequency components are preserved in the noise-smoothing process by combining spatially adaptive constraints. By experimentation, interpolated images using the conventional algorithms are compared with the proposed adaptive fusion-based algorithm. Experimental results show that the proposed algorithm has the advantage of preserving directional high-frequency components and suppressing undesirable artifacts such as noise.

8.1 Introduction

High-resolution (HR) image restoration has many applications in image processing. One important application of HR image restoration is the reconstruction of an uncorrupted image from a noisy-blurred image [andrews77]. The other is image

M.A. Abidi et al., *Optimization Techniques in Computer Vision*, Advances in Computer Vision and Pattern Recognition, DOI 10.1007/978-3-319-46364-3_8

interpolation associated with an increase in the spatial resolution of a single or a set of image frames [jain89, lim90]. In addition, application areas of HR image interpolation includes, but is not limited to, digital high-definition television (HDTV), aerial photography, medical imaging, surveillance video, and remote sensing [schowengerdt97].

In this chapter, image interpolation algorithms that can increase the resolution of an image are examined. By introducing image fusion and adaptive regularization algorithms, the proposed algorithm can restore HR image frames from LR video. Originally, the objective of image fusion was to combine information from multiple images of the same scene. As a result of image fusion, a single image, which is more suitable for human and machine perception or further image-processing tasks, can be obtained [zhang99]. Data fusion algorithms are usually used in applications ranging from earth resource monitoring, weather forecasting, and vehicular traffic control to military target classification and tracking [klein99]. By utilizing the properties of image fusion mentioned above, we can not only increase the resolution of LR image frames but can also construct a general framework based on image fusion solving the problem of multiframe image interpolation.

Many algorithms have been proposed to improve the resolution of images. Conventional interpolation algorithms, such as zero-order or nearest neighbor, bilinear, cubic B-spline, and the DFT-based interpolation, can be classified by basis functions, and they focus on just enlargement of an image [lim90, unser91, parker83]. Those algorithms have been developed under the assumption that there is no mixture among adjacent pixels in the imaging sensor, no motion blur due to finite shutter speed of the camera, no isotropic blur due to out-of-focus, and no aliasing in the process of subsampling. Since one or more of these algorithms are not true in general low-resolution imaging systems, restoring the original high-resolution image by using the conventional interpolation algorithms is almost impossible.

In order to improve the performance of the previously mentioned algorithms, a spatially adaptive cubic interpolation method has been proposed in [hong96]. Although this method can preserve a number of directional edges in the interpolation process, restoring original high-frequency components that have been lost in the subsampling process is not as easy. As an alternative, multiframe interpolation techniques which use sub-pixel motion information have been proposed in [kim90, patti94, patti97, hong97, tom96, schultz96, shin99, hardie97, hardie98, patti95, elad97, tekalp95, shah99, kang97, shin98, shin00].

It is well known that image interpolation is an ill-posed problem. More specifically, the subsampling process is regarded as a general image degradation process. Then the regularized image interpolation is to find the inverse solution defined by the image degradation model subject to a priori constraint [hong97, tom96]. Since the conventional regularized interpolation methods used isotropic smoothness as a priori constraints, their interpolation performance for images with various edges is limited.

A different approach to the high-resolution image interpolation problem has been proposed by Schultz and Stevenson. They addressed a method for nonlinear single-channel image expansion which preserves the discontinuities of the original image based on the maximum a posteriori (MAP) estimation technique [schultz94]. They also proposed a video superresolution algorithm, which used the MAP estimator with edge-preserving Huber-Markov random field (HMRF) prior [schultz96]. A similar approach to the superresolution problem has also been proposed by Hardie, Barnard, and Armstrong. They simultaneously estimate image registration parameters and the high-resolution image [hardie97, hardie98]. However, all of these methods have similar Gibbs priors which represent nonadaptive smoothness constraints.

A spatial image sequence interpolation algorithm using projections onto convex sets (POCS) theory has been studied in [patti94, patti97, shin99, patti95, trussel84, youla82]. In spite of several shortcomings, such as nonuniqueness of solution and computational complexity, POCS-based algorithms, in general, have a simple structure to implement. For regularized interpolation, deterministic or statistical information about the fidelity to the HR image and statistical information about the smoothness prior are incorporated to obtain the feasible solution between constraint sets.

On the other hand, a hybrid algorithm combining several different optimization methods, such as maximum likelihood (ML), MAP, and POCS approaches, was presented by Elad and Feuer [elad97]. They used the regularization weight matrix with locally adaptive smoothness to obtain a higher resolution image.

A video interpolation algorithm utilizing temporal motion information has been proposed by Shah and Zakhor [shah99]. This algorithm used a set of candidate motion estimates and chrominance components to compute accurate motion vectors. In [kang97], modified edge-based line average (ELA) techniques for the scanning rate conversion were proposed.

The common drawback of the abovementioned interpolation algorithms is that there is no direct effort to restore the high-frequency details lost in the subsampling process. One reasonable approach to solve this problem is fusion of temporarily adjacent image frames together with the existing multiframe image interpolation algorithm.

Restored HR image frames with high-frequency components along the direction of the edges can be obtained from LR image frames. In this chapter, incorporation of the data fusion technique into the interpolation algorithm is described.

8.2 Mathematical Model for the Subsampling Process

In this section, we present an observation model of the LR video formation system. The continuous-discrete model that includes the temporal sampling, sensor limitations, sampling, and noise effect is described. Then the model is transformed into the discrete-discrete type using ideally sampled input HR image frames.

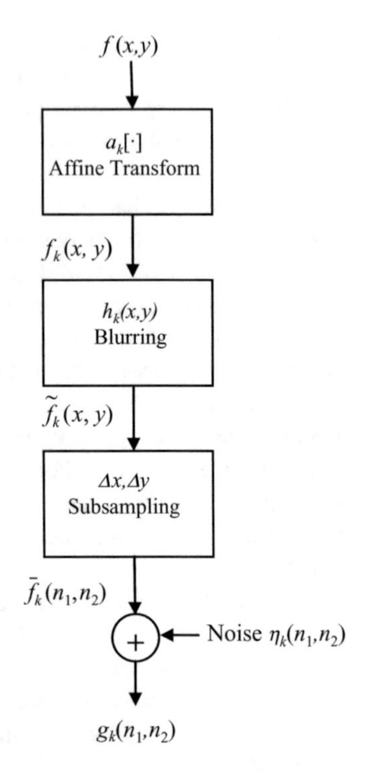

Fig. 8.1 Observation model of LR video formation system

8.2.1 Continuous-Discrete Model

The LR video formation process is depicted in Fig. 8.1. The continuous HR image is denoted by $f(x, y)$ in the continuous two-dimensional (2D) coordinate system (x, y). Here we assume that the HR image goes through as an affine transformation, blurring, and subsampling process. The affine transformation, denoted by $a_k[\bullet]$, includes three basic transformations: translation, scaling, and rotation. According to the transformation, the kth observed image frame in a sequence can be expressed as

$$f_k(x,y) = a_k[f(x,y)]_{d_k,\theta_k,s_k}, \text{ for } k = 1,2,\ldots,L, \tag{8.1}$$

where the parameters d_k, θ_k, and s_k, respectively, represent a translation vector $\{d_x, d_y\}$, a rotation angle, and the 2D scaling factor $\{s_x, s_y\}$ at the kth frame.

The blurring effect of the LR photodetectors is modeled by the second block in Fig. 8.1 as

$$\tilde{f}_k(x,y) = f_k(x,y)**h_k(x,y), \tag{8.2}$$

where ** represents the 2D convolution. Note that the blur is assumed to be linear space-invariant (LSI) and temporally variant. This type of blur results from imperfect optical systems such as out-of-focus and imaging sensors with limited resolution. In this chapter, only the sensor blurs are considered for simplicity.

The third step of the observation model is subsampling with constant sampling intervals, $(\Delta x, \Delta y)$, on the 2D rectangular sampling grid. The resulting subsampled version of $f_k(x, y)$ is given as

$$\bar{f}_k(n_1, n_2) = \sum_{n_1=1}^{N_1} \sum_{n_2=1}^{N_2} \bar{f}_k(x, y)\, \delta(x - n_1 \Delta x, y - n_2 \Delta y), \tag{8.3}$$

where δ represents the Dirac delta function.

Finally, the sampled image is corrupted by additive noise as

$$g(n_1, n_2) = \bar{f}_k(n_1, n_2) + \eta_k(n_1, n_2), \quad \text{for} \quad n_1 = 1, 2, \ldots, N_1,\ n_2 = 1, 2, \ldots, N_2. \tag{8.4}$$

8.2.2 Discrete-Discrete Model

The relationship has been established between the continuous HR image and the sampled LR image by using Eqs. (8.1) through (8.4). The HR interpolation, or equivalently restoration problem, is then posed as the reconstruction of the HR image from one or more observed LR image frames. Assume that the continuous HR image $f(x, y)$ is sampled over the Nyquist rate so that the sampled version of the HR continuous image has no aliasing. By ideal sampling at the Nyquist rate, the discrete HR image $f(m_1, m_2)$ of size $M_1 \times M_2$ can be obtained.

The discrete HR image goes through the same processes as shown in Fig. 8.1 except that f_k, g_k, h_k and η_k are replaced by the corresponding discrete arrays sampled on the 2D rectangular grid.

The discrete version of the affine transformation yields

$$f_k(m_1, m_2) = A_k[f(m_1, m_2)]_{d_k, \theta_k, s_k}, \quad \text{for} \quad m_1 = 1, \ldots, M_1, m_2 = 1, \ldots, M_2, m_1 = 1, \ldots, M_1, \text{and}\, k = 1, \ldots, L. \tag{8.5}$$

The discrete versions of blurring and subsampling are respectively given as

$$\tilde{f}_k(m_1, m_2) = f_k(m_1, m_2) ** h_k(m_1, m_2) \tag{8.6}$$

and

$$\bar{f}_k(n_1, n_2) = \tilde{f}_k\left(n_1 \frac{M_1}{N_1}, n_2 \frac{M_2}{N_2}\right), \quad \text{for} \quad n_1 = 1, \ldots, N_1, n_2 = 1, \ldots, N_2, \text{ and } k = 1, \ldots, L. \tag{8.7}$$

Finally, the noise corrupted subsampled discrete images are given as

$$g_k(n_1, n_2) = \bar{f}_k(n_1, n_2) + \eta_k(n_1, n_2). \tag{8.8}$$

As a result, the set of L observed LR image frames, $\{g_k(n_1, n_2)\}_{k=1}^{L}$, can be obtained. For notational simplicity, it is convenient to use the matrix-vector representation as

$$g_k = H_k f_k + \eta_k \quad \text{for} \quad k = 1, \ \ldots, \ L, \tag{8.9}$$

where f_k represents the row-ordered vector of the HR image of size $N^2 \times 1$, $H_k = S_k L_k A_k$, A_k denotes the $M^2 \times M^2$ affine transformation matrix, L_k represents the blurring operator of size $M^2 \times M^2$, and S_k represents the subsampling matrix of size $N^2 \times M^2$. g_k and η_k respectively represent the kth observed LR image of size $N^2 \times 1$ and additive Gaussian noise. In this chapter, the observed LR image g_k in Eq. (8.9) is assumed to be the output of the kth imperfect LR sensor.

8.3 Multiframe Regularized Image Interpolation Algorithm

In this section, a framework of multiframe image interpolation using data fusion and steerable constraints is proposed. The proposed interpolation algorithm is summarized in Fig. 8.2. As shown in the figure, the multiframe image interpolation algorithm is composed of two levels of data fusion. In level-1 fusion, sensor data

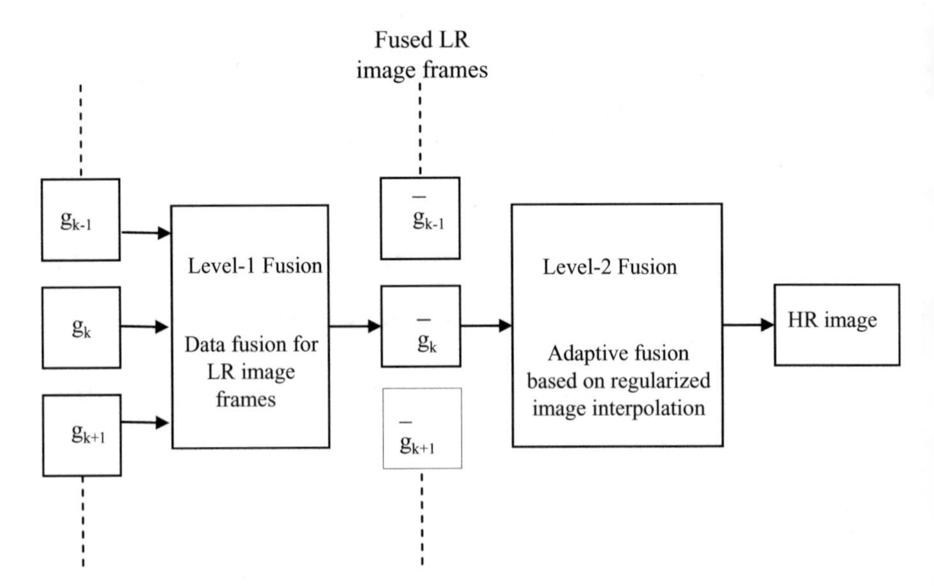

Fig. 8.2 The structure of the proposed multiframe image interpolation algorithm

from the LR imaging sensor is acquired and combined to obtain LR image frames with more information. Here, data from different sensors or different channels within a common sensor are combined.

Next, the level-2 fusion processing fuses data compatibility and spatially adaptive smoothness constraints to regularize the ill-posed interpolation problem. In general, the level-2 fusion performs either mixing or voting images depending on the customized purpose. In this chapter, level-2 fusion incorporates a spatially adaptive constraint to restore/interpolate the HR image which is then suitable for the human visual system. As a result, the HR image is restored from several LR sensor images.

8.3.1 Data Fusion for LR Image Frames: Level-1 Fusion

A brief review of the data fusion algorithm, which combines two or more LR image frames to provide a more feasible input image, is presented in this subsection. In order to perform the level-1 fusion, sensor data is registered by data-alignment processing. Although the objective of this algorithm is image interpolation with respect to the same scene, the image frames from a single sensor may include arbitrary motion due to moving objects. Therefore, data alignment is performed through spatiotemporal reference adjustment, coordinate system selection, and transformations that establish a common space-time reference for fusion processing [klein99]. Further details for the specific image interpolation can be found in [shin00].

Data fusion can be implemented at either the signal, the pixel, the feature, or the symbolic level of representation. Here, we use pixel-level fusion which merges multiple images on a pixel-by-pixel basis to improve the performance of general image processing tasks [kim90, hall92]. The reason for using this fusion concept in image interpolation is that multiple image frames present higher resolution and more information from the scene.

There are two approaches to combine LR image frames according to the processing domain. Spatial domain processing uses gradient and local variance information [shin00]. On the other hand, transformed domain processing uses the discrete wavelet transform (DWT), discrete wavelet frame (DWF) [shin00b], or a steerable pyramid. By performing level-1 fusion to be discussed in this subsection, fused LR image frames can be constructed as $\{\overline{g}_k(n_1, n_2)\}_{k=1}^{L}$. After the data fusion process, regularization-based adaptive fusion is performed, as described in the following subsection.

8.3.2 Regularization-Based Adaptive Fusion: Level-2 Fusion

Level-2 fusion incorporates a voting fusion into the existing iterative regularization structure which fuses data compatibility and smoothness of the solution to solve the ill-posed problem.

It is well known that the existing regularized image interpolation algorithms do not satisfy the human visual system [shin98]. Human visual characteristics have been partly revealed by psychophysical experiments. According to these experiments, the human visual system is sensitive to noise in flat regions and becomes less sensitive to noise at sharp transitions in image intensity, i.e., the human visual system is less sensitive to noise in edges than in flat regions [anderson76]. Based upon experimental results, various methods to subjectively improve the quality of the restored image have been proposed in [katsaggelos89, katsaggelos91].

In this subsection, a novel adaptive, efficient fusion algorithm which is adequate for the human visual system is introduced. The steerable orientation analysis and its application to spatially adaptive constraints are also presented.

8.3.3 Regularized Image Restoration

In solving Eq. (8.9), the regularized image restoration algorithm is used to find the estimate $\hat{x}$ which satisfies the following optimization problem [kang97]:

$$\hat{x} = \underset{x}{\text{argmin}} f(x), \tag{8.10}$$

where

$$f(x) = \sum_{k=1}^{L} f_k(x) = \sum_{k=1}^{L} \left[\|\overline{g}_k - H_k x\|^2 + \lambda \|Cx\|^2 \right], \tag{8.11}$$

C is an $M^2 \times M^2$ matrix which represents a high-pass filter and $\|Cx\|^2$ represents a stabilizing functional whose minimization suppresses high-frequency components due to noise amplification. λ represents the regularization parameter which controls the fidelity to the given data and smoothness of the restored image.

The cost function given in Eq. (8.11) can be minimized when x satisfies the following equation:

$$\sum_{k=1}^{L} [H_k^{\mathrm{T}} H_k + \lambda C^{\mathrm{T}} C] x = \sum_{k=1}^{L} H_k^{\mathrm{T}} \overline{g}_k. \tag{8.12}$$

The solution for Eq. (8.12) can be obtained if the entire set of both L fused LR image frames $\overline{g}_k$ and the corresponding subsampling matrices H_k is available at the same time. This approach has the advantage of providing more feasible solutions with the cost of heavy computation. Therefore, it can be used when processing time is not critical.

For solving Eq. (8.12), the successive approximation procedure at the $(l+1)$th iteration step is given as

$$x^{l+1} = x^l + \beta\left\{\sum_{k=1}^{L} H_k^{\mathrm{T}}\overline{g}_k - \sum_{k=1}^{L}\left[H_k^{\mathrm{T}}H_k + \lambda C^{\mathrm{T}}C\right]x^l\right\}, \tag{8.13}$$

where $\overline{g}_k$ represents the fused LR image frame from the set of P consecutive LR image frames, $\left\{g_{k-\frac{P-1}{2}}, \ldots, g_k, \ldots, g_{h+\frac{P+1}{2}}\right\}$, and β the relaxation parameter that controls the convergence.

8.3.4 Orientation Analysis Based on the Steerable Filter

This subsection provides an orientation analysis algorithm that can be used to find the exact orientation of an edge, which will then be used to form a spatially adaptive regularization. Orientation-specific filters are useful in many image processing and early vision tasks. Freeman and Adelson developed the concept of steerable filters, in which a filter of arbitrary orientation is synthesized as a linear combination of a set of basis filters [freeman91]. The quadrature pair filters include, in general, the pth order derivative of Gaussian filters and their Hilbert transforms.

In this chapter, two different sets of steerable filters are used: (a) the second derivative of a Gaussian filter and (b) its Hilbert transformed filter. For the Gaussian filter, the following basis filters are used:

$$\begin{aligned} G_{2a} &= 0.92132\left(2x_1^2 - 1\right)\mathrm{e}^{-\left(x^2+y^2\right)}, \\ G_{2b} &= 1.84264xy\mathrm{e}^{-\left(x^2+y^2\right)}, \text{ and} \\ G_{2c} &= 0.92132\left(2y_1^2 - 1\right)\mathrm{e}^{-\left(x^2+y^2\right)}. \end{aligned} \tag{8.14}$$

The corresponding steering functions are

$$\begin{aligned} K_a(\theta) &= \cos^2(\theta), \\ K_b(\theta) &= -2\cos(\theta)\sin(\theta), \text{ and} \\ K_c(\theta) &= \sin^2(\theta). \end{aligned} \tag{8.15}$$

A synthesized filter with orientation θ can be represented as

$$G_2^{\theta} = \left(K_a(\theta)G_{2a} + K_b(\theta)G_{2b} + K_c(\theta)G_{2c}\right). \tag{8.16}$$

A quadrature pair is formed by using the Hilbert transformation of the Gaussian filters. A normalized numerical approximation of the Hilbert transform results in the following four basis filters:

$$\begin{aligned} H_{2a} &= 0.97780(-2.254x + y^3)\mathrm{e}^{-\left(x^2+y^2\right)}, \\ H_{2b} &= 0.97780(-0.7515x + x^2)y\mathrm{e}^{-\left(x^2+y^2\right)}, \\ H_{2c} &= 0.97780(-0.7515x + y^3)x\mathrm{e}^{-\left(x^2+y^2\right)}, \quad \text{and} \\ H_{2d} &= 0.97780(-2.254y + y^3)\mathrm{e}^{-\left(x^2+y^2\right)}. \end{aligned} \tag{8.17}$$

The corresponding steering functions are

$$\begin{aligned} K_a(\theta) &= \cos^3(\theta), \\ K_b(\theta) &= -3\cos^2(\theta)\sin(\theta), \\ K_c(\theta) &= 3\cos(\theta)\sin^2(\theta), \quad \text{and} \\ K_d(\theta) &= -\sin^3(\theta). \end{aligned} \tag{8.18}$$

The resulting filter steered with the orientation θ is

$$H^\theta = (K_a(\theta)H_{2a} + K_b(\theta)H_{2b} + K_c(\theta)H_{2c} + K_d(\theta)H_{2d}). \tag{8.19}$$

By choosing the nth derivative of a Gaussian and its Hilbert transform as band-pass filters, we have that

$$E_n(\theta) = \left[G_n^\theta\right]^2 + \left[H_n^\theta\right]^2. \tag{8.20}$$

Expressing the steered filter response in terms of the basis filters and steering weights, the estimated energy in Eq. (8.20) can be rewritten as

$$E_n(\theta) = C_1 + C_2\cos(2\theta) + C_3\cos^2(2\theta) + (\text{higher order terms}). \tag{8.21}$$

The dominant orientation (one that maximizes the output energy) θ_d is given as

$$\theta_d = \frac{\arg[C_2, C_3]}{2}, \tag{8.22}$$

and the orientation strength is

$$S_\theta = \sqrt{C_2^2 + C_3^2}. \tag{8.23}$$

8.3.5 Spatially Adaptive Fusion

As an alternative to the nonadaptive processing in Eq. (8.13), we propose a spatially adaptive fusion algorithm by using a set of P different high-pass filters, C_i, for $i = 1, \ldots, P$, which selectively suppresses high-frequency components along the corresponding edge direction. The assumption is made that the number of edge directions is equal to five, that is $P = 5$, and each pixel in an image can be classified into either a monotone, horizontal edge, vertical edge, or two diagonal edges by the steerable orientation analysis. By applying this adaptive fusion algorithm, an HR image can be interpolated from LR image frames, and five directional edges are preserved in the HR image simultaneously.

To apply the proposed spatially adaptive fusion algorithm into the existing regularization, the isotropic high-pass filter C in Eq. (8.13) is replaced by a set of P different high-pass filters, C_i, for $i = 1, \ldots, P$, and the $(l+1)$st regularized iteration step can be given as

$$x^{l+1} = x^l + \beta\left\{\sum_{k=1}^{L} H_k^{\mathrm{T}} g_k - \left[\sum_{k=1}^{L} H_k^{\mathrm{T}} H_k + \sum \lambda_i I_i C_i^{\mathrm{T}} C_i\right] x^l\right\}, \tag{8.24}$$

where I_j represents a diagonal matrix with diagonal elements of either zero or one. The properties of I_i can simply be summarized as

$$I_i I_j = 0, \quad \text{for} \quad i \neq j, \text{and} \sum_{i=1}^{M} I_i = I, \tag{8.25}$$

where I represents an $N^2 \times 1$ identity matrix. More specifically, the diagonal element in I_j, which has a one-to-one correspondence with each pixel in the image, is equal to one if it is on the corresponding edge, or zero otherwise.

In order to incorporate adaptive smoothness constraints to this algorithm, the edge orientation of each pixel must be determined and classified. At the same time, we should define the spatially adaptive high-pass filters $\{C_i\}_{i=1}^{P}$ corresponding to the direction of dominant orientation at each pixel. When $P = 5$, a typical set of directional constraints C_i are defined as follows:

$$C_1 = \frac{1}{16}\begin{bmatrix} 0 & 0 & -1 & 0 & 0 \\ 0 & -1 & -2 & -1 & 0 \\ -1 & -2 & 6 & -2 & -1 \\ 0 & -1 & -2 & -1 & 0 \\ 0 & 0 & -1 & 0 & 0 \end{bmatrix}, \quad C_2 = \frac{1}{6}\begin{bmatrix} 0 & 0 & 0 & 0 & 0 \\ 0 & -1 & 2 & -1 & 0 \\ 0 & -1 & 2 & -1 & 0 \\ 0 & -1 & 2 & -1 & 0 \\ 0 & 0 & 0 & 0 & 0 \end{bmatrix},$$

$$C_3 = \frac{1}{6}\begin{bmatrix} 0 & 0 & 0 & 0 & 0 \\ 0 & -1 & -1 & -1 & 0 \\ 0 & 2 & 2 & 2 & 0 \\ 0 & -1 & -1 & -1 & 0 \\ 0 & 0 & 0 & 0 & 0 \end{bmatrix}, \quad C_4 = \frac{1}{16}\begin{bmatrix} 0 & 0 & -1 & 0 & 0 \\ -1 & 2 & -1 & -1 & 0 \\ 0 & -1 & 2 & -1 & 0 \\ 0 & 0 & -1 & 2 & -1 \\ 0 & 0 & 0 & 0 & 0 \end{bmatrix},$$

$$\text{and } C_5 = \frac{1}{6}\begin{bmatrix} 0 & 0 & 0 & 0 & 0 \\ 0 & 0 & -1 & 2 & -1 \\ 0 & -1 & 2 & -1 & 0 \\ -1 & 2 & -1 & 0 & 0 \\ 0 & 0 & 0 & 0 & 0 \end{bmatrix}. \tag{8.26}$$

The orientation analysis algorithm using a steerable filter determines the direction θ_d and magnitude S_d of the dominant orientation at each pixel as given in Eqs. (8.22) and (8.23). According to the direction of dominant orientation at each pixel, the current image x^l is convolved with the corresponding high-pass filter C_i in Eq. (8.26). As a result, directional smoothness constraints can be fused. Figure 8.3 gives an overview of the proposed adaptive fusion processing based on regularization.

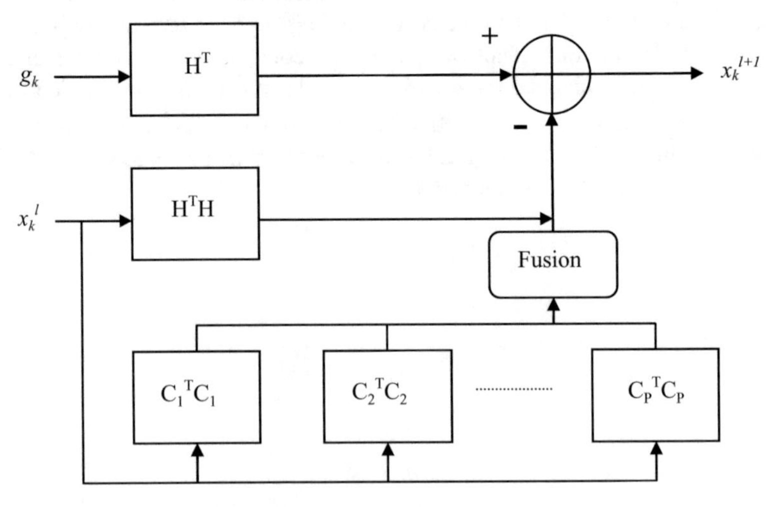

Fig. 8.3 Adaptive fusion processing based on regularization

In order to demonstrate the performance of the proposed algorithm, 16 64×64 image frames can first be made by subsampling a 256×256 Lena image at 16 differently shifted positions. Here we assume that 64×64 image frames represent the LR image and the 256×256 image represents the HR image. After subsampling, white Gaussian noise is added with a 20 dB signal-to-noise ratio (SNR). In this experiment, we assume that the LR image frames have only global translational motion and suffer from 4×4 uniform blur. In other words, the affine transform matrix A_k has only translational motion components.

Figure 8.4a represents the 256×256 HR Lena image, and Fig. 8.4b represents the 16 64×64 LR images. Figures 8.5a, b show the 4×4 interpolated image of the

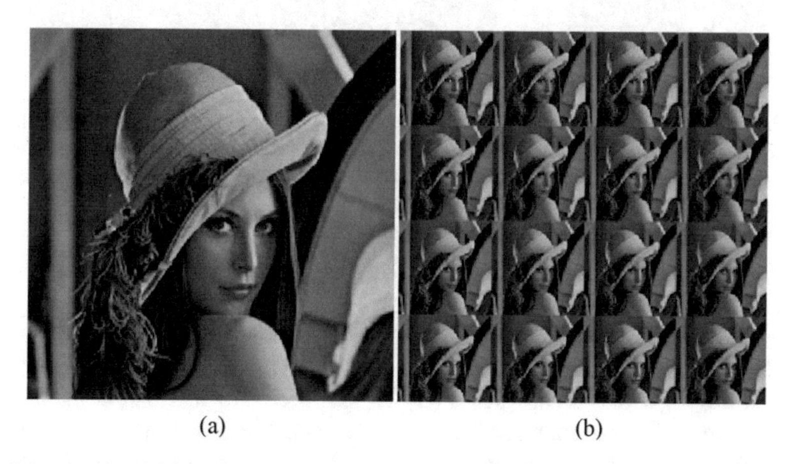

(a) (b)

Fig. 8.4 (**a**) Original HR Lena image and (**b**) 16 synthetically subsampled LR image frames with factor of 1/4 and 20 dB additive Gaussian noise. (**a**) Original image; (**b**) subsampled image

(a) (b)

Fig. 8.5 Simulated HR images by interpolating the *top*, *left* image frame of Fig. 8.4b using (**a**) zero-order and (**b**) bilinear interpolation algorithms. (**a**) Zero-order interpolated image (PSNR = 23.10|dB|); (**b**) bilinear interpolated image (PSNR = 21.46|dB|)

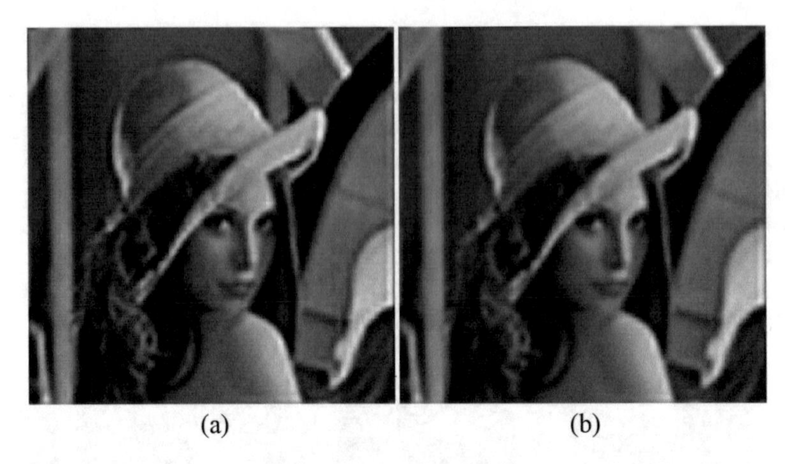

Fig. 8.6 Simulated HR images obtained by interpolating 16 image frames of Fig. 8.4b using the conventional regularization methods with different regularization parameters. (**a**) Nonadaptively regularized interpolated image (PSNR = 25.66|dB|, $\lambda = 0.01$); (**b**) nonadaptively regularized interpolated image (PSNR = 25.31|dB|, $\lambda = 1.0$)

first frame (top, left image frame of Fig. 8.4b) obtained by using the zero-order and bilinear interpolation methods, respectively.

Figures 8.6a, b respectively show interpolated images by using the existing regularized interpolation algorithm given in Eq. (8.13). For both interpolated images, the algorithm terminates after ten iterations. Regularization parameters, $\lambda = 0.01$, and 1.0 are used, respectively. As shown in Fig. 8.6, a large λ can suppress noise effectively but blurs the image. On the contrary, a small λ cannot efficiently suppress noise amplification.

Figures 8.7a, b show interpolated images gained by using the adaptive fusion algorithm with and without level-1 fusion, respectively. For both interpolated images, the algorithm terminates after ten iterations, and the set of adaptive constraints in Eq. (8.26) is used. In Fig. 8.7b, noise remains in some regions because of level-1 fusion. Actually, when level-1 fusion combines higher-frequency components from severely noisy LR image frames, noise is also considered to be high-frequency components.

In this chapter we proposed a general framework of multichannel image interpolation using data fusion and steerable filters. The proposed framework is composed of level-1 and level-2 fusions. Level-1 fusion processing is performed at the pixel level and provides enhanced LR images so that HR image frames can be obtained in the main processing of the level-2 fusion. Level-2 fusion is implemented at the feature level, and features at this level provide the spatially adaptive constraints in the fusion process.

For example, edge features in the LR image frame are classified and fused with data compatibility which can restore high-resolution images. As a result, the level-2 fusion processing image is obtained at the intersection of two uncertainty sets such as the data compatibility and smoothness constraint sets. The space variant

Fig. 8.7 Simulated HR images obtained by the proposed regularization-based adaptive fusion algorithm (**a**) without and (**b**) with level-1 fusion. (**a**) Adaptively regularized interpolated image (PSNR = 25.60|dB|, $\lambda_1 = 1.0$, $\lambda_{2...5} = 0.1$); (**b**) Adaptively regularized interpolated image using level-1 fusion (PSNR = 25.63|dB|, $\lambda_1 = 1.5$, $\lambda_{2...5} = 0.5$)

regularization parameter λ can be adopted at this level. Steerable constraints along the orientation of the edge are also used in the level-2 processing.

The regularization-based fusion algorithm can be divided into two classes: nonadaptive and adaptive. The adaptive algorithm provides higher quality than the nonadaptive version in the sense of high-frequency details along the direction of edges. The proposed framework presents the optimal solution to HR video with edge-preserving constraints from LR video. By applying the data fusion concept within this algorithm, the regularized image interpolation algorithm and orientation analysis results using steerable filters are successfully combined.

References

[andrews77] H.C. Andrews, B.R. Hunt, *Digital Image Restoration* (Prentice-Hall, Englewood Cliffs, 1977)

[jain89] A.K. Jain, *Fundamentals of Digital Image Processing* (Prentice-Hall, Englewood Cliffs, 1989)

[lim90] J.S. Lim, *Two-Dimensional Signal and Image Processing* (Prentice-Hall, Englewood Cliffs, 1990)

[schowengerdt97] R.A. Schowengerdt, *Remote Sensing: Models and Methods for Image Processing*, 2nd edn. (Academic, New York, 1997)

[zhang99] Z. Zhang, R.C. Blum, A categorization of multiscale-decomposition-based image fusion schemes with a performance study for a digital camera application. Proc. IEEE **87**, 1315–1326 (1999)

[klein99] L.A. Klein, *Sensor and Data Fusion Concepts and Applications* (SPIE Optical Engineering Press, 1999)

[unser91] M. Unser, Fast b-spline transforms for continuous image representation and interpolation. IEEE Trans. Pattern Anal. Mach. Intell. **13**, 277–285 (1991)

[parker83] J.A. Parker, R.V. Kenyon, D.E. Troxel, Comparison of interpolating methods for image resampling. IEEE Trans. Med. Imaging **2**, 31–39 (1983)

[hong96] K.P. Hong, J.K. Paik, H.J. Kim, C.H. Lee, An edge-preserving image interpolation system for a digital camcoder. IEEE Trans. Consum. Electron. **42**, 279–284 (1996)

[kim90] S.P. Kim, H.K. Bose, H.M. Valenzuela, Recursive reconstruction of high-resolution image from noisy undersampled frames. IEEE Trans. Acoust. **38**, 1013–1027 (1990)

[patti94] A. Patti, M.I. Sezan, and A.M. Tekalp, High-resolution image reconstruction from a low-resolution image sequence in the presence of time varying motion blur, Proc. Int. Conf. Image Processing (1994)

[patti97] A. Patti, M.I. Sezan, A.M. Tekalp, Superresolution video reconstruction with arbitrary sampling lattices and nonzero aperture time. IEEE Trans. Image Process. **6**, 1064–1076 (1997)

[hong97] M.C. Hong, M.G. Kang, A.K. Katsaggelos, An iterative weighted regularized algorithm for improving the resolution of video sequences. Proc. Int. Conf. Image Process. **2**, 474–477 (1997)

[tom96] B.C. Tom and A.K. Katsaggelos, An iterative algorithm for improving the resolution of video sequences, Proc. SPIE Visual Comm. Image Proc., 1430–1438 (1996)

[schultz96] R.R. Schultz, R.L. Stevenson, Extraction of high-resolution frames form video sequences. IEEE Trans. Image Process. **5**, 996–1011 (1996)

[shin99] J.H. Shin, J.H. Jung, and J.K. Paik, Spatial interpolation of image sequences using truncated projections onto convex sets, IEICE Trans. Fund. Electron. Comm. Comput. Sci. (1999)

[hardie97] R.C. Hardie, K.J. Barnard, E.E. Armstrong, Joint map registration and high-resolution image estimation using a sequence of undersampled images. IEEE Trans. Image Process. **6**, 1621–1633 (1997)

[hardie98] R.C. Hardie, K.J. Barnard, J.G. Bognar, E.E. Armstrong, E.A. Watson, High-resolution image reconstruction from a sequence of rotate and translated frames and its application to an infrared imaging system. Opt. Eng. **37**, 247–260 (1998)

[patti95] A.J. Patti, M.I. Sezan, and A.M. Tekalp, High-resolution standards conversion of low resolution video, Proc. IEEE Int. Conf. Acoust. Speech Signal. Process., 2197–2200 (1995)

[elad97] M. Elad, A. Feuer, Restoration of a single superresolution image from several blurred, noisy, and undersampled measured images. IEEE Trans. Image Process. **6**, 1646–1658 (1997)

[tekalp95] A.M. Tekalp, *Digital video processing* (Prentice-Hall, Englewood Cliff, 1995)

[shah99] N.R. Shah, A. Zakhor, Resolution enhancement of color video sequences. IEEE Trans. Image Process. **8**, 879–885 (1999)

[kang97] M.G. Kang, A.K. Katsaggelos, Simultaneous multichannel image restoration and estimation of the regularization parameters. IEEE Trans. Image Process. **6**, 774–778 (1997)

[shin98] J.H. Shin, J.H. Jung, J.K. Paik, Regularized iterative image interpolation and its application to spatially scalable coding. IEEE Trans. Consum. Electron. **44**, 1042–1047 (1998)

[shin00] J.H. Shin, J.S. Yoon, J.K. Paik, Image fusion-based adaptive regularization for image expansion. Proc. SPIE Image, Video Comm. Process. **3974**, 1040–1051 (2000)

[shin00b] J.H. Shin, J.H. Jung, J.K. Paik, M.A. Abidi, Adaptive image sequence resolution enhancement using multiscale decomposition based image fusion. Proc. SPIE Visual Comm. Image Proc. **3**, 1589–1600 (2000)

[schultz94] R.R. Schultz, R.L. Stevenson, A bayesian approach to image expansion for improved definition. IEEE Trans. Image Process. **3**, 233–242 (1994)

[trussel84] H.J. Trussell, M.R. Civanlar, The feasible solution in signal restoration. IEEE Trans. Acoust. **32**, 201–212 (1984) [youla82] D.C. Youla, H. Webb, Image restoration by the method of convex projections: part 1-theory. IEEE Trans. Med. Imaging **MI-1**, 81–94 (1982)

[hall92] D.L. Hall, *Mathematical Techniques in Multisensor Data Fusion* (Artech House, 1992)

[anderson76] G.L. Anderson, A.N. Netravali, Image restoration based on a subjective criterion. IEEE Trans. Syst. Man Cybern. **SMC-6**, 845–853 (1976)

[katsaggelos89] A.K. Katsaggelos, Iterative image restoration algorithms. Opt. Eng. **28**, 735–748 (1989)

[katsaggelos91] A.K. Katsaggelos, J. Biemond, R.W. Schafer, R.M. Mersereau, A regularized iterative image restoration algorithms. IEEE Trans. Signal Process. **39**(4), 914–929 (1991)

[freeman91] W.T. Freeman, E.H. Adelson, The design and use of steerable filters. IEEE Trans. Pattern Anal. Mach. Intell. **13**, 891–906 (1991)

Part IV

Chapter 9
Enhancement of Compressed Video

Abstract In this chapter, a modified, regularized image restoration algorithm useful in reducing blocking artifacts in predictive-coded (P) pictures of compressed video, based on the corresponding image degradation model, is presented. Since most video coding standards adopt a hybrid structure of macroblock-based motion compensation (MC) and block discrete cosine transform (BDCT), the blocking artifacts occur at both the block boundary and block interior, and the degradation process due to quantization is generated on just differential images. Based on observation, a new degradation model is needed for differential images and the corresponding restoration algorithm, which directly processes the differential images before reconstructing decoded images. For further removal of both kinds of blocking artifacts, the restored differential image must satisfy two constraints: directional discontinuities on the block boundary and on the block interior. These constraints have been used for defining convex sets for restoring differential images. In-depth analysis of differential domain processing is presented in the appendix and serves as the theoretical basis for justifying differential domain image processing. Experimental results also show significant improvement over conventional methods in the sense of both objective and subjective criteria.

9.1 Introduction

As the demand for video communication has grown, many efficient image compression techniques have been developed and standardized. Narrowband transmission channels for video communication and limited amount of video storage necessitate efficient video compression techniques. Most video coding standards, such as H.261, H.263, MPEG-1, and MPEG-2, adopt a hybrid structure of macroblock-based MC and BDCT for compressing video data [mitchell96, itu-t96]. In the abovementioned video coding processes, the macroblock-based MC causes discontinuities inside the DCT blocks, while the BDCT causes blocking artifacts on block boundaries [joung99, joung00]. On the other hand, in still images the blocking artifact occurs on only block boundaries, which results from independent processing of blocks without considering between-block pixel correlation.

M.A. Abidi et al., *Optimization Techniques in Computer Vision*, Advances in Computer Vision and Pattern Recognition, DOI 10.1007/978-3-319-46364-3_9

In order to enhance the quality of coded still images, various techniques to remove blocking artifacts have been proposed. These can be summarized as (a) low-pass filtering of the boundary region between blocks [reeve84], (b) iterative image restoration based on the theory of projections onto convex sets (POCS) [yang95, rosenholtz92] or regularization [yang93], (c) adaptive constrained least square (CLS) filtering method [kim98], and (d) the prediction of transform coefficients using the mean squared difference of slope (MSDS) [minami95] or boundary orthonormal function [jeon95]. The low-pass filtering approach requires neither additional information to be transmitted nor significant computational overhead. The POCS-based and regularization-based methods iteratively reconstruct images with prior knowledge about the smoothness of the original image and boundedness of quantized transform coefficients. The adaptive CLS filtering method classifies the edge direction of each block by using a part of the DCT coefficients and carries out the directional CLS filtering on the corresponding block. The transform coefficient prediction method finds the optimal set of transform coefficients that minimizes the prespecified function for the block-discontinuity measure.

A common drawback of the abovementioned methods is that they only consider discontinuities on block boundaries caused by the BDCT without considering discontinuities inside each block caused by motion estimation (ME) and motion compensation (MC). Figure 9.1 shows the process where discontinuities occur on both the boundary and interior of 8×8 DCT blocks.

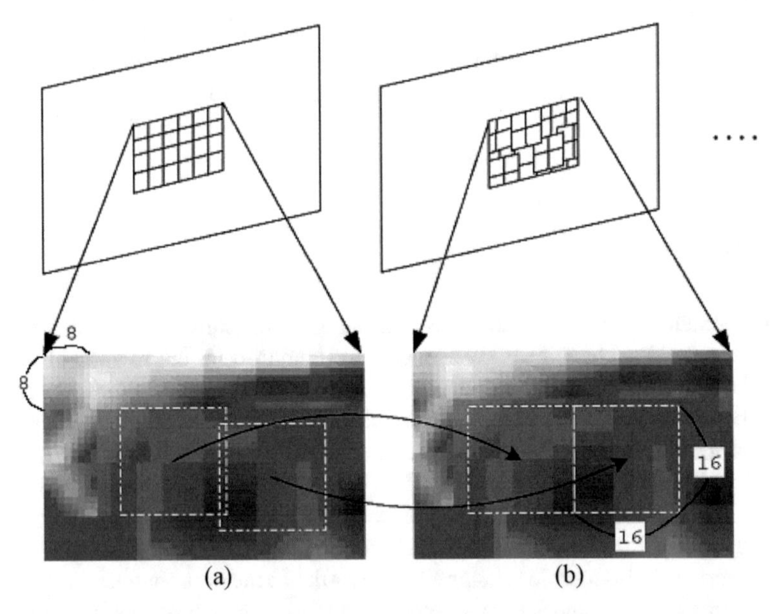

Fig. 9.1 Blocking artifacts on block boundary and block interior: (**a**) the previous reconstructed image with discontinuities on 8×8 DCT block boundary and (**b**) the motion-compensated image from (**a**) with discontinuities on both boundary and interior of blocks

Because the motion-compensated image, as shown in Fig. 9.1b, has been reconstructed from the previous image shown in Fig. 9.1a, discontinuities appear inside the macroblock with any nonzero motion vector. Moreover, if motion vectors of the two adjacent macroblocks are not the same, discontinuities also appear on the macroblock boundary. Consequently, blocking artifacts exist on both the boundary and interior of the 8×8 DCT blocks in a motion-compensated image.

In order to remove block discontinuities in compressed video with motion compensation, the existing still or intra-coded image-based algorithms should be properly modified and extended. For that purpose, the regularized iterative image restoration algorithm is particularly suitable because (a) this process minimizes a user-defined objective function, which represents error between the desired solution and the current estimate; (b) this algorithm can incorporate various types of prior knowledge about the desired solution, which is usually implemented in the form of constraints; and (c) this algorithm can also be implemented in real time or almost real time by preconditioning the iteration [shin99].

The general formulation of regularized optimization can be expressed as

$$\arg \min_{x} f(x) = g(x) + \lambda h(x), \tag{9.1}$$

where x represents the lexicographically ordered vector of the image. The first term in the objective function $g(x) = \|y - Hx\|^2$ represents the data compatibility term, where y and H respectively represent the degraded blocky image and the block matrix that models degradation. The second term $h(x)$ represents the regularizing term that incorporates prior knowledge about the solution, and λ is the regularization parameter that controls the ratio of minimization between $g(x)$ and $h(x)$ [katsaggelos89, kang95].

For the purpose of removing blocking artifacts, the mathematical model of the degradation matrix H must be defined. Because the blocking artifact generally occurs in a spatially variant manner, the degradation cannot be modeled as a block-Toeplitz matrix. Furthermore, it is impossible to have a fixed form matrix for the degradation because there is no one-to-one correspondence between the input original image and the output blocky image due to the nature of quantization. In order to make the problem manageable with regularized optimization, Kim and Paik approximated the blocking artifact generating process as a simple low-pass filtering [kim98, kim00]. The basic idea for this approximation is that the relatively high-frequency DCT coefficients tend to be removed due to coarse quantization. They also let the regularizing term compensate the approximation error.

In this chapter we propose an image enhancement algorithm for compressed video based on the approximated degradation model in [kim98]. The major contribution of this work is to propose a novel, regularized iterative image restoration algorithm that (a) directly improves differential images in the processing of motion-compensated video coding and (b) efficiently removes discontinuities due to motion-compensated coding on both the boundary and interior of blocks.

Because the proposed restoration algorithm does not reconstruct each image frame for enhancement purposes, there is neither loss of information during each compression-reconstruction process nor significant computational overhead.

In order to deal with two different types of blocking artifacts, multiple constraints for the regularizing term are proposed. The first is a smoothness constraint on the block boundary. Because the motion-compensated image has discontinuities on the block boundary, the differential image must have discontinuities in the opposite direction at the same position. The second is a smoothness constraint inside the block, which assumes that the differential image must have discontinuities in the opposite direction inside the block. These constraints are used to define convex sets for iteration of the POCS.

9.2 Image Degradation Model

Still image compression such as JPEG and an intra (I)-picture coding in MPEG is performed mainly by reducing the spatial redundancy. Randomly scattered energy of the image can be compacted into the low-frequency band by the orthogonal transform, and most high-frequency components with small magnitudes may be neglected by quantization. Because the quantized zero coefficients have little influence on the length of the bit stream due to run-length coding (RLC), efficient image compression is achieved. Recently, DCT has been the most widely used orthogonal transform in image coding. The decompression process is composed of inverse quantization and inverse DCT. Therefore, the degraded image obtained from the entire compression and decompression process is modeled as

$$y = Hx \tag{9.2}$$

and $H = C^{-1}D^{-1}QC$, where y and x, respectively, represent the degraded and original images. The degradation matrix is represented as

$$H = C^{-1}D^{-1}QC, \tag{9.3}$$

where C and C^{-1}, respectively, represent the block-based forward and the inverse DCT matrices, and Q and D^{-1}, respectively, the corresponding quantization and the inverse quantization matrices. The RLC process is omitted in the degradation model since it is a lossless process. More detail in mathematical expressions for vectors x and y and matrices C, D, Q, D, and H can be found in previous works [kim98, kim00].

In the case of video, the inter-frame compression for predictive-coded (P) pictures is different from still image coding. Motion estimation and compensation is additionally used to reduce temporal redundancy.

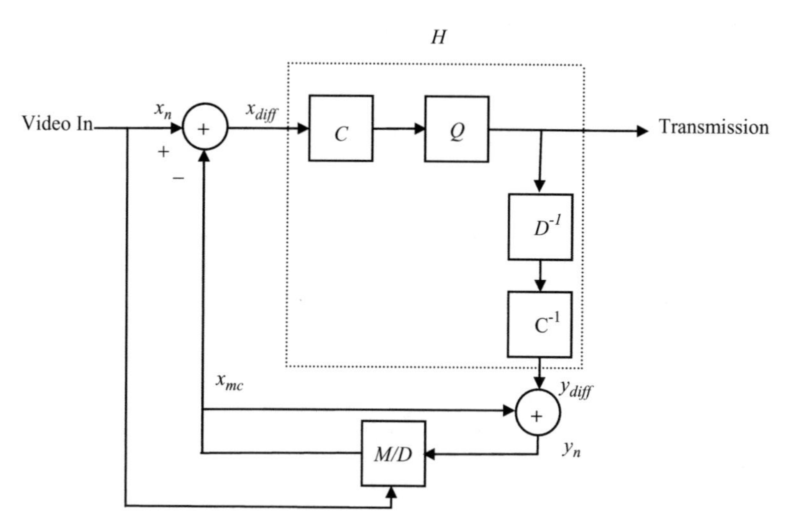

Fig. 9.2 Block diagram of the degradation model for inter-frame coded images

Figure 9.2 shows the inter-frame encoding process of a general video compression system, where M/D represents the motion compensation and frame delay process.

The motion-compensated image from the previous decompressed image is subtracted from the current input image, and then the corresponding differential image is compressed by the DCT and quantization processes. If the nth incoming picture is of type P, the transmitted bit stream, denoted by R^P, is described as

$$R^P = QC(x_n - P(y_{n-1})), \tag{9.4}$$

where $P(\cdot)$ represents the motion compensation operation for P pictures. In the decoder, the bit stream is decoded by the inverse quantization and the inverse DCT.

The decompressed differential image is then added to the motion-compensated image. According to the previously indicated compression-decompression process, the nth decompressed P picture can be represented as

$$y_n = C^{-1}D^{-1}R^P + P(y_{n-1}). \tag{9.5}$$

From Eqs. (9.4) and (9.5), the degradation model of P pictures is defined as

$$y_n = H(x_n - Py_{n-1}) + Py_{n-1}, \text{ for } n = 1, \ldots, M, \tag{9.6}$$

where M represents the number of P pictures per intra (I)-picture and the lossless RLC coding process is omitted. In Eq. (9.6), the motion-compensated version of the previous image $P(y_{n-1})$ can be completely characterized in the receiver part of an image communication system, so we replace this factor with x_{mc} for notational

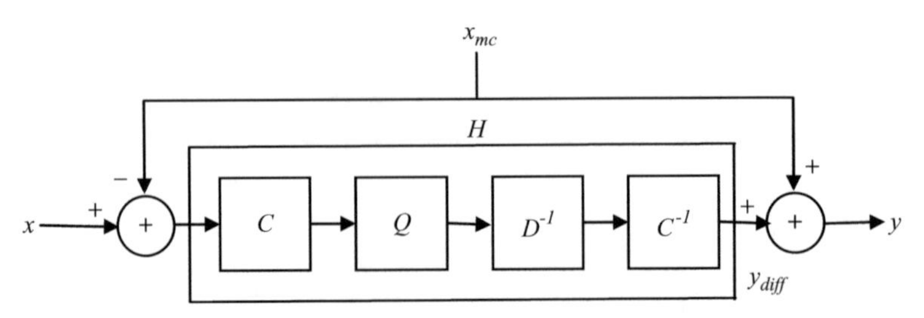

Fig. 9.3 Block diagram of a part of the degradation model for inter-frame coded images. The degradation process occurs on differential images only

simplicity. From Eq. (9.6), the pure degradation model for *P* pictures without considering the lossless operations can be represented as

$$y_{\text{diff}} = Hx_{\text{diff}}, \tag{9.7}$$

where y_{diff} and x_{diff} represent the decompressed differential image $y - x_{\text{mc}}$ and the original differential image $x - x_{\text{mc}}$, respectively. The subscript n, which represents the frame number, is omitted for simplicity.

Figure 9.3 represents the block diagram of a part of the degradation model for inter-frame coded images. The normal degradation of the compressed image is caused by just H, which includes the DCT and the quantization processes. As shown in Fig. 9.3 and Eq. (9.7), we can easily know that the degradation in a *P* picture occurs in the differential image only and the reconstructed blocky image is just obtained by adding the degraded differential image to the motion-compensated image. Image restoration on differential images rather than reconstructed images is therefore more suitable.

9.3 Definition of Convex Sets Based on Analysis of Differential Images

In this section, important properties of differential images which will be used to formulate the objective function of regularization, or to define convex sets for the POCS method, are discussed.

We assume that each pixel has an intensity value in the range [0, 255]. On the other hand, in the differential image x_{diff}, each pixel has an intensity range of [−255, 255]. Therefore, in order to apply the POCS method to differential images, the Hilbert space H includes all $N^2 \times 1$ dimensional differential image vectors, whose elements take any integer value between −255 and 255.

In most image processing areas, such as image restoration, edge detection, and image indexing, frequency domain properties play an important role in analyzing and processing images [li99, wang98]. In order to apply the POCS method to restore BDCT-based compressed images, many researchers have defined a convex set, where BDCT coefficients should lie in a specific range [yang95, rosenholtz92]. The major advantages in using the BDCT coefficient constraint are twofold: (a) there is no additional computation, and (b) they include useful information about the original image.

We note that the quantization constraint on the DCT coefficient can, however, be defined only on differential images because the BDCT coefficients in P pictures are computed from differential images. It is therefore more appropriate to perform the restoration process on differential images rather than reconstructed images.

A convex set, denoted by C_1, represents the set of images whose N^2 DCT coefficients remain in their original quantization interval, such as

$$\begin{aligned} C_1 &= \left\{x_{\text{diff}} : \left(C_{y\text{diff}}\right)_n - \frac{q_n}{2} < \left(C_{x\text{figg}}\right)_n \leq (Cy_{\text{diff}})_n + \frac{q_n}{2}, \forall_n \in \tau\right\} \\ &= \left\{x_{\text{diff}} : \left(QC_{y\text{diff}}\right)_n = \left(QC_{x\text{figg}}\right)_n, \forall_n \in \tau\right\}, \end{aligned} \tag{9.8}$$

where τ and q_n, respectively, represent the set of indices of the BDCT coefficients and the quantization interval for the nth component. The convex set implies that the quantized BDCT coefficients of the projected image onto C_1 are the same as those of y_{diff}. It is straightforward to show that C_1 is closed and convex [yang93].

Another important characteristic of general images is incorporated by the smoothness constraint. Special modification is necessary to implement the smoothness constraint onto x_{mc} because x_{diff} is obtained by subtracting the input image x from x_{mc}. As shown in Fig. 9.1, the motion-compensated image x_{mc} obtained from the previously reconstructed image has discontinuities on both the boundary and interior of each block. If a block size of 8×8 is assumed, that the previous frame is the I picture and that the motion vector of the macroblock containing the (i, j)th block is (a, b), then discontinuities should appear on the a'th column and b'th raw in the (i, j)th block interior, where

$$a' = (8 - a) \bmod 8, \text{ and } b' = (8 - b) \bmod 8. \tag{9.9}$$

If motion vectors of the two adjacent macroblocks are not the same, discontinuities also occur on the macroblock boundary. Therefore, x_{mc} has discontinuities on both the block boundary and block interior. Consequently, we know that x_{diff} must have discontinuity in the opposite direction to that in x_{mc} on the same position so that the reconstructed image can be preserved as smooth. The previously mentioned properties are used in defining smoothness constraints of the differential image.

The convex set for the smoothness constraint between adjacent blocks can be defined as follows. Let F represent the difference between the adjacent columns at the block boundaries of the image x, such as

$$x = \left[x_1^{\text{T}}, x_2^{\text{T}}, \ \ldots \ , x_N^{\text{T}}\right]^{\text{T}}, \tag{9.10}$$

and

$$Fx = \begin{bmatrix} x_8 - x_9 \\ x_{16} - x_{17} \\ \vdots \\ x_{N-8} - x_{N-7} \end{bmatrix}, \tag{9.11}$$

where x_i denotes the $N \times 1$ vector representing the ith column of the $N \times N$ image x. Then, set C_2 captures the intensity variations between columns on block boundaries and is defined as

$$C_2 = \{x_{\text{diff}} : -Fx_{\text{mc}} - \varepsilon < Fx_{\text{diff}} \leq -Fx_{\text{mc}} + \varepsilon\}, \tag{9.12}$$

where ε represents an upper bound of the error residual. The minus signs in Eq. (9.12) represent the direction of discontinuity. In similar fashion, we can define set $C_2^{'}$, which captures the intensity variations between the rows of the block boundaries. C_2 and $C_2^{'}$ serve as convex sets for differential images that have discontinuity in the opposite direction to the previous motion-compensated image on the block boundary. These sets are also closed and convex [yang93].

In order to define new constraint sets C_3 and $C_3^{'}$ for the block interior, let

$$F'x = \begin{bmatrix} D^1 \\ D^2 \\ \vdots \\ D^{\frac{N}{8}} \end{bmatrix}, \text{ and } D^i = \begin{bmatrix} d^{i1} \\ d^{i2} \\ \vdots \\ d^{i\frac{N}{8}} \end{bmatrix}, \tag{9.13}$$

where $d^{i,j}$ represents an 8×1 vector, whose elements have intensity variations between the columns in the (i,j)th block interior. Namely, if the (i,j)th block has motion vector (a,b),

$$d^{i,j} = \begin{bmatrix} x(8i + a', 8j) - x(8i + a' + 1, 8j) \\ x(8i + a', 8j + 1) - x(8i + a' + 1, 8j + 1) \\ \vdots \\ x(8i + a', 8j + 7) - x(8i + a' + 1, 8j + 7) \end{bmatrix}, \tag{9.14}$$

where $x(u_1, u_2)$ represents a pixel intensity value at the position (u_1, u_2). Then, set C_3 captures the intensity variations between columns in the block and is defined as

$$C_3 = \{x_{\text{diff}} : -F'x_{\text{mc}} - \varepsilon < F'x_{\text{diff}} \leq -F'x_{\text{mc}} + \varepsilon\}. \tag{9.15}$$

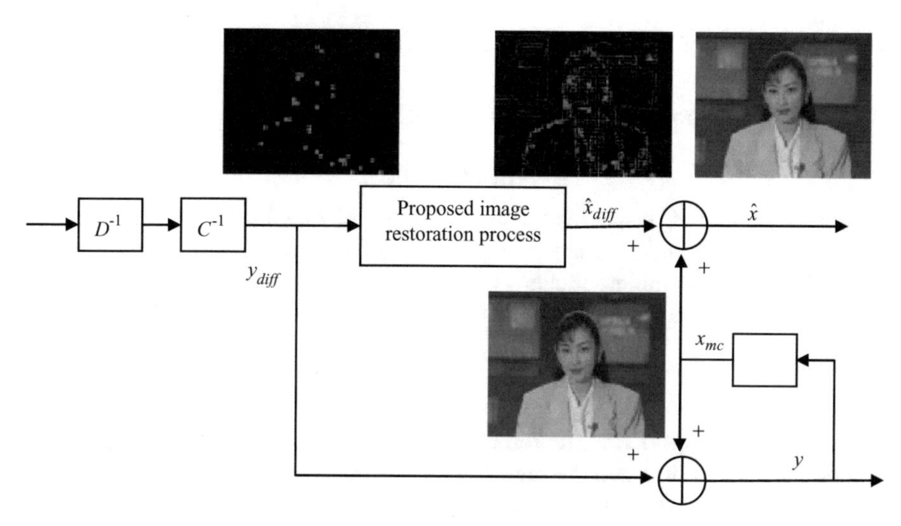

Fig. 9.4 Modified decoding process with the proposed restoration algorithm

In the similar fashion, set C_3' captures the intensity variations between rows in the block. Those convex sets reflect the fact that x_{diff} must have discontinuity in the opposite direction to the previous motion-compensated image at the interior of the block. C_3 and C_3' are also closed and convex.

9.4 Modified Regularized Image Restoration

In this section an iterative image restoration algorithm for enhancing inter-frame coded images is proposed based on the degradation model described in the previous section. The proposed restoration process is performed only on differential images. Figure 9.4 shows the modified decoding process with the proposed restoration process. As shown in the figure, the finally reconstructed image $\hat{x}$ is obtained by adding the differential image $\hat{x}_{\text{diff}}$ and the motion-compensated image x_{mc}. Theoretical justification for restoration in the differential image domain is given in the following section.

In Sect. 9.4.1 conventional regularized iterative image restoration algorithms are briefly reviewed, and modification and extension of those algorithms to inter-frame coded images are discussed. In Sects. 9.4.2 and 9.4.3, a block classification method is proposed for spatially adaptive restoration, and the proposed projection operator is described based on the constraint sets given in Sect. 9.3, respectively.

9.4.1 Spatially Adaptive Image Restoration

The degradation model for a compressed still image is given as

$$y = Hx + \eta, \tag{9.16}$$

where y, H, and x, respectively, represent the blocky image, the degradation operator, and the original image. η represents additive noise, which approximates errors from the incomplete model. A general image restoration process based on regularized optimization is to find $\hat{x}$ which minimizes the regularized functional, as

$$f(\hat{x}) = \|y - H\hat{x}\|^2 + \lambda\|A\hat{x}\|^2, \tag{9.17}$$

where A and λ, respectively, represent a high-pass filter for incorporating a priori smoothness constraints and the regularization parameter that controls the compatibility to the original image and smoothness of the restored image. The iterative process for the solution that minimizes Eq. (9.17) can be given as

$$x^k = x^{k-1} + \beta\{H^{\mathrm{T}}y - (H^{\mathrm{T}}H + \lambda A^{\mathrm{T}}A)x^{k-1}\}, \tag{9.18}$$

where x^k and β represent the restored image after the kth iteration and the step length which controls the convergence rate, respectively.

When H represents the image degradation for the blocking artifact due to quantization of BDCT coefficients, $H^{\mathrm{T}}H$ in Eq. (9.18) can equivalently be reduced to H and $H^{\mathrm{T}}y$ to y because repeated processes by the same quantizer provide the same result. The regularizing term, denoted by $\lambda A^{\mathrm{T}}A$, can also be substituted by low-pass filtering in the form of a constraint at each iteration step as

$$x^k = L\{x^{k-1} + \beta(y - Hx^{k-1})\}, \tag{9.19}$$

where L represents a spatially adaptive low-pass filter.

According to the degradation model given in Sect. 9.2, the objective function of regularized image restoration for inter-frame coded images in an image sequence can be rewritten as

$$f(\hat{x}_{\mathrm{diff}}) = \|y - H\hat{x}_{\mathrm{diff}}\|^2 + \lambda\|A\hat{x}_{\mathrm{diff}}\|^2. \tag{9.20}$$

Note that the solution $\hat{x}$ in Eq. (9.17) is replaced by the restored differential image $\hat{x}_{\mathrm{diff}}$. Correspondingly, Eqs. (9.18) and (9.19) are respectively rewritten as

$$x^k{}_{\mathrm{diff}} = x^{k-1}{}_{\mathrm{diff}} + \beta\{H^{\mathrm{T}}y_{\mathrm{diff}} - (H^{\mathrm{T}}H + \lambda A^{\mathrm{T}}A)x^{k-1}{}_{\mathrm{diff}}\}, \tag{9.21}$$

and

$$x^k{}_{\mathrm{diff}} = L\{x^{k-1}{}_{\mathrm{diff}} + \beta(y_{\mathrm{diff}} - Hx^{k-1}{}_{\mathrm{diff}})\}, \tag{9.22}$$

where x^k_{diff} represents the restored differential image after the kth iteration step. Although the low-pass filter, denoted by L, removes the blocking artifact inherent to differential images, the blocking artifact still remains when the block-free differential image is added to the previous blocky motion-compensated image. Therefore, it is necessary to add an additional constraint that the differential image must have discontinuity in the opposite direction to that of the previous motion-compensated image. More specifically, let P be an operator that produces a differential image with discontinuity in the opposite direction to that of the previous motion-compensated one, then each iteration step in the differential image can be represented as

$$x^k{}_{\text{diff}} = P \cdot L\left\{x^{k-1}{}_{\text{diff}} + \beta\left(y_{\text{diff}} - Hx^{k-1}{}_{\text{diff}}\right)\right\}. \tag{9.23}$$

Finally, the reconstructed image at each step is given as

$$x^k = x_{\text{mc}} + x^k_{\text{diff}}. \tag{9.24}$$

Projection operator P, shown in Eq. (9.23), will be explained more precisely in Sect. 9.4.3.

9.4.2 Block-Based Edge Classification

In this subsection we present a simple block classification algorithm, using a portion of the DCT coefficients, which is suitable for implementing the block-adaptive directional low-pass filters used in Eqs. (9.19) and (9.23).

Generally the block size for the DCT is 8×8, and only $C(0,1)$ and $C(1,0)$, which represent the amount of the most significant vertical and horizontal edge components, respectively, are used. Using those coefficients, each block is classified as monotone, vertical, horizontal, 45°, or 135° blocks. This simple edge classification algorithm is summarized in Fig. 9.5.

9.4.3 Projection Operator

It is well known that previously defined sets C_1, C_2, $C_2^{'}$, C_3, and $C_3^{'}$ are all convex and intersection of any combination of those convex sets is not empty [yang93]. Let P_i and $P_i^{'}$ represent projections onto convex sets C_i and $C_i^{'}$, respectively, and

$$T_i = I + \lambda_i(P_i - I), \text{ for } 0 < \lambda_i \leq 2, \tag{9.25}$$

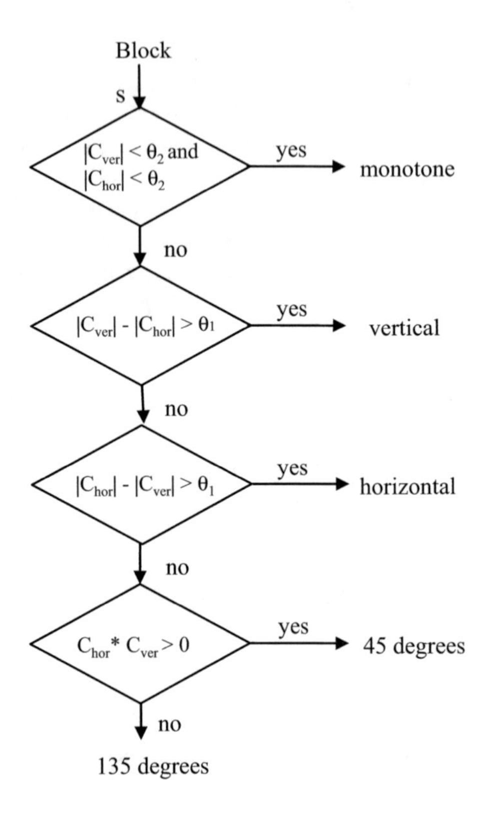

Fig. 9.5 Block-based edge classification algorithm

where I represents the identity matrix. Then the projection operator in Eq. (9.23) is defined as

$$P = T_1 T'_3 T_3 T'_2 T_2. \tag{9.26}$$

The complete projection P in Eq. (9.26) implies that the projection onto the quantization constraint C_1 is performed after the projections onto C_2 and C_3.

9.5 Analysis of Restoration in the Differential Image Domain

In this section the difference between restoration in the differential image domain and in the reconstructed image domain is discussed.

The mathematical representation for the restoration process can be represented as

$$x^k{}_{\text{diff}} = P \cdot L\left\{x^{k-1}{}_{\text{diff}} + \beta\left(y_{\text{diff}} - Hx^{k-1}{}_{\text{diff}}\right)\right\}, \quad x = x_{\text{mc}} + x^k_{\text{diff}}, \tag{9.27}$$

where x^k_{diff} and x^k, respectively, represent the restored differential image and the reconstructed image at the kth iteration step. P, L, β, H, and x_{mc}, respectively, represent the projection operator, spatially adaptive low-pass filter, step length, the degradation operator for P pictures, and the motion-compensated image.

In the remainder of this subsection, these nonlinear projection operators P will be omitted because this variable does not affect the comparison result. Let $x, \bar{x}^k$, and e^k be the restoration result after the kth iteration on the differential image domain, the same result on the reconstructed image domain, and the difference between two restoration results, that is $\bar{x}^k - x^k$, respectively.

The first iteration results on the differential and the reconstructed image domains are given as

$$\begin{aligned} x^1 &= x_{\text{mc}} + x^1_{\text{diff}} \\ &= x_{\text{mc}} + L\{\, y_{\text{diff}} + \beta\,(y_{\text{diff}} - Hy_{\text{diff}})\ \} \end{aligned} \tag{9.28}$$

and

$$\bar{x}^1 = L\{y + \beta(y - Hx)\}. \tag{9.29}$$

Because Hy_{diff} can be replaced by y_{diff}, Eqs. (9.28) and (9.29) can be simplified as

$$x^1 = x_{\text{mc}} + Ly_{\text{diff}} \tag{9.30}$$

and

$$\bar{x}^1 = Ly. \tag{9.31}$$

The difference after the first iteration is then given as

$$e^1 = \bar{x}^1 - x^1 = L(y - y_{\text{diff}}) - x_{\text{mc}} = Lx_{\text{mc}} = Lx_{\text{mc}} - x_{\text{mc}}, \tag{9.32}$$

where L is assumed to be linear. The second iteration results can be computed in the same manner as

$$x^2 = x_{\text{mc}} + L\{Ly_{\text{diff}} + \beta(y_{\text{diff}} - HLy_{\text{diff}})\}, \tag{9.33}$$

$$\bar{x}^2 = L\{Ly - \beta(y - HLy)\}, \tag{9.34}$$

and

$$\begin{aligned} e^2 &= L^2(y - y_{\text{diff}}) + \beta L(y - y_{\text{diff}}) - x_{\text{mc}} - \beta LHLy + \beta LHLy_{\text{diff}} \\ &= L^2 x_{\text{mc}} + \beta Lx_{\text{mc}} - x_{\text{mc}} - \beta LHLy + \beta LHLy_{\text{diff}}. \end{aligned} \tag{9.35}$$

The last two terms in Eq. (9.35), that is, $(\beta LHLy - \beta LHLy_{\text{diff}})$, cannot be simplified to $\beta LHLx_{\text{mc}}$ if H is nonlinear and the distributive property does not hold. As seen in Eq. (9.35), the difference e^k becomes extremely complex with a large number of iterations. In order to simplify the expression for e^k, we assume that H is linear.

By neglecting the projection operator, Eq. (9.27) can be rewritten as

$$x_{\text{diff}}^{k+1} = L(I - \beta H)x_{\text{diff}}^{k} + \beta L y_{\text{diff}}, \tag{9.36}$$

which can be represented as an ignition equation as

$$x_{\text{diff}}^{k+1} + \frac{\beta L y_{\text{diff}}}{L(I - \beta H) - I} = L(I - \beta H)\left(x_{\text{diff}}^{k} + \frac{\beta L y_{\text{diff}}}{L(I - \beta H) - I}\right), \tag{9.37}$$

where I represents the identity matrix. From Eq. (9.37), as the term $x_{\text{diff}}^{k+1} + \frac{\beta L y_{\text{diff}}}{L(I-\beta H)-I}$ is the geometric progression with a common rate $L(I - \beta H)$ and initial value $y_{\text{diff}}^{k+1} + \frac{\beta L y_{\text{diff}}}{L(I-\beta H)-I}$, the kth general terms can be represented as

$$x_{\text{diff}}^{k} + \frac{\beta L y_{\text{diff}}}{L(I - \beta H) - I} = \frac{[L(I - \beta H)]^{k-1} - I}{L(I - \beta H) - I}\left(y_{\text{diff}} + \frac{\beta L y_{\text{diff}}}{L(I - \beta H) - I}\right), \tag{9.38}$$

$$x_{\text{diff}}^{k} = \frac{[L(I - \beta H)]^{k-1} - I}{L(I - \beta H) - I}\left(y_{\text{diff}} + \frac{\beta L y_{\text{diff}}}{L(I - \beta H) - I}\right) - \frac{\beta L y_{\text{diff}}}{L(I - \beta H) - I}, \tag{9.39}$$

and

$$\begin{aligned} x_{\text{diff}}^{k} = x_{\text{mc}} &+ \frac{[L(I - \beta H)]^{k-1} - I}{L(I - \beta H) - I}\left(y_{\text{diff}} + \frac{\beta L y_{\text{diff}}}{L(I - \beta H) - I}\right) \\ &- \frac{\beta L y_{\text{diff}}}{L(I - \beta H) - I}. \end{aligned} \tag{9.40}$$

In the same manner, the kth general term of $\overline{x}^k$ is represented as

$$\overline{x}^k = \frac{[L(I - \beta H)]^{k-1} - I}{L(I - \beta H) - I}\left(y + \frac{\beta L y}{L(I - \beta H) - I}\right) - \frac{\beta L y}{L(I - \beta H) - I}. \tag{9.41}$$

From Eqs. (9.40) and (9.41), the kth difference can easily be computed as

$$\begin{aligned} e^k &= \frac{[L(I - \beta H)]^{k-1} - I}{L(I - \beta H) - I}\left(y - y_{\text{diff}} + \frac{\beta L(y - y_{\text{diff}})}{L(I - \beta H) - I}\right) - \frac{\beta L(y - y_{\text{diff}})}{L(I - \beta H) - I} - x_{\text{mc}} \\ &= \frac{[L(I - \beta H)]^{k-1} - I}{L(I - \beta H) - I}\left(x_{\text{mc}} + \frac{\beta L x_{\text{mc}}}{L(I - \beta H) - I}\right) - \frac{\beta L x_{\text{mc}}}{L(I - \beta H) - I} - x_{\text{mc}}. \end{aligned} \tag{9.42}$$

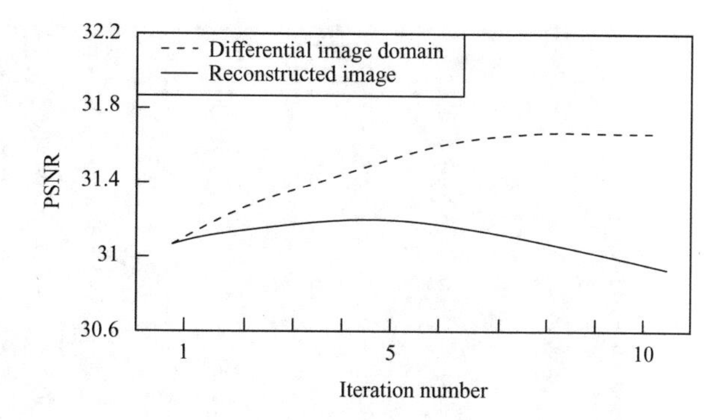

Fig. 9.6 PSNR variation about the first P picture of the Akiyo sequence

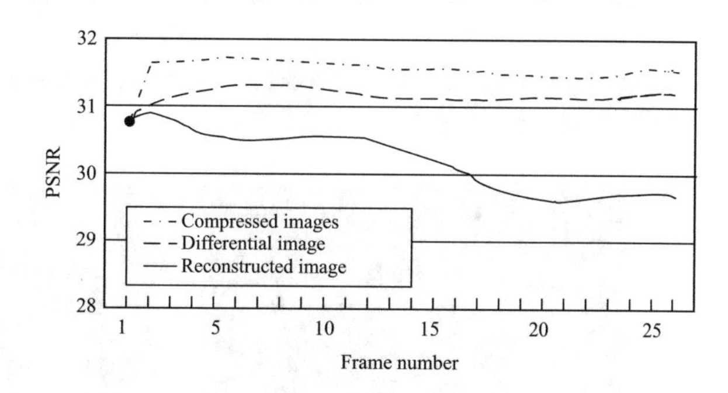

Fig. 9.7 PSNRs of processed images in the different domains

From Eq. (9.42), one can see that the restoration on the differential image domain outperforms the restoration on the reconstructed image domain and that the difference between two restoration results increases as the iteration progresses.

The simulation result of comparing the restoration performances in the differential and the reconstructed domains are shown in Figs. 9.6 and 9.7, respectively. Figure 9.6 indicates that PSNR decreases as iteration continues, when the proposed algorithm is applied on the reconstructed image domain. Figure 9.7 shows the variation in PSNR for 30 frames when the proposed method is applied on the differential image domain.

For experimental purposes, the CIF Akiyo and the QCIF Foreman sequences with quantization levels of 30 and 25, respectively, are used. For image coding, the basic H. 263 was used without additional options. Figure 9.8a, b shows the original second frame and the macroblock-based motion-compensated image from the first I picture, denoted by x and x_{mc}, respectively. Figure 9.8a, b shows the decompressed differential image and the reconstructed image. In Fig. 9.8b, one can easily see that there exist two different types of discontinuities.

At each iteration, every block is classified into one of five different edge blocks as explained in Sect. 9.4. After selecting a directional low-pass filter according to the classified edge information, the proposed adaptive iterative image restoration given in Eq. (9.23) for the differential image is performed. Two parameters for block edge classification, denoted by θ_1 and θ_2, are set equal to $2q$, where q represents the quantization level. The step length $\beta = 0.1$ is chosen

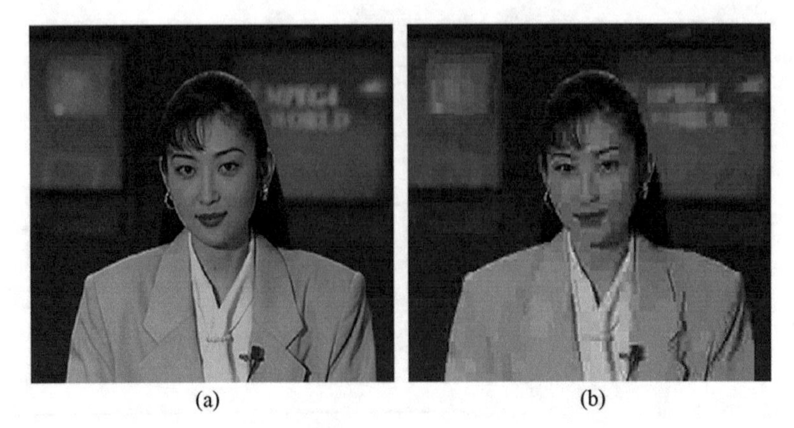

Fig. 9.8 (**a**) The second frame of the original Akiyo sequence and (**b**) the corresponding motion-compensated image from the first *I* picture

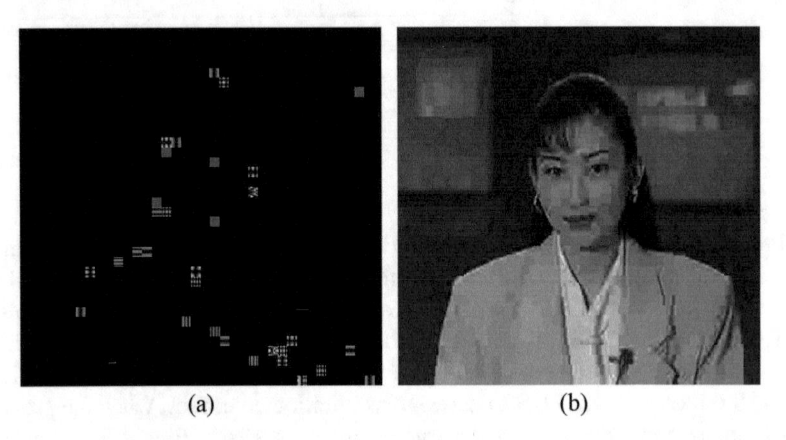

Fig. 9.9 (**a**) Decompressed differential image and (**b**) the reconstructed image from inter-frame coding

experimentally, and all λ's in Eq. (9.25) have the value 0.8. The differential image restored by the proposed algorithm is shown in Fig. 9.9a, which is obtained after ten iterations. For clearer comparison, absolute values multiplied by a factor of ten are taken for displaying the differential images shown in Figs. 9.8a and 9.9a. Figure 9.9b is the finally reconstructed image that is obtained by adding the restored differential image and the previously motion-compensated image. In Fig. 9.9b the two different types of blocking artifacts are significantly reduced, and the PSNR has increased by over 0.7 dB.

Figures 9.11 and 9.12 show the PSNR of 30 frames of the Akiyo and the Foreman sequences, respectively. The first frame of each sequence is the *I* picture and not processed by the restoration algorithm. As shown in the figures, the first *P* picture, that is, the second frame of the sequence, is enhanced by 0.7 dB in the sense of PSNR. This result shows that the proposed degradation model for the first *P* picture is applicable and that two different smoothness constraint operators play a

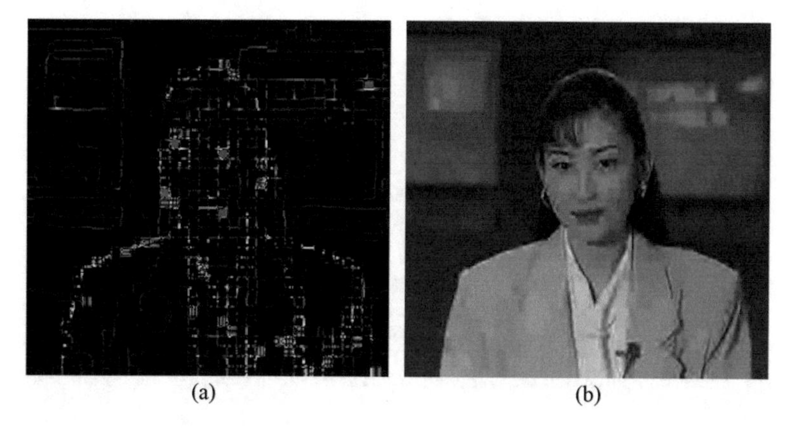

Fig. 9.10 (**a**) Restored differential image generated by the proposed algorithm and (**b**) the finally restored image

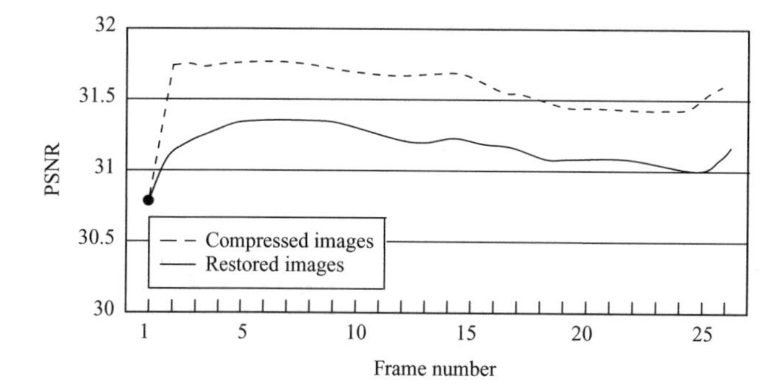

Fig. 9.11 PSNR comparison of the compressed and the restored frame of the Akiyo sequence

role in removing discontinuities on both the boundary and interior of each block. Because blocking artifacts at the interior of the block propagate into the following *P* pictures, the restoration quality of *P* frames tends to degrade as the frame number increases. In order to solve the error propagation problem, motion vectors from the nearest *I* picture need to be stored. If the stored motion vectors are applied to restore the following *P* pictures, the restoration result will be as good as the first *P* picture.

In this chapter information is provided which indicates that the blocking artifact in inter-frame coded images is caused by both the 8×8 DCT and 16×16 macroblock-based MC, while that of the intra-coded images is caused by the 8×8 DCT only. Based on this observation, a new degradation model is proposed for differential images and the corresponding restoration algorithm that utilizes additional constraints and convex sets for both inter- and intra-block discontinuities. The proposed algorithm provides significantly enhanced reconstructed images compared with conventional methods, which do not consider the degradation

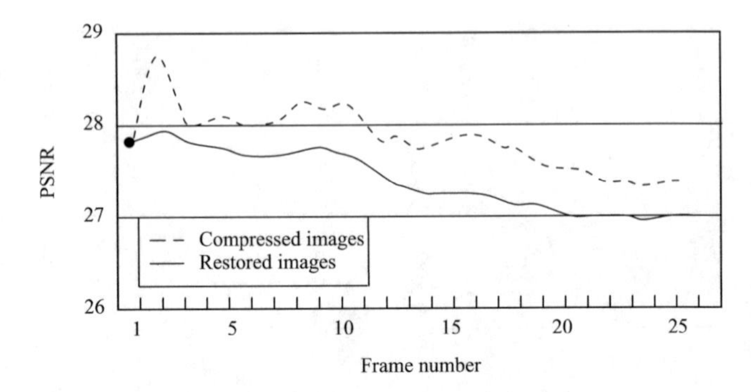

Fig. 9.12 PSNR comparison of the compressed and the restored frames of the Foreman sequence

process of inter-frame coded images. The proposed method may be used as a post-processor in the decoder of moving image compression systems, such as digital VCR and digital HDTV systems.

In the experiments previously restored frames were not used for the current restoration process. If the previously restored frames are used in class A video sequences, higher-quality images can be obtained with faster convergence.

References

[mitchell96] J.L. Mitchell, W.B. Pennebaker, C.E. Fogg, D.J. LeGall, *MPEG video compression standard* (Chapman and Hall, New York, 1996)

[itu-t96] ITU-T Recommendation H. 263: Video coding for low rate communication, ITU-T (1996)

[joung99] S.C. Joung, J.K. Paik, Modified regularized image restoration for postprocessing inter-frame coded images. Proc. Int. Conf. Image Process. **3**, 474–478 (1999)

[joung00] S.C. Joung, S.J. Kim, J.K. Paik, Postprocessing of inter-frame coded images based on convex projection and regularization. Proc. SPIE Image Video Comm. Process. **3974**, 396–404 (2000)

[reeve84] H. Reeve, J.S. Lim, Reduction of blocking effects in image coding. Opt. Eng. **23**(1), 34–37 (1984)

[yang95] Y. Yang, N.P. Galantsanos, A.K. Katsaggelos, Projection-based spatially adaptive reconstruction of block-transform compressed images. IEEE Trans. Image Process. **4** (7), 896–908 (1995)

[rosenholtz92] R. Rosenholtz, A. Zakhor, Iterative procedures for reduction of blocking effects in transform image coding. IEEE Trans. Circuits Syst. Video Technol. **2**(1), 91–94 (1992)

[yang93] Y. Yang, N.P. Galantsanos, A.K. Katsaggelos, Regularized reconstruction to reduce blocking artifacts of block discrete cosine transform compressed images. IEEE Trans. Circuits Syst. Video Technol. **3**(6), 421–432 (1993)

[kim98] T.K. Kim, J.K. Paik, Fast image restoration for reducing block artifacts based on adaptive constrained optimization. J Vis Commun Image Represent **9**(3), 234–242 (1998)

[minami95] S. Minami, A. Zakhor, An optimization approach for removing blocking effects in transform coding. IEEE Trans. Circuits Syst. Video Technol. **5**(2), 74–82 (1995) [jeon95] B. Jeon, J. Jeong, and J. Jo, Blocking artifacts reduction in image coding based on minimum block boundary discontinuity, Proc. Vis. Commun. Image Process., 198–209 (1995)

[shin99] J.H. Shin, J.S. Yoon, J.K. Paik, M.A. Abidi, Fast superresolution for image sequence using motion adaptive relaxation parameters. Proc. Int. Conf. Image Process. **3**, 676–680 (1999)

[katsaggelos89] A.K. Katsaggelos, Iterative image restoration algorithms. Opt. Eng. **28**(7), 735–748 (1989)

[kang95] M.G. Kang, A.K. Katsaggelos, General choice of the regularization functional in regularized image restoration. IEEE Trans. Image Process. **4**(5), 594–602 (1995)

[kim00] T.K. Kim, J.K. Paik, C.S. Won, Y.S. Choe, J. Jeong, J.Y. Nam, Blocking effect reduction of compressed images using classification-based constrained optimization. Signal Process. Image Commun. **15**(10), 869–877 (2000)

[li99] C.T. Li and D.C. Lou, Edge detection based on the multiresolution Fourier transform, IEEE Workshop on Signal Processing Systems, 686–693 (1999)

[wang98] C. Wang, K.L. Chan, and S.Z. Li, Spatial-frequency analysis for color image indexing and retrieval, Proc. Int. Conf. Control, Automation, Robotics and Vision, 1461–1465 (1998)

Chapter 10
Volumetric Description of Three-Dimensional Objects for Object Recognition

Abstract Three-dimensional (3D) object representation and shape reconstruction play an important role in the field of computer vision. In this chapter, a parametric model is proposed as a coarse description of the volumetric part of 3D objects for object recognition. The parameterized part-based superquadric model is used as a volumetric parameter model. This superquadric model has the advantage that it can describe various primitive shapes using a finite number of parameters including translation, rotation, and global deformation. Shape recovery is performed by least square minimization of the objective function for all range points belonging to a single part of the object. The set of superquadric parameters representing each single part of the 3D object can then be obtained.

10.1 Introduction

In the field of computer vision, 3D object representation is a key technique for object recognition. Most representation methods are classified by feature types, such as points, contours, surfaces, and volumetric features. Since 3D objects can be completely described by volume information, volumetric description plays an important role in many 3D object recognition systems. The volumetric representation includes voxels,[1] octrees,[2] and generalized cylinders (GC)[3] [besl85]. Most conventional 3D shape representation methods, however, use local features such as the surface normal and surface curvature, which are sensitive to either noise or erroneous data.

Recently, superquadric-based representation methods have attracted considerable interest since the first work in this area was proposed by Pentland in 1986. Superquadrics are represented by using a set of geometrical parameters and can generate various kinds of 3D shape primitives including ellipsoids, cylinders,

---

[1] A voxel represents a small volume element on the discretized 3D space.

[2] The octree represents hierarchical representation of volume occupancy.

[3] A generalized cylinder represents a 3D volume by sweeping a planar cross section along a space-curve spine.

M.A. Abidi et al., *Optimization Techniques in Computer Vision*, Advances in Computer Vision and Pattern Recognition, DOI 10.1007/978-3-319-46364-3_10

parallelepipeds, pyramids, cones, and round-edged shapes. The human visual system intuitively perceives 3D objects by decomposing an object into parts based on the recognition by components (RBC) theory [biederman85]. Superquadrics are versatile for part primitives since they can be deformed and glued together into realistic-looking models [hwang91, leonardis95, solina90, wu94].

Bajcsy and Solina suggested an iterative technique for estimating the parameters, which minimizes an error-of-fit function by the Levenberg-Marquardt method for nonlinear least squares [solina90]. The superquadric model is also used in the segmentation of range images [leonardis95].

In this chapter we show the feasibility of part-based superquadric models for 3D object representation. The representation procedure will be viewed as superquadric parameter estimation, which can be performed by an optimization process. For this purpose, the 3D input range image is first segmented into smooth surface patches by using surface curvature information. Then the 3D object is decomposed into several volumetric primitives using these segmented surfaces to recover the part-based superquadric model.

10.2 Parametric Model Representation Using Superquadrics

Superquadrics are well known and often used to describe 3D object surfaces in computer graphics [solina90]. Superquadrics as a family of parametric shapes can describe a wide variety of complex and realistic 3D shapes effectively with a compact and finite number of parameters. As the values of the parameters vary, superquadrics can assume many different shapes such as spheres, ellipsoids, rectangular blocks, pyramids, cones, and cylinders. These basic shapes can then be deformed to further increase the number of shape primitives. Typical deformation includes tapering and bending. GCs, proposed by Binford, are more powerful in describing 3D objects than superquadrics, but are too complex for general applications.

10.3 Definition and Properties of Superquadrics

A superquadric surface is defined by the following 3D vector:

$$r(\eta, \omega) = \begin{bmatrix} x(\eta, \omega) \\ y(\eta, \omega) \\ z(\eta, \omega) \end{bmatrix} = \begin{bmatrix} a_1 \cos^{\varepsilon_1}\eta \cos^{\varepsilon_2}\omega \\ a_2 \cos^{\varepsilon_1}\eta \cos^{\varepsilon_2}\omega \\ a_3 \sin^{\varepsilon_1}\eta \end{bmatrix}, \tag{10.1}$$

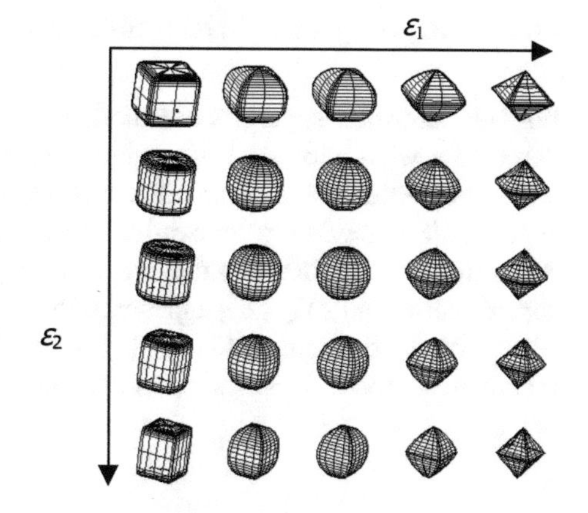

Fig. 10.1 Different superquadric shapes varying by parameters ε_1 and ε_2. From *left* to *right* and from *top* to *bottom*, ε_1 and ε_2, respectively, take values: 0.2, 0.8, 1.0, 1.5, and 2.0 ($a_1 = a_2 = a_3 = 30$)

where $\eta \in \left[-\frac{\pi}{2}, \frac{\pi}{2}\right]$ represents the north–south (xz) parameter, such as latitude in the spherical coordinate system; $\omega \in [-\pi, \pi]$ the east–west (xy) parameter, such as longitude; ε_1 the squareness parameter in the xz direction; and ε_2 the squareness parameter in the xy direction. Parameters a_1, a_2, and a_3 determine the size of the superquadric along the x, y, and z directions, respectively. Figure 10.1 shows a family of differently shaped superquadrics with the same size parameters.

10.4 Superquadric Inside-Outside Function

By eliminating η and ω in Eq. (10.1) by using the equality $\sin^2\alpha + \cos^2\alpha = 1$, the following implicit equation is obtained:

$$\left(\left(\frac{x}{a_1}\right)^{2/\varepsilon_2} + \left(\frac{y}{a_2}\right)^{2/\varepsilon_2}\right)^{\varepsilon_2/\varepsilon_1} + \left(\frac{z}{a_3}\right)^{2/\varepsilon_1} = 1. \tag{10.2}$$

Based on this superquadric surface equation, the superquadric inside-outside function is defined as

$$F(x, y, z) = \left(\left(\frac{x}{a_1}\right)^{2/\varepsilon_2} + \left(\frac{y}{a_2}\right)^{2/\varepsilon_2}\right)^{\varepsilon_2/\varepsilon_1} + \left(\frac{z}{a_3}\right)^{2/\varepsilon_1}. \tag{10.3}$$

This equation determines where a given point $[x, y, x]^{\mathrm{T}}$ lies on, above, or below the superquadric surface. That is, if $F(x, y, z) = 1$, the corresponding point (x, y, z) lies on the superquadric surface. If $F(x, y, z) > 1$, the point lies above the superquadric surface and vice versa [solina90].

10.5 Superquadric Deformation

For nondeformed superquadric models as shown in Fig. 10.1, it is difficult to describe a wide variety of 3D object shapes existing in the real world. By using additional deformation parameters such as tapering, bending, twisting, and cavity, superquadric modeling capabilities can be enhanced [hwang91, solina90]. Assuming a coordinate value (x, y, z) represents a point on a nondeformed superquadric and (X, Y, Z) represents the corresponding point after deformation, we can derive the equation $X = D(x)$, where D represents the deformation function that relates x and X. Translation and/or rotation can be performed after deformation.

10.6 Tapering Deformation

Tapering deformation is utilized to elongate the object shape along the z axis. In this deformation, it is assumed that tapering is performed along the z axis only and that the tapering rate is linear with respect to z. By using this tapering, an object can have a wedge-type shape. Based on these assumptions, tapering deformation is given by

$$X = \left(\frac{K_x}{a_3}z + 1\right)x, \quad Y = \left(\frac{K_y}{a_3}z + 1\right)y, \quad Z = z, \tag{10.4}$$

where X and Y represent the transformed coordinates of the primitives after tapering. Tapering parameters K_x and K_y in the x and y coordinates have constraints $0 \leq K_x, K_y \leq 1.0$, to avoid invalid tapering. If $K_x, K_y > 1.0$, invalid deformation occurs as shown in Fig. 10.2.

At the initial value estimation stage of recovering the superquadric model from the deformed object, no deformation is assumed to occur. Thus, the values of the tapering parameters are initialized with zeros, and $K_x = K_y = 0$.

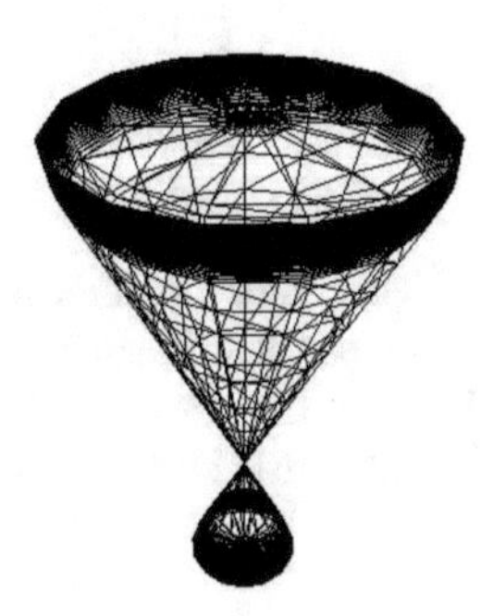

Fig. 10.2 Invalid tapering deformation in the case where K_x, $K_y > 1.0$

10.7 Bending Deformation

A simple bending operation that corresponds to a circular section is used, as shown in Fig. 10.3. The bending operation is applied along the z axis in the positive x direction. This process requires only one parameter of the curvature κ of the circular section to describe the bending feature.

Figure 10.3 shows bending deformation in the xz plane. The Y axis is perpendicular to this plane, projecting into the paper. The shaded area delimits the original primitive, O represents the center of bending curvature, and θ the bending angle. The equations describing the bending deformation are given by

$$X = k^{-1} - \left(k^{-1} - x\right)\cos\theta, \quad Y = y, \quad Z = \left(k^{-1} - x\right)\sin\theta, \tag{10.5}$$

where θ represents the bending angle, which can be computed from the curvature parameter k. The inverse transformation is given by

$$x = k^{-1} - \sqrt{z^2 + \left(k^{-1} - x\right)^2}, \quad y = Y, \quad z = k^{-1}\theta = k^{-1}\tan^{-1}\frac{Z}{k^{-1} - X}. \tag{10.6}$$

The superquadric model for a block and a cylinder deformed by tapering and bending is shown in Fig. 10.4.

In Fig. 10.4, size parameters $a_1 = a_2 = 30$ and $a_3 = 60$ are used. As shown in this figure, the z axis can be placed along the longer side of elongated objects and perpendicular to flat, rotationally symmetric objects.

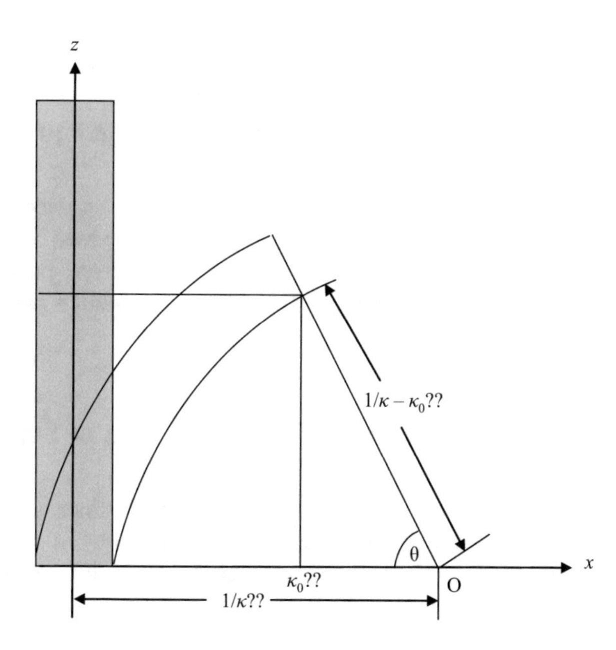

Fig. 10.3 Bending deformation in the xz plane

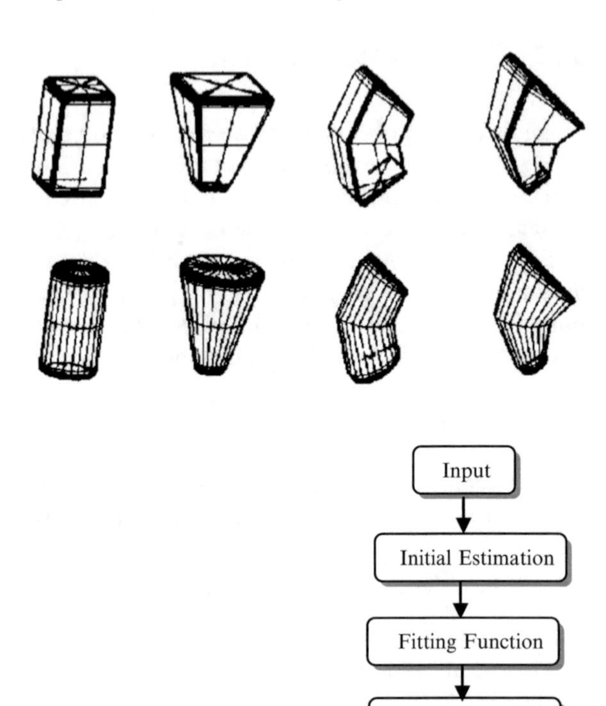

Fig. 10.4 Typical deformed superquadrics with deformation parameters, $K_x = K_y = 0.5$ and $k = \frac{\pi}{240}$

Fig. 10.5 Recovery process for nondeformed superquadric models

10.8 Recovery of Nondeformed Superquadric Models

In this section, recovery of nondeformed superquadric models in an arbitrary position is described. The recovery method is based on the least square minimization of a fitting function derived from the inside-outside superquadric function given in Eq. (10.3). Figure 10.5 shows the overall recovery process.

10.9 Homogeneous Transformation of Superquadrics

The superquadric inside-outside function defines superquadric surfaces in an object-centered or canonical coordinate system. On the other hand, 3D input data is expressed in the world coordinate system with a homogeneous transformation T^{-1} as shown in Fig. 10.6.

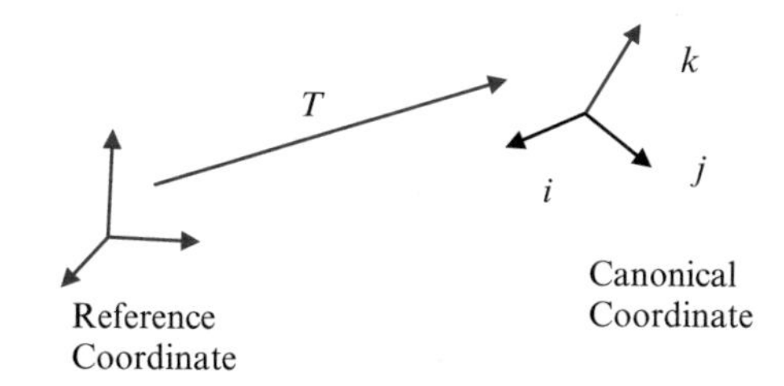

Fig. 10.6 Homogeneous coordinate transformation using a 4 × 4 matrix

Coordinate position and orientation with respect to 3D space are defined by four vectors as shown in Fig. 10.6. Those vectors represent directions of the x, y, and z axes, such as $\bar{i} = \begin{bmatrix} i_x & i_y & i_z \end{bmatrix}^{-1}, \bar{j} = \begin{bmatrix} j_x & j_y & j_z \end{bmatrix}^{-1}, \bar{k} = \begin{bmatrix} k_x & k_y & k_z \end{bmatrix}^{-1}$, and $\bar{p} = \begin{bmatrix} p_x & p_y & p_z \end{bmatrix}^{-1}$ that represents the origin. Homogeneous coordinate transformation T is given by the following matrix:

$$T = \begin{bmatrix} i_x & j_x & k_x & p_x \\ i_y & j_y & k_y & p_y \\ i_z & j_z & k_z & p_z \\ 0 & 0 & 0 & 1 \end{bmatrix}, \quad \text{where} \quad \begin{matrix} i_x^2 + i_y^2 + i_z^2 = 1 \\ j_x^2 + j_y^2 + j_z^2 = 1 \\ k_x^2 + k_y^2 + k_z^2 = 1 \end{matrix}. \tag{10.7}$$

If a point (x_s, y_s, z_s) is on the superquadric surface represented in the object-centered coordinate system, (x_w, y_w, z_w) can be obtained by using the homogeneous coordinate transformation, such as

$$\begin{bmatrix} x_w \\ y_w \\ z_w \\ 1 \end{bmatrix} = T \begin{bmatrix} x_s \\ y_s \\ z_s \\ 1 \end{bmatrix} = \begin{bmatrix} i_x & j_x & k_x & p_x \\ i_y & j_y & k_y & p_y \\ i_z & j_z & k_z & p_z \\ 0 & 0 & 0 & 1 \end{bmatrix} \begin{bmatrix} x_s \\ y_s \\ z_s \\ 1 \end{bmatrix}. \tag{10.8}$$

The homogeneous coordinate transformation matrix T can be defined using the set of Euler angles $\{\varphi, \theta, \psi\}$ as

$$T = \begin{bmatrix} \cos\phi\ \cos\theta\ \cos\psi - \sin\phi\ \sin\psi & -\cos\phi\ \cos\theta\ \sin\psi - \sin\theta\ \cos\varphi & \cos\phi\ \sin\theta & p_x \\ \sin\phi\ \cos\theta\ \cos\psi - \cos\phi\ \sin\psi & -\sin\phi\ \cos\theta\ \sin\varphi + \cos\phi\ \cos\psi & \sin\phi\ \sin\theta & p_y \\ -\sin\theta\ \cos\psi & \sin\theta\ \sin\psi & \cos\theta & p_z \\ 0 & 0 & 0 & 1 \end{bmatrix}. \tag{10.9}$$

By inverting the homogeneous transformation matrix, (x_s, y_s, z_s) can be calculated using Euler angles, translation vectors (p_x, p_y, p_z), and the world coordinate. Then, by substituting this equation into Eq. (10.3), the general inside-outside function for superquadrics can be obtained as

$$F(x_w, y_w, z_w) = F(x_w, y_w, z_w; \Lambda), \tag{10.10}$$

where

$$\Lambda = \begin{cases} \{a_1, a_2, a_3, \varepsilon_1, \varepsilon_2, \varphi, \theta, \psi, p_x, p_y, p_z\} & \text{for regular superquadrics,} \\ \{a_1, a_2, a_3, \varepsilon_1, \varepsilon_2, \varphi, \theta, \psi, p_x, p_y, p_z, K_x, K_y, \alpha, k\} & \text{for deformed superquadrics.} \end{cases} \quad (10.11)$$

The superquadric model can thus be defined using 11 parameters. If 3D range data and the expanded superquadric inside-outside function are given, these parameters can be extracted using nonlinear least square minimization (LSM) methods such as the Gauss-Newton or Levenberg-Marquardt algorithm.

10.10 Superquadric Parameter Estimation

Superquadric representation aims to recover superquadric parameters from data set of real-world objects. This representation is obtained by a data fitting process, which is well known to be an inverse problem. Most superquadric representation methods try to estimate an optimal set of parameters by minimizing an objective function. This objective function serves as an error metric to measure how well the recovered model fits the data. A commonly used objective function is defined based on the radial Euclidean distance as

$$G(\Lambda) = \sum_{i=1}^{N} d^2, \quad (10.12)$$

where d represents the Euclidean distance between an arbitrary point and the superquadric as depicted in Fig. 10.7. This distance is measured along the line segment that connects the center of a canonical coordinate and the point of interest, (x_s, y_s, z_s). Let $\mathbf{r}_o$ and $\mathbf{r}_s$ be vectors pointing from the origin to the point (x_s, y_s, z_s) and to the surface of the superquadric, respectively. Then we can compute the radial distance using the following relationships:

$$d = |\mathbf{r}_o - \mathbf{r}_s| = |\mathbf{r}_o| \left|1 - F^{-\frac{\varepsilon_1}{2}}(x_s y_s, z_s)\right| = |\mathbf{r}_s| \left|F^{\frac{\varepsilon_1}{2}}(x_s y_s, z_s) - 1\right|. \quad (10.13)$$

The superquadric representation algorithm is summarized as follows:

Algorithm 10.1: Superquadric Representation
Given N, 3D surface points, (x_{wi}, y_{wi}, z_{wi}), for $i = 1, \ldots, N$, estimate the set of superquadric parameters given in Eq. (10.11) as follows:

Step 1. Estimate initial superquadric parameters.
Step 2. Perform the Levenberg-Marquardt algorithm to iteratively refine the superquadric parameters.

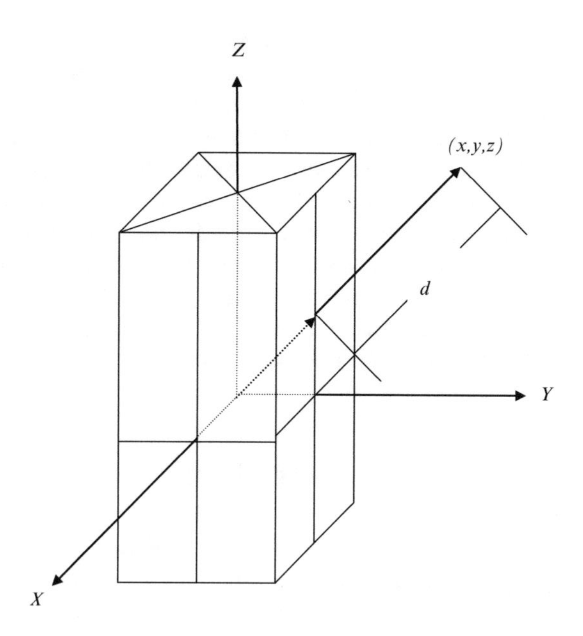

Fig. 10.7 Radial Euclidean distance of superquadrics

The initial superquadric parameters can be evaluated using the method in [solina90]. Initial parameter estimation and the Levenberg-Marquardt algorithm are described in more detail in the following subsections.

10.11 Initial Values Estimation

In order to guarantee convergence for the iterative superquadric model recovery, reasonably correct estimation of the object's position, orientation, and size are required. Unfortunately, the parameter estimation is usually inaccurate because only the visible sides of the object can be used. ε_1 and ε_2 are first set to unity so that the initial model is an ellipsoid. Then, the initial position of the ellipsoid is set to the center of gravity among all N range points as

$$\overline{x} = p_x = \frac{1}{N}\sum_{i=1}^{N} x_{wi}, \quad \overline{y} = p_y = \frac{1}{N}\sum_{i=1}^{N} y_{wi}, \quad \text{and } \overline{z} = p_z = \frac{1}{N}\sum_{i=1}^{N} z_{wi}. \tag{10.14}$$

The initial orientation of ellipsoid (φ, θ, ψ) can be estimated by computing the following central moments matrix:

$$M = \frac{1}{N}\sum_{i=1}^{N} \begin{bmatrix} (y_i - \overline{y})^2 + (z_i - \overline{z})^2 & -(y_i - \overline{y})(x_i - \overline{x}) & -(z_i - \overline{z})(x_i - \overline{x}) \\ -(x_i - \overline{x})(y_i - \overline{y}) & (x_i - \overline{x})^2 + (z_i - \overline{z})^2 & -(z_i - \overline{z})(y_i - \overline{y}) \\ -(x_i - \overline{x})(z_i - \overline{z})^2 & -(y_i - \overline{y})(z_i - \overline{z}) & (x_i - \overline{x})^2 + (y_i - \overline{y})^2 \end{bmatrix}. \tag{10.15}$$

Eigenvector e_1, which corresponds to the smallest eigenvalue λ_1, represents the minimum-inertia line, and eigenvector e_3, corresponding to the largest eigenvalue, represents the maximum-inertia line. The minimum-inertia line is also called the principal axis. In order to obtain eigenvectors, Jacobi's method can be used [press92]. Under the assumption that bent and/or tapered objects along their longest side, the decision was made to orient the object-centered coordinate system so that the z axis aligns to the longest side for the elongated object. This makes the z axis coincide with the axis of rotational symmetry. From the transformation matrix T, initial values of the Euler angle (φ, θ, ψ) can be obtained. Initial size parameters of the ellipsoid are simply the distance between the outmost range points along each coordinate axis of the new object-centered coordinate system.

10.12 The Levenberg-Marquardt Algorithm

Based on the objective function defined in Eqs. (10.12) and (10.13), the superquadric representation turns into a nonlinear least square fitting problem. The corresponding numerical optimization problem is generally solved by the Levenberg-Marquardt algorithm due to its efficiency and stable performance [press92]. To concentrate on the optimization aspect of the superquadric representation, we assume a set of unstructured 3D data has been scanned from the surface of a single object.

Due to self-occlusion, not all sides of an object are visible at the same time. Thus, the general viewpoint has to be considered to provide enough information for the entire object. By introducing a fitting function as

$$R = \sqrt{a_1 a_2 a_3}(F - 1), \tag{10.16}$$

a new objective function is defined as

$$\min \sum_{i=1}^{N} \left\{ R\left(x_{wi}, y_{wi}, z_{wi}; \Lambda\right\} \right.^{2}. \tag{10.17}$$

Since R is a nonlinear function composed of 11 parameters (15 parameters for a deformed superquadric), minimization must proceed iteratively until the sum of the least squares in Eq. (10.12) stops decreasing, or the changes become smaller than a sufficiently small value. The Levenberg-Marquardt method is used for nonlinear least square minimization [press92]. Constraints $a_1, a_2, a_3 > 0$ and $0.1 < \varepsilon_1, \varepsilon_2 < 1.0$ are used in order to prevent ambiguity in the superquadric shape description.

Example 10.1

Figure 10.8 shows two sets of synthetic 3D data of a cylinder collected from two different views. Table 10.1 shows the ground truth values used to create the cylinder and recovered superquadric parameters.

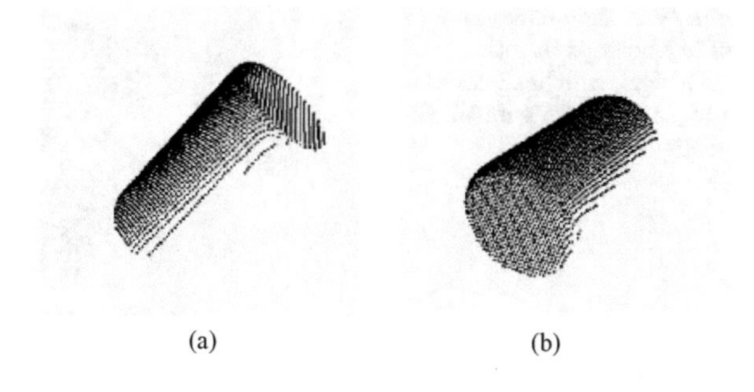

Fig. 10.8 Synthetic range data of a cylinder from two views. (**a**) 3D data from view 1 with 3194 points and (**b**) 3D data from view 2 with 3106 points

Table 10.1 Recovered superquadric parameters of the objects shown in Fig. 10.8 scanned from two views

	a_1	a_2	a_3	ε_1	ε_2	e_{FT}
GT	30.000	30.000	60.000	0.100	1.000	0.000
RP1	29.245	27.182	58.381	0.100	1.021	0.354
RP2	28.861	29.613	58.809	0.100	1.003	0.312

GT represents the ground truth values, *RP1* recovered parameters from view 1, *RP2* recovered parameters from view 2. *GT* ground truth and e_{FT} fitting error

The fitting error is calculated as

$$e_{\mathrm{FT}} = \frac{1}{N}\sum_{i=1}^{N} d_i^2, \tag{10.18}$$

where N represents the number of data points and d_{i} represents the radial Euclidean distance between the i-th point of the input data and the corresponding recovered superquadric surface.

Example 10.2
Figure 10.9 shows two 3D calibrated real range images of a wooden block and a cylinder. The corresponding parameters of the recovered superquadric models are summarized in Tables 10.2 and 10.3, respectively.

Both Tables 10.2 and 10.3 show that the superquadric parameters have been recovered very well compared with the ground truth values. The algorithm converges after 30 iterations (a few seconds) for most data.

10.13 Superquadric Model of 3D Objects

In general, simple 3D objects can be decomposed into combinations of several volumetric part primitives such as spheres, cylinders, and blocks. Furthermore, the correctly estimated shape of an object in terms of its parts can both distinguish among different classes of objects and provide an efficient indexing mechanism for recognition systems. Therefore, we need to decompose the 3D input range data into

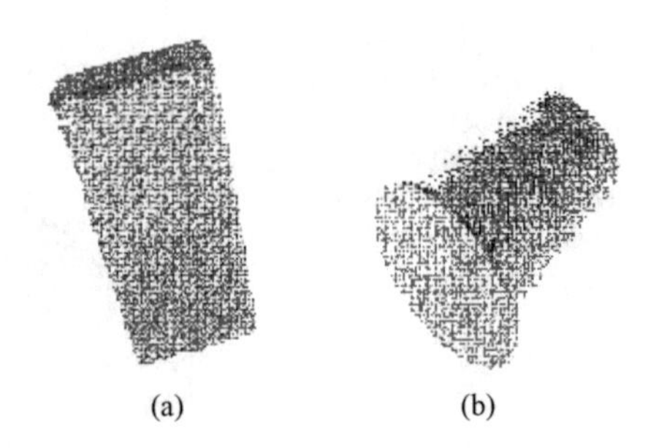

Fig. 10.9 Real range data of regular superquadrics: (**a**) a block with 5322 points and (**b**) a cylinder with 4675 points

Table 10.2 Recovered superquadric parameters of the object in Fig. 10.9a

	a_1	a_2	a_3	ε_1	ε_2	e_{FT}
GT	29.000	56.000	116.000	0.100	0.100	0.000
RP	23.263	60.833	111.232	0.100	0.100	3.423

Table 10.3 Recovered superquadric parameters of the object in Fig. 10.9b

	a_1	a_2	a_3	ε_1	ε_2	e_{FT}
GT	52.000	52.000	50.000	0.100	1.000	0.000
RP	51.794	53.728	51.819	0.100	1.000	1.344

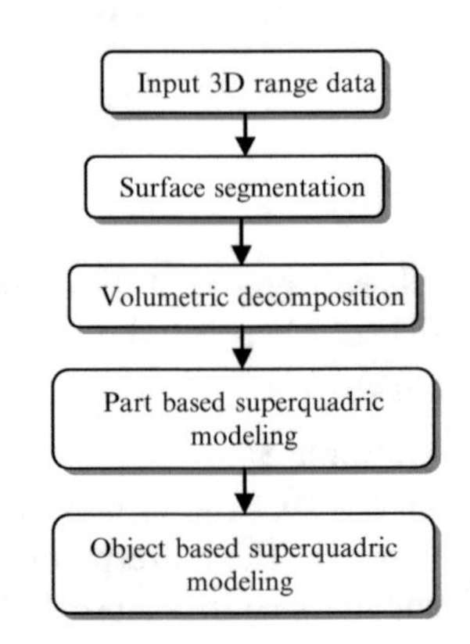

Fig. 10.10 3D object representation using a superquadric model

volumetric primitives in terms of geons [raja92, wu94] and then construct the 3D object model using the part-based superquadric model recovery. In this section, simple volumetric primitives such as a sphere, a cylinder, and a block are used without any deformation. In order to decompose 3D objects into these volumetric primitives, the 3D input range image has to be segmented into smooth surface patches, which have the same surface curvature information. These patches are classified as a plane, a cylindrical surface, or a spherical surface according to the surface curvature characteristics. After classification, these segmented patches are grouped to make the volumetric fitting. Figure 10.10 shows this overall procedure of 3D object representation using a superquadric model. We assume that at least two sides of a volumetric primitive are visible at the same time in an input image. Just one side of a cube, for example, does not provide enough information on the extent of the whole object.

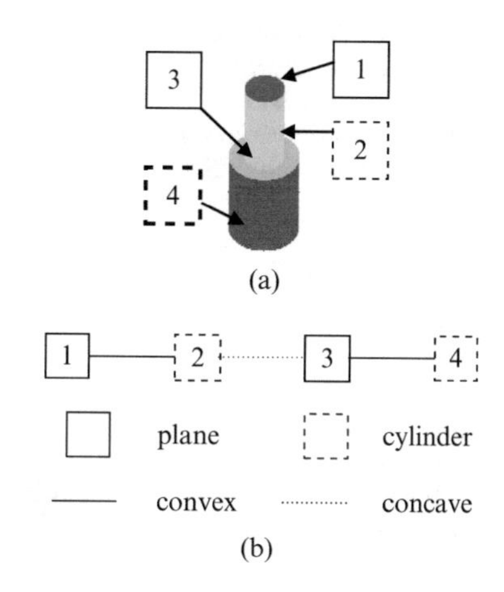

Fig. 10.11 Surface segmentation and surface adjacency graph of an object. (**a**) Surface segmentation of an object. (**b**) Surface adjacency graph of an object

10.14 Surface Segmentation of 3D Object

Given range data for a 3D object, we have to segment the object into the smooth surface patches by using curvature-based surface segmentation. As visibly invariant surface characteristics, the Gaussian and mean curvatures are used. Those curvatures are referred to as collective surface curvatures. Note that they are also invariant to translation and rotation of object surfaces. In this subsection, three fundamental surface types, planar, cylindrical, and spherical surfaces, are introduced.

After surface segmentation, a 3D object is decomposed into several surface patches that have the same curvature characteristics. Finally, a relationship between neighboring surfaces is established by using a surface adjacency graph (SAG) as shown in Fig. 10.11. In the SAG, a node indicates each surface and a link shows junction type between two surfaces. Figure 10.11a demonstrates the result of surface segmentation, where the box drawn with straight lines indicates a planar surface while the box drawn with dotted line shows a cylindrical surface. As shown in Fig. 10.11b, there is a concave junction depicted by a dotted line between surfaces 2 and 3. We assume that the SAG has no concave junctions.

10.15 Volumetric Decomposition

By using the segmented surface patches and SAG of the input 3D object, the superquadric part model can be constructed. Blocks, cylinders, and spheres are used as volumetric primitives for the superquadric part model. A block primitive consists of six planar surfaces; however, the number of visible surfaces is either three or two (if occluded), according to viewpoint. A cylinder has two planes

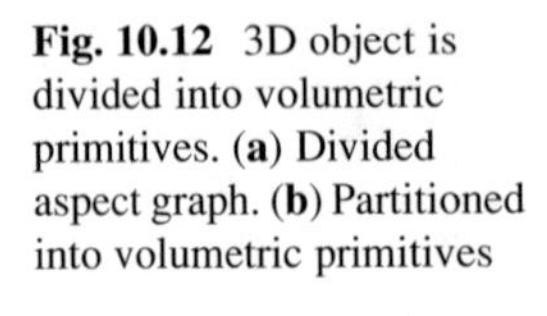

Fig. 10.12 3D object is divided into volumetric primitives. (**a**) Divided aspect graph. (**b**) Partitioned into volumetric primitives

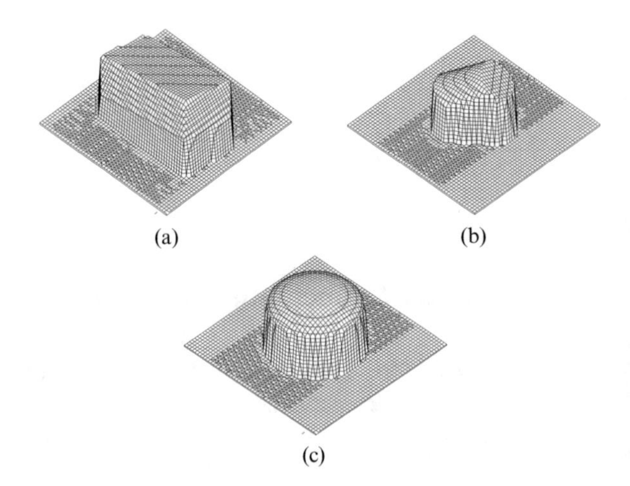

Fig. 10.13 Synthetic range images used in this experiment ($200 \times 200 \times 256$). (**a**) Block. (**b**) Cylinder. (**c**) Sphere

and a cylindrical surface, but the maximum number of visible surfaces is two, while a sphere has only one visible surface regardless of viewpoint. When there is a concave junction in the SAG, this junction should be broken to rebuild the volumetric primitives, based on our assumption. Therefore, the concave junction in Fig. 10.11b is drawn broken in Fig. 10.12a, dividing the object into two volumetric cylinder primitives as shown in Fig. 10.12b. In this way, the 3D object can be decomposed into finite sets of volumetric primitives for use with superquadric part models.

Noise-free $200 \times 200 \times 256$ synthetic range images and a $240 \times 240 \times 256$ real range image, which were obtained from MSU's Technical Arts 100X Range Scanner, were used for the experiment. The synthetic range images were taken from a CAD modeling system which allowed exact control over the elongation and other shape attributes. We assumed that the 3D input object consists mainly of blocks, cylinders, and spheres. Figure 10.13 shows the volumetric primitives used in this experiment, where the block in Fig. 10.13a has dimensions of $40 \times 40 \times 160$ pixels, the cylinder in Fig. 10.13b has a radius of 15 and a height of 70, and the sphere in Fig. 10.13c has a radius of 30 pixels.

Figure 10.13 shows the results of shape recovery by using the nondeformed superquadric model recovery. Figure 10.14a–c, respectively, shows recovery

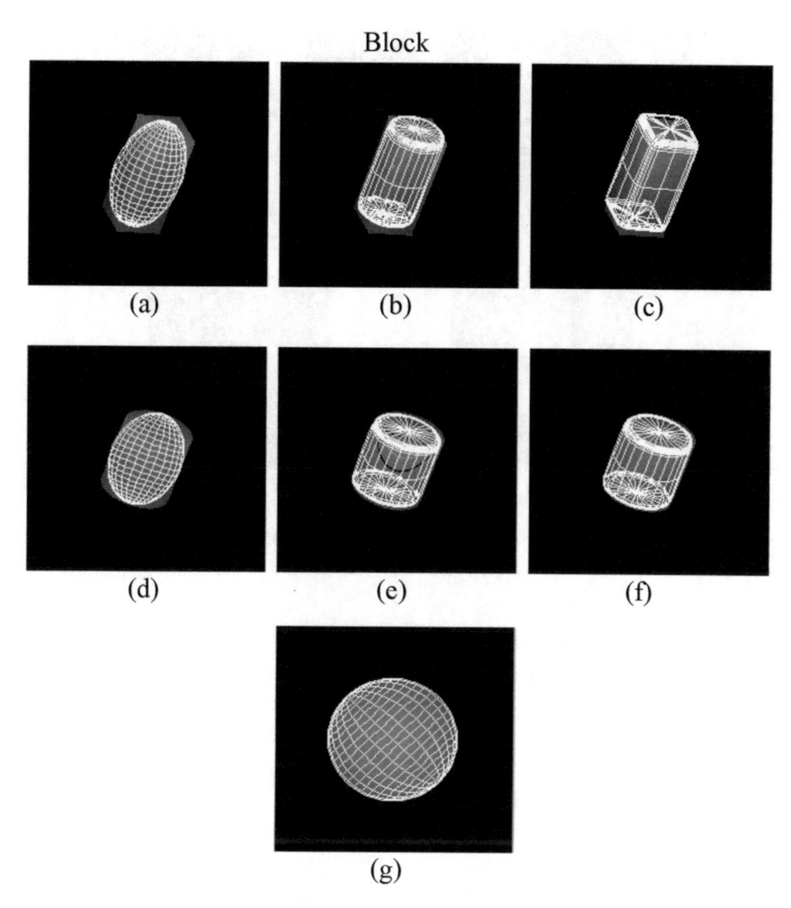

Fig. 10.14 Parametric shape recovery of simple volumetric primitives. (**a**) One iteration. (**b**) Five iterations. (**c**) 25 iterations. (**d**) One iteration. (**e**) Five iterations. (**f**) 25 iterations. (**g**) One iteration

results of the block after 1, 5, and 26 iterations. Figure 10.14d–f shows the same results for the cylinder and Fig. 10.14g the results for the sphere. To recover a superquadric model from an input range image, we used the nonlinear iterative optimization method of Levenberg-Marquardt with 50 iterations.

The volumetric primitives used in this experiment are simple part objects, which are symmetric with respect to the cross section without deformation. In addition, these primitives should be elongated along one or more axis directions. If objects have the same size parameters along the x, y, and z axes, that is, a_x, a_y, and a_z, and/or are rotationally asymmetric with respect to the cross section, the shape cannot be recovered. Bending and linear tapering are two deformations commonly applied to superquadrics, which affect the symmetry attributes of the objects as a whole. However, those operations do not affect the symmetry attributes of the cross section.

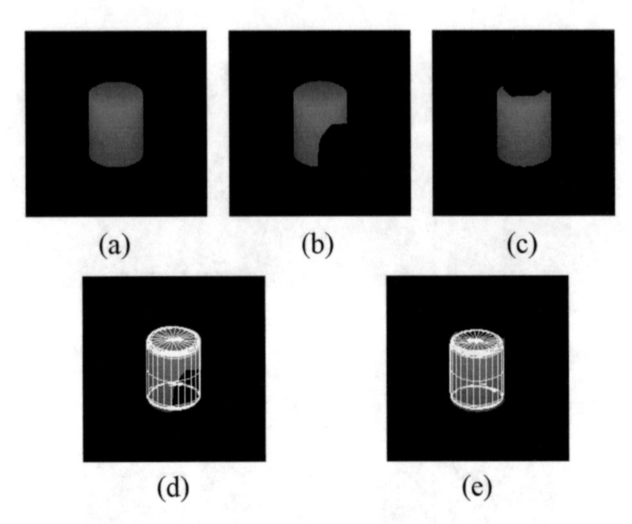

Fig. 10.15 Superquadric recovery with data loss. (**a**) Original cylinder, (**b**) bottom loss, (**c**) top loss, (**d**) recovered results of bottom loss, (**e**) recovered results of top loss

In order to demonstrate robustness of superquadric shape recovery, experimental results for input objects with a certain amount of data loss are presented. Composite objects can be constructed by combining several volumetric primitives and initially assuming there is data loss due to occlusions. The object without defects is shown in Fig. 10.15a with 5055 range data points. Figure 10.15b, c includes data loss at the bottom of cylinder (3846 data points) and at the top (4415 points, respectively). As shown in Fig. 10.15d, e, these cylinders can successfully be recovered in spite of the defects.

We assume that the 3D object consists of a number of simple volumetric primitives. Therefore, the 3D input object is segmented into smooth surface patches by using surface curvature information, which results in volumetric segmentation decomposing the object into part-based superquadric models. The experimental results of this segmentation procedure are shown in Fig. 10.16. Figure 10.17 shows the results of superquadric model recovery. The results for a real object are shown in Figs. 10.18 and 10.19.

In this chapter, a method for 3D object representation by using part-based superquadric models is presented. The superquadric model consists of geometric parameter representations and recovers 3D shapes by extracting the superquadric parameters. Superquadric-based shape recovery is performed by least square minimization of the predefined objective function for all range points belonging to a single-part volumetric primitive. For the purpose of iterative minimization, rough estimation of initial position, orientation, and size of the part has to be computed. As shown by the experimental results, estimation of superquadric parameters is highly dependent on the viewpoint. Estimation accuracy may become very low for objects with rough surfaces or noise. This part-based superquadric model representation provides rough descriptions of 3D objects and makes possible object

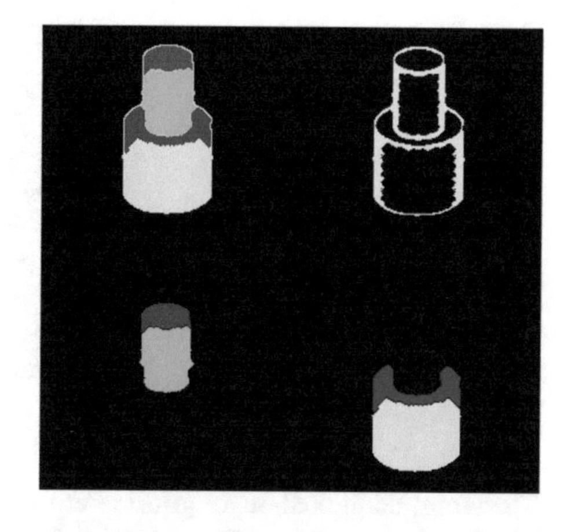

Fig. 10.16 Surface and volumetric segmentation of the two-cylinder object

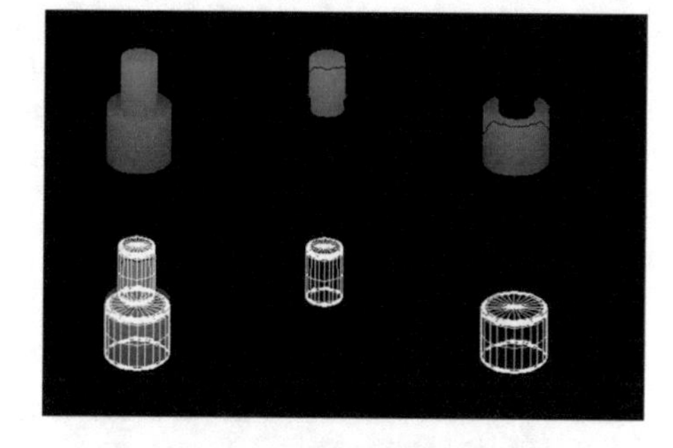

Fig. 10.17 Recovered superquadric model of the two-cylinder object

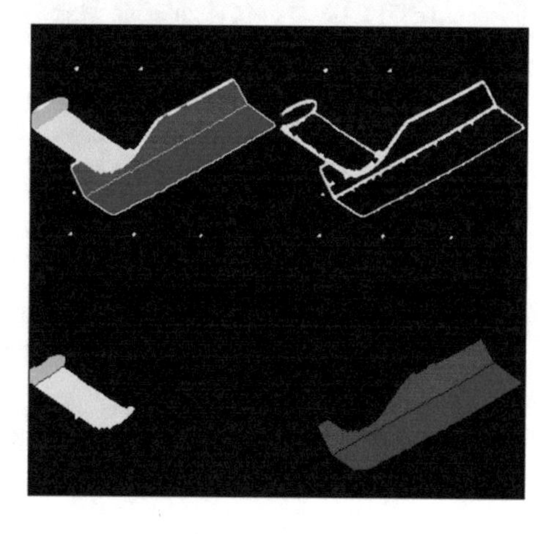

Fig. 10.18 Surface and volumetric segmentation of a real object

avoidance, grasping, and shape classification. By decomposing a complicated object into finite sets of part-based superquadric models, we can represent the object as a combination of multiple simple part objects. For more complicated 3D objects, part-based superquadric models should be combined with deformation.

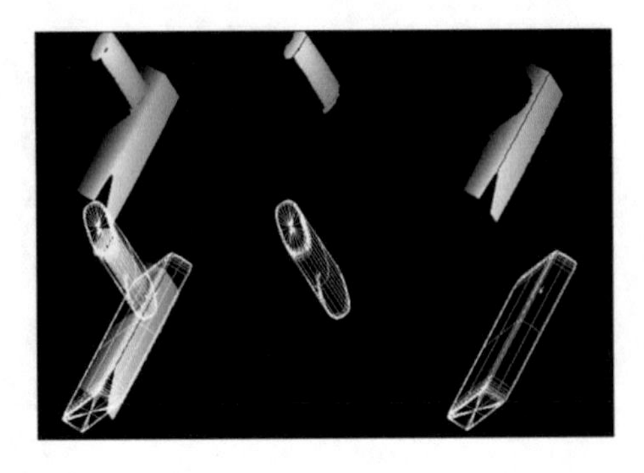

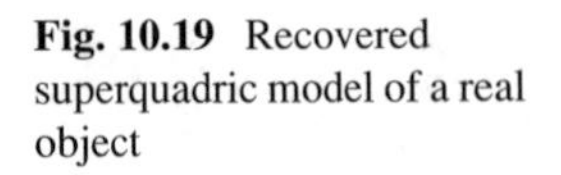

Fig. 10.19 Recovered superquadric model of a real object

Therefore, additional superquadric deformation parameters such as bending and tapering are necessary. For a precise 3D object recognition, the relationship of combinations between part-based volumetric primitives must be considered.

Based on the experimental results, part-based superquadric models for 3D object representation have shown their effectiveness in facilitating object recognition or indexing.

References

[besl85] P.J. Besl, R.C. Jain, Three-dimensional object recognition. ACM Comput. Surv. **17**(1), 75–145 (1985)

[biederman85] I. Biederman, Human image understanding: recent research and theory. Comput. Vis. Graph. Image Process. **32**, 29–73 (1985)

[hwang91] S.C. Hwang, H.S. Yang, 3D object representation using the CSG tree and superquadrics. KITE J. Electr. Eng. **2**(1), 76–83 (1991)

[leonardis95] A. Leonardis, A. Gupta, R. Bajacy, Segmentation of range images as the search for geometric parametric models. Int. J. Comput. Vis. **14**, 253–277 (1995)

[press92] W.H. Press, B.P. Flannery, S.A. Teukolsky, W.T. Vettering, *Numerical Recipes in C* (Cambridge University Press, Cambridge, England, 1992)

[raja92] N.S. Raja, A.K. Jain, Recognizing geons from superquadrics fitted to range data. Image Vis. Comput. **10**(3), 179–190 (1992)

[solina90] F. Solina, R. Bajcsy, Recovery of parametric models from range images: the case for superquadrics with global deformation. IEEE Trans. Pattern Anal. Mach. Intell. **12**(2), 131–147 (1990)

[wu94] K. Wu, M.D. Levine, Recovering parametric geons from multiview range data, in *Proceedings of IEEE Conference Computer Vision and Pattern Recognition*, pp. 159–166, June, 1994

Chapter 11
Regularized 3D Image Smoothing

Abstract This chapter discusses a new surface smoothing method based on area decreasing flow, which can be used for preprocessing raw range data or postprocessing reconstructed surfaces. Although surface area minimization is mathematically equivalent to the mean curvature flow, area decreasing flow is far more efficient for smoothing the discrete surface on which the mean curvature is difficult to estimate. A general framework of regularization based on area decreasing flow is proposed and applied to smoothing range data and arbitrary triangle mesh. Crease edges are preserved by adaptively changing the regularization parameter. The edge strength of each vertex on a triangle mesh is computed by fusing the tensor voting and the orientation check of the normal vector field inside a geodesic window. Experimental results show that the proposed algorithm provides successful smoothing for both raw range data and surface meshes.

11.1 Introduction

As three-dimensional (3D) modeling techniques cover wider application areas, 3D image processing attracts more attention. Examples include geospatial science, photorealistic modeling of real environments, and reverse engineering of small parts and entire factories, to name a few.

So far, the most precise techniques to digitize surfaces of real 3D objects employ laser range finders. The acquired range signals are, however, corrupted by noise. Smoothing the corrupted surfaces in the 3D image is one of the most basic operations for noise suppression in real range data. This process can be applied to range data acquisition in a preprocessing step.

Although laser range finders produce the most accurate 3D data available, their employment for 3D modeling fields is limited for several reasons such as safety restrictions, physical properties of 3D objects, and runtime requirements. As an alternative to laser range finders, passive techniques like stereo vision can be used. 3D surfaces are computed based on a triangulation of corresponding points that have to be recognized in the images when a stereo vision technique is applied. The resulting reconstructed 3D surfaces are often corrupted by mismatches of pixels in

M.A. Abidi et al., *Optimization Techniques in Computer Vision*, Advances in Computer Vision and Pattern Recognition, DOI 10.1007/978-3-319-46364-3_11

the stereo image pairs. After 3D surfaces are reconstructed, smoothing can also be applied as postprocessing to reduce such reconstruction errors.

Furthermore, successful surface smoothing can greatly improve the visual appearance of a 3D object and at the same time can feed improved data to successive processes, such as matching, surface segmentation, and mesh simplification.

If we consider that a range value of a range image is equivalent to an intensity value of a two-dimensional (2D) image, 2D low-pass filtering can serve as a simple surface smoothing method. For an arbitrary surface, 2D image processing algorithms such as spatial or frequency domain low-pass filterings, however, cannot provide promising results because they do not take 3D parameters into consideration. For this reason, the surface smoothing problem has been tackled in the literature using different approaches, including regularization, surface fairing, and surface evolution using the level set method.

Regularization has been used for surface interpolation from sparse range data and for restoring noisy surfaces. Regularization performs smoothing operations by minimizing an energy function that includes data compatibility and smoothing terms. The result of minimization is a trade-off between remaining close to the given observed data and avoiding a bumpy, coarse surface. Different definitions of the smoothing term have been proposed in the literature. Blake and Zissermen [blake87] introduced the membrane and plate model where smoothing terms are the gradient variation and square Laplacian, respectively. Stevenson and Delp [stevenson92] defined the smoothing term as the sum of squared principal curvatures. Yi and Chelberg [yi95] proposed a simple first-order smoothing term, which tends to minimize the surface area. Since these methods assume that the surface is a height map, they cannot be applied to smooth arbitrarily defined surfaces. Another category of surface smoothing methods, which is known as discrete surface fairing, directly process the surface mesh. Taubin [taubin95] applied a weighted Laplacian smoothing on the surface mesh. Shrinkage is avoided by alternating the scale factors of opposite signs. Vollmer et al. [vollmer99] proposed another modified Laplacian smoothing. In [ohtake00], Ohtake et al. showed that Laplacian smoothing tends to develop unnatural deformations. They applied the mean curvature flow to smooth the surface and the Laplacian flow to improve the mesh regularity.

Mean curvature flow originated from generating minimal surfaces in mathematics and material sciences [sethian98] and has been applied to implicit surface smoothing by Whitaker [whitaker98] and Zhao et al. [zhao00]. The surface, represented by a zero crossing field in the volumetric grid, is deformed according to the magnitude of the mean curvature using the level set methods.

Mean curvature flow is mathematically equivalent to surface area minimization, as shown in Sect. 11.2. However, direct area decreasing flow better fits the discrete surface smoothing than mean curvature flow, because mean curvature is not well defined on a discrete surface where the area can be explicitly computed.

In this chapter, a new regularized 3D image smoothing method based on locally adaptive minimization of the surface area is proposed. The approach is first applied to range image smoothing. Since range values are optimally estimated along the ray of measurement, there is no overlapping data problem. The method is then extended to process arbitrary surfaces represented by the triangle mesh. The position of each vertex is adjusted along the surface normal to minimize the simplex area. The optimal, unique magnitude of the adjustment can be obtained. Crease edges are preserved by adaptive smoothing according to the edge strength of each vertex on a triangle mesh, which is computed by fusing the tensor voting and the orientation check of the normal vector field inside a geodesic window.

This chapter is organized as follows. In Sect. 11.2, the regularized energy function is formulated for surface smoothing based on area decreasing flow. In Sect. 11.3, we derive the area decreasing stabilizer for calibrated range data and show how to preserve the edges. The stabilizer for the triangle mesh is presented in Sect. 11.4 along with the discussion of adaptive edge preservation based on the tensor voting approach. Experimental results are shown in Sect. 11.5, and Sect. 11.6 concludes the chapter.

11.2 Regularization Based on Area Decreasing Flow

The 3D image smoothing problem corresponds to a constrained optimization problem. Regularization is the most widely used method to solve practical optimization problems with one or more constraints. Major advantages of regularization include (a) simple and intuitive formulation of the objective or energy function and (b) flexibility in incorporating one or more constraints into the optimization process.

General n-dimensional regularization involves minimizing the energy function

$$f(x) = g(x) + \lambda h(x),\ x \in R^n, \tag{11.1}$$

where minimization of $g(x)$ involves the compatibility of the solution to the original observation, and minimization of $h(x)$ incorporates prior knowledge. λ is called the regularization parameter which determines the weight of minimization between $g(x)$ and $h(x)$. When regularization is applied to surface smoothing, x is replaced by X, a parameterized surface which represents a differentiable map from an open set $U \subset R^2$ into R^3, that is, $X : U \subset R^2 \to R^3$. Given a bounded domain $D \subset U$ and a differentiable function $l : \overline{D} \subset R$, where $\overline{D}$ represents the union of the domain D and its boundary, the variation of $X(\overline{D})$ along normal n is given as

$$\phi : \overline{D} \times (-\varepsilon, \varepsilon) \to R^3, \tag{11.2}$$

and

$$\phi(u, v, t) = X(u, v) + tl(u, v)n(u, v), \tag{11.3}$$

where $(u, v) \in \overline{D}$ and $t \in (-\varepsilon, \varepsilon)$.

For the map $X^t(u,v) = \phi(u,v,t)$, the data compatibility term $g(x)$ is chosen to be the distance between the original surface and the smoothed surface, such as

$$g(X) = \|X - X^t\|^2, \tag{11.4}$$

where $\|\cdot\|$ denotes a norm. The smoothness is assumed to be the prior knowledge of the surface. From the frequency analysis point of view, the smoothing or stabilizing term $h(X)$ should reflect the high-frequency energy. From the observation that noisy surfaces usually have larger areas than smooth surfaces, we choose $h(X)$ to be the surface area A_s. This is closely related to the mean curvature flow that has been applied to 3D image processing [ohtake00, whitaker98, zhao00, desbrun99].

The area $A_s(t)$ of $X^t(\overline{D})$ is obtained as

$$A_s(t) = \int_{\overline{D}} \sqrt{1 - 4tlH + R}\sqrt{EG - F^2}dudv, \tag{11.5}$$

where $\lim_{t\to 0}(R/t) = 0$, H represents the mean curvature, and E, F, and G are the coefficients of the first fundamental form. The derivative of $A_s(t)$ at $t = 0$ is

$$A_s'(0) = -\int_{\overline{D}} 2lH\sqrt{EG - F^2}dudv. \tag{11.6}$$

The area is always decreasing if the normal variation is set as $l = H$, which is called the mean curvature flow. This flow will generate a minimal surface whose mean curvature vanishes everywhere. The details of minimal surface theory can be found in [docarmo76]. A minimal surface may take the form of a plane, catenoid, Enneper's surface, Scherk's surface, etc. Under mean curvature flow, a smooth cylinder can deform into a catenoid or two planes. To apply the mean curvature flow to a regularized surface smoothing, the compatibility term $g(X)$ should be used to generate variation near the original surface. The curvature on discrete surfaces, such as for the range data and surface meshes, is difficult to compute because it is defined on an infinitesimal area. For this reason, direct surface area minimization is better than mean curvature flow for smoothing discrete surfaces. The triangle mesh is used to represent surfaces because most types of surface meshes can be easily translated into a triangle mesh.

11.3 Range Data Smoothing

The viewpoint invariant reconstruction of surfaces from sparse range images by regularization, which is essentially an interpolation problem, has been studied in [blake87, stevenson92, yi95, vaidya98]. Reconstruction from a single-view, dense range image, on the other hand, is a smoothing or restoration problem.

In Sect. 11.3.1, the area decreasing flow is applied to range data smoothing. Edge preservation by adaptive regularization is discussed in Sect. 11.3.2.

11.3.1 Area Decreasing Stabilizer for Range Data

In previous works [blake87, stevenson92, yi95], the surface was considered as a graph $z(x, y)$ and represented as z_{ij} over a rectangular grid. Given the observed data c_{ij} on the same rectangular grid, the viewpoint invariant surface reconstruction can be performed by minimizing the regularized energy function as

$$f = \sum_{ij} \left(z_{ij} - c_{ij}\right)^2 / \sigma_{ij}^2 + \lambda h, \tag{11.7}$$

where $1/\sigma_{ij}$ denotes the confidence of the measurement. In practice, $1/\sigma_{ij}$ approximates the surface slant $\cos \varsigma$ with respect to the incident laser. The larger the angle ς between the surface normal and the direction of measurement, the smaller the confidence becomes. Because $\left(z_{ij} - c_{ij}\right)/\sigma_{ij}$ is also the perpendicular distance between the estimated and the real surfaces, this distance is viewpoint invariant. The stabilizing function h can take several different forms. For example, a first-order term is used in [yi95], while a second-order term is employed in [stevenson92].

Estimating the elevation z_{ij} is feasible for sparse data. But in dense range images from a range scanner with a spherical coordinate system, $z(x, y)$ is no longer a graph, and estimating the elevation may result in overlap in range measurement. Therefore, we would like to estimate the range r_{ij} instead of z_{ij} so that all refinement takes place along the line of measurement.

For the Perceptron range scanner [perceptron93], the range value of each pixel z_{ij} is converted to (x_{ij}, y_{ij}, z_{ij}) in Cartesian coordinates. We adopt the calibration model described in [hoover96]:

$$\begin{cases} x_{ij} = dx + r \sin\alpha \\ y_{ij} = dy + r \cos\alpha \sin\beta \\ z_{ij} = dz - r \cos\alpha \cos\beta \end{cases}, \tag{11.8}$$

$$\begin{cases} \alpha = \alpha_0 + H_0(\mathrm{col}/2 - j)/N_0 \\ \beta = \beta_0 + V_0(\mathrm{row}/2 - i)/M_0 \end{cases}, \tag{11.9}$$

$$\begin{cases} r_1 = (dz - p_2)/\delta \\ r_2 = \sqrt{dx^2 + (p + dy)^2}/\delta \\ r = \left(R_{ij} + r_0 - r_1 - r_2\right)/\delta \end{cases}, \tag{11.10}$$

and

$$\begin{cases} dx = (p_2 + dy)\tan\alpha \\ dy = dz\tan(\theta + 0.5\beta) \\ dz = -p_1(1.0 - \cos\alpha)/\tan\gamma \end{cases}, \qquad (11.11)$$

where $\{p_1, p_2, \gamma, \theta, \alpha_0, \beta_0, H_0, V_0, r_0, \delta\}$ represents the set of calibration parameters of the scanner, and (M_0, N_0) refers to the image size. For estimating r, we can use the parameterization

$$X(\alpha, \beta) = (r\sin\alpha, r\cos\alpha\sin\beta, -r\cos\alpha\cos\beta) \qquad (11.12)$$

and ignore small values denoted by dx, dy, and dz in the analysis. The coefficients of the first fundamental form, which will be used shortly in the computation of the surface area, are given in the basis of $\{X_\alpha, X_\beta\}$ as

$$\begin{cases} E = X_\alpha \cdot X_\beta = r^2 + r_\alpha^2 \\ F = X_\alpha \cdot X_\beta = r_\alpha r_\beta \\ G = X_\beta \cdot X_\beta = r^2\cos^2\alpha + r_\beta^2 \end{cases}, \qquad (11.13)$$

where

$$X_\alpha = \frac{\partial X}{\partial \alpha}, \; X_\beta = \frac{\partial X}{\partial \beta}, \; r_\alpha = \frac{\partial r}{\partial \alpha}, \text{ and } r_\beta = \frac{\partial r}{\partial \beta}. \qquad (11.14)$$

c represents the observed value of r. Range data smoothing can then be performed by minimizing the following energy function:

$$f = \sum_{ij} \left(r_{ij} - c_{ij}\right)^2 / \sigma_{ij}^2 + h. \qquad (11.15)$$

We let the stabilizing function h be the surface area, which can be calculated as

$$h = A_s = \iint_{\overline{D}} \sqrt{EG - F^2}\, d\alpha d\beta, \qquad (11.16)$$

where D represents the domain of (α, β).

The stabilizing function used by Yi [yi95] has the same effect of minimizing surface area, but the assumption is made that the surface is a graph in Cartesian coordinates. By using the height map $z(x, y)$, the coefficients of the first fundamental form are obtained as

$$\begin{cases} E = 1 + z_x^2 \\ F = z_x z_y \\ G = 1 + z_y^2 \end{cases}. \qquad (11.17)$$

Accordingly, the stabilizing function is obtained as

$$h = \iint_D \sqrt{1 + z_x^2 + z_y^2} dxdy, \tag{11.18}$$

which is similar to the function used in [yi95]. As Eq. (11.16) is not easily minimized due to the square root operation, we instead minimize

$$h = \iint_D (EG - F^2) d\alpha d\beta. \tag{11.19}$$

The following theorem justifies that minimizations of Eqs. (11.16) and (11.19) gave the same solution.

Theorem 12.1 *The minimization of an area integration of the square root of a function is equivalent to that of the same area integration of the function without the square root.*

Proof We define

$$\xi_{iN_0+j} = \sqrt{E_{ij}G_{ij} - F_{ij}^2} \text{ and } \eta_{iN_0+1} = 1, \text{ for } i < M_0, \ j < N_0, \text{and} \tag{11.20}$$

$$\xi_k = \eta_k = 0, \text{ for } k > M_0 N_0. \tag{11.21}$$

Using the Cauchy-Schwarz inequality, we know that

$$\sum_{j=1}^{\infty} |\xi_j \eta_j| \leq \sqrt{\sum_{k=1}^{\infty} |\xi_k|^2} \sqrt{\sum_{m=1}^{\infty} |\eta_m|^2}, \tag{11.22}$$

where $\sum_{j=1}^{\infty} |\xi_j|^2 < \infty$ and $\sum_{j=1}^{\infty} |\eta_j|^2 < \infty$, because only a finite number of terms are nonzero. This then yields

$$\sum_{i,j}^{M_0 N_0} \sqrt{E_{ij}G_{ij} - F_{ij}^2} \leq \sqrt{M_0 N_0 \sum_{i,j}^{M_0 N_0} \left(E_{ij}G_{ij} - F_{ij}^2\right)}, \tag{11.23}$$

which shows that the minimizer of Eq. (11.16) implies that of Eq. (11.19). □

From Eqs. (11.13), (11.15), and (11.19), the finally regularized energy function is given as

$$f = \sum_{ij} \left(r_{ij} - c_{ij}\right)^2 / \sigma_{ij}^2 + \sum_{ij} \lambda_{ij} \left(r_{ij}^4 \cos^2\alpha + r_{ij}^2 r_\beta^2 + r_\alpha^2 r_{ij}^2 \cos^2\alpha\right). \tag{11.24}$$

Among various optimization methods, the simple gradient descent method is adopted to minimize Eq. (11.24), because convergence can easily be controlled with an adaptively varying regularization parameter. The estimation r'_{ij} of each measurement r_{ij} is given as

$$r'_{ij} = r_{ij} - w\frac{\partial f}{\partial r_{ij}}, \tag{11.25}$$

where w represents the iteration step size and

$$\begin{aligned}\frac{\partial f}{\partial r_{ij}} &= 2\left(r_{ij} - c_{ij}\right)/\sigma_{ij}^2 + \lambda_{ij}\Big\{4r_{ij}^3\cos^2\alpha + \Big[2r_{ij}\left(r_{i+1,j} - r_{ij}\right)^2 - 2r_{ij}^2\left(r_{i+1,j} - r_{ij}\right)\\ &+2r_{i-1,j}^2\left(r_{ij} - r_{i-1,j}\right)\Big]\left(\tfrac{1}{d\beta}\right)^2\\ &+\Big[2r_{ij}\left(r_{i,j+1} - r_{ij}\right)^2 - 2r_{ij}^2\left(r_{i,j+1} - r_{ij}\right) + 2r_{i,j-1}^2\left(r_{ij} - r_{i,j-1}\right)\Big]\left(\tfrac{\cos\alpha}{d\alpha}\right)^2\Big\}.\end{aligned} \tag{11.26}$$

In calculating the derivative of f in Eq. (11.26), the following forward difference approximations were used:

$$r_\alpha = \frac{r_{i,j+1} - r_{ij}}{d\alpha} \text{ and } r_\beta = \frac{r_{i+1,j} - r_{ij}}{d\beta}. \tag{11.27}$$

Alternatively, the central difference approximation can also be used for r_α and r_β, such as

$$r_\alpha = \frac{r_{i,j+1} - r_{i,j-1}}{2d\alpha} \text{ and } r_\beta = \frac{r_{i+1,j} - r_{i-1,j}}{2d\beta}. \tag{11.28}$$

The derivative of f is then differently obtained as

$$\begin{aligned}\frac{\partial f}{\partial r_{ij}} &= 2\left(r_{ij} - c_{ij}\right)/\sigma_{ij}^2 + \lambda_{ij}\Big\{4r_{ij}^3\cos^2\alpha + \Big[2r_{ij}\left(r_{i+1,j} - r_{i-1,j}\right)^2\\ &- 2r_{i+1,j}^2\left(r_{i+2,j} - r_{ij}\right) + 2r_{i-1,j}^2\left(r_{ij} - r_{i-2,j}\right)\Big] \times \left(\tfrac{1}{2d\beta}\right)^2\\ &+\Big[2r_{ij}\left(r_{i,j+1} - r_{i,j-1}\right)^2 - 2r_{i,j+1}^2\left(r_{i,j+2} - r_{ij}\right) + 2r_{i,j-1}^2\left(r_{ij} - r_{i,j-2}\right)\Big]\left(\frac{\cos\alpha}{2d\alpha}\right)^2\Big\}.\end{aligned} \tag{11.29}$$

11.3.2 Edge Preservation for Range Data

Incorporation of the regularizing term in the regularized energy function, as shown in Eq. (11.24), tends to suppress local change in the range image. Although the smoothing function is good for suppressing undesired noise, it also degrades important feature information such as edges, corners, and segment boundaries. Using an additional energy term to preserve discontinuity [blake87], however, generally makes the minimization very difficult. Instead, we use the results of a 2D edge detection operation to adaptively change the weight of the regularization parameter so that edges are preserved during the regularization. Although there are various simple edge enhancement filters, we use the optimal edge enhancer, which guarantees both good detection and localization [canny86].

Let $J_x(i,j)$ and $J_y(i,j)$ be the gradient component of the Gaussian-filtered version of r_{ij} in the horizontal and the vertical directions, respectively. Then the edge strength image can be obtained as

$$\mathrm{e}_{ij} = \sqrt{J_x^2(i,j) + J_y^2(i,j)}. \tag{11.30}$$

The regularizing term in Eq. (11.24) can then be adaptively weighted as in [katsaggelos89] using

$$\lambda_{ij} = \frac{\rho}{1 + \kappa \mathrm{e}_{ij}^2}, \text{ for } 0 < \kappa < 1, \tag{11.31}$$

where k represents a parameter that determines the sensitivity of edge strength, and ρ is a prespecified constant. The selection of ρ generally depends on the desired data compatibility as well as the level of noise. Here we introduce a method to approximately estimate ρ, which is determined in such a way that the average relative adjustment (ARA) from the observed range value is in the same range as that produced by other popular techniques, such as median filtering. For a sample range image, the ARA was found in our experiments to be 0.44 % after twice applying a 3×3 median filtering. ρ is selected to make the ARA smaller than that of the median-filtered version. For example, the ARA was found to be 0.25 % after 50 iterations using the regularization method for the same image. Median filtering is still applied twice before using the regularization method, because the regularization method does not suppress impulsive (salt and pepper) noise effectively. For the same sample image, the ARA was found to be 0.50 % after median filtering twice and 50 iterations of regularization. This is comparable to the ARA produced using only median filtering.

11.4 Surface Mesh Smoothing

In order to apply the proposed regularized range image smoothing algorithm, the accurate calibration model of the scanner must be known. However, for some scanners, such as the RIEGL system LMS-Z210 [riegl00] used in this work, the calibration parameters are used by the manufacturer and not released. It is also desirable to smooth an arbitrary surface instead of single-view range data. In Sect. 11.4.1, we extend the area decreasing flow to process arbitrary surfaces represented by the triangle mesh. Adaptive smoothing is discussed in Sect. 11.4.2, where edge strength is computed based on the tensor voting approach.

11.4.1 Area Decreasing Stabilizer for Surface Mesh

For an umbrella neighborhood [kobbelt98] with I triangles on a triangle mesh, the position of the center vertex v is adjusted along the normal direction n, as shown in Fig. 11.1. The superscript k represents the kth adjustment. The original center vertex position is denoted as $v^{(0)}$.

In the kth adjustment, the center vertex moves from $v^{(k)}$ to $v^{(k+1)}$ by the length of l 1 in the direction of $n^{(k)}$ such as

$$v^{(k+1)} = v^{(k)} + l \cdot n^{(k)}. \tag{11.32}$$

An adjustment is made to minimize the area of the umbrella and at the same time be compatible to the original measurement. The local energy function is defined as

$$f(l) = \sum_{i=1}^{I} 4\left[S_i^{(k+1)}\right]^2 + \lambda \left\| \Delta v^{(k+1)} \right\|^2, \tag{11.33}$$

where $S_i^{(k+1)}$ i is the area of the triangle $v_{i1}v_{i2}v^{(k+1)}$, denoted by T_i, and $\Delta v^{(k+1)}$is defined as

$$\Delta v^{(k+1)} = v^{(k+1)} - v^0 = \Delta v^{(k)} + l \cdot n^{(k)}. \tag{11.34}$$

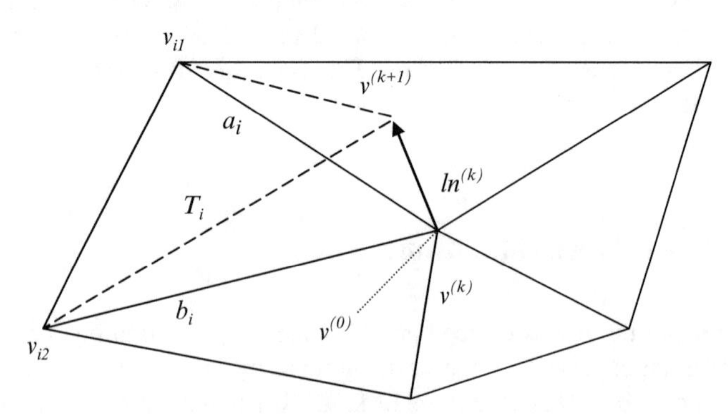

Fig. 11.1 Umbrella operation. In the kth iteration, the vertex moves from $v^{(k)}$ to $v^{(k+1)}$ by $l \times n^{(k)}$

For computational convenience, $\sum S_i$ is replaced by $\sum S_i^2$. Similar to the justification shown in Theorem 12.1, the replacement can be justified using the Cauchy-Schwarz inequality, such as

$$\sum_{i=1}^{I} S_i \leq \sqrt{I \sum_{i=1}^{I} S_i^2}, \tag{11.35}$$

where $S_i^{(k+1)}$ is computed as

$$S_i^{(k+1)} = \frac{1}{2} \left\| \left(v_{i1} - v^{(k+1)} \right) \wedge \left(v_{i2} - v^{(k+1)} \right) \right\|. \tag{11.36}$$

For notational simplicity, we define

$$a_i = v_{i1} - v^{(k)} \text{ and } b_i = v_{i2} - v^{(k)}. \tag{11.37}$$

The energy function defined in Eq. (11.30) can then be rewritten as

$$\begin{aligned} f(l) &= \sum_{i=1}^{I} \left\| \left(a_i - l n^{(k)} \right) \wedge \left(b_i - l n^{(k)} \right) \right\|^2 + \lambda \left\| \Delta v^{(k+1)} \right\|^2 \\ &= l^2 \left\{ \sum_{i=1}^{I} \left\{ \|a_i - b_i\|^2 - \left[(a_i - b_i) \cdot n^{(k)} \right]^2 \right\} + \lambda \right\} \\ &\quad + 2l \left\{ \sum_{i=1}^{I} \left[\left(a_i \cdot n^{(k)} \right) b_i - \left(b_i \cdot n^{(k)} \right) a_i \right] \cdot (a_i - b_i) + \lambda \left(\Delta v^{(k)} \cdot n^{(k)} \right) \right\} \\ &\quad + \sum_{i=1}^{I} \|a_i\|^2 \|b_i\|^2 - (a_i \cdot b_i)^2 + \lambda \left\| \Delta v^{(k)} \right\|^2. \end{aligned} \tag{11.38}$$

The optimum value of l, which minimizes f, can be obtained by solving $\frac{\partial f}{\partial l} = 0$, that is,

$$l = \frac{A - \lambda \Delta v^{(k)} \cdot n^{(k)}}{B + \lambda} \tag{11.39}$$

where

$$A = \sum_{i=1}^{I} \left\{ \left(b_i \cdot n^{(k)} \right) a_i - \left(a_i \cdot n^{(k)} \right) b_i \right\} \cdot \left(a_i - b_i \right), \tag{11.40}$$

and

$$B = \sum_{i=1}^{I} \|a_i - b_i\|^2 - \left\{ (a_i - b_i) \cdot n^{(k)} \right\}^2. \tag{11.41}$$

The surface is iteratively deformed in the sense of minimizing the area of the umbrella according to Eq. (11.32).

11.4.2 Edge Preservation for Surface Mesh

Similar to the process used for 2D image, edge detection on a triangle mesh is performed by operation in a local window that is usually called the neighborhood. From a small neighborhood, such as an umbrella, it is difficult to determine if the vertex is from noise or near an edge. Window size is a trade-off between mesh resolution and noise level on the surface. More specifically, higher mesh resolution requires a smaller window, while a larger window is necessary for strong noise. The irregular connections on the triangle mesh, however, make window selection not as straightforward as with 2D images. We propose a new 3D edge detection algorithm using the geodesic window instead of the neighborhood defined by the Euclidean measure as used in previous works. The geodesic window is a small surface patch whose boundary has the same geodesic distance to the center vertex. Details for computing the geodesic distance on a triangle mesh can be found in [kimmel98] and [sun01].

Medioni et al. [medioni00] introduced a tensor voting approach that can signal the presence of a salient structure, a discontinuity, or an outlier at any location. In this work, tensor voting and the orientation check are combined inside the geodesic window to detect the crease edges on a triangle mesh.

The tensor voting method for detecting crease edges can simply be considered as the eigen analysis of the surface normal vector field. For a certain vertex q, the votes are cast by the neighboring triangles, as shown in Fig. 11.2a. The voted tensor cast

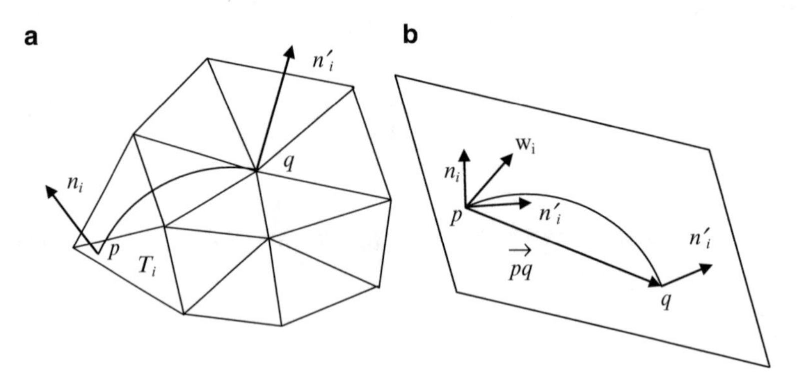

Fig. 11.2 Voting process. (**a**) n_i' is the voted normal by T_i's normal n_i. (**b**) The normal n_i at p is transported through the arc pq producing the voted normal n_I' at q

by the triangle T_i at vertex q is $\mu_i n_i' n_i'^{\mathrm{T}}$, where n_i' is the voted normal by T_i's normal n_i and μ_i is the weight of the vote. The new tensor collected at q is

$$T = \sum_{i=1}^{M} \mu_i n_i' n_i'^{\mathrm{T}}, \tag{11.42}$$

where M is the number of triangles inside the geodesic window of q and $M > I$. n_i' is obtained by transporting n_i through a sector of arc connecting p and q where p represents the centroid of T_i. Figure 11.2b illustrates the voting process. The arc is on the plane defined by two vectors n_i and $\vec{pq}$. The normals at two terminals of the arc are n_i, and $n_i' \cdot n_i'$ is computed as

$$n_i' = 2(n_i \cdot w_i) w_i - n_i, \tag{11.43}$$

where

$$w_i = \frac{(\vec{pq} \wedge n_i) \wedge \vec{pq}}{\|(\vec{pq} \wedge n_i) \wedge \vec{pq}\|}. \tag{11.44}$$

The weight μ_i exponentially decreases according to the geodesic distance d between p and q, such as

$$\mu_i = \mathrm{e}^{-(d/\tau)^2}, \tag{11.45}$$

where τ controls the decaying speed and depends on the scale of the input triangle mesh.

The singular value decomposition is applied to the new tensor T as

$$T = \begin{bmatrix} \mathrm{e}_1 & \mathrm{e}_2 & \mathrm{e}_3 \end{bmatrix} \begin{bmatrix} \nu_1 & 0 & 0 \\ 0 & \nu_2 & 0 \\ 0 & 0 & \nu_3 \end{bmatrix} \begin{bmatrix} \mathrm{e}_1^{\mathrm{T}} \\ \mathrm{e}_2^{\mathrm{T}} \\ \mathrm{e}_3^{\mathrm{T}} \end{bmatrix}, \tag{11.46}$$

where $\nu_1 \geq \nu_2 \geq \nu_3$.

T can be rewritten as

$$\begin{aligned} T = {} & (\nu_1 - \nu_2)\mathrm{e}_1\mathrm{e}_1^{\mathrm{T}} + (\nu_2 - \nu_3)\left(\mathrm{e}_1\mathrm{e}_1^{\mathrm{T}} + \mathrm{e}_2\mathrm{e}_2^{\mathrm{T}}\right) \\ & + \nu_3\left(\mathrm{e}_1\mathrm{e}_1^{\mathrm{T}} + \mathrm{e}_2\mathrm{e}_2^{\mathrm{T}} + \mathrm{e}_3\mathrm{e}_3^{\mathrm{T}}\right), \end{aligned} \tag{11.47}$$

where $\mathrm{e}_1\mathrm{e}_1^{\mathrm{T}}$ describes a stick, $\mathrm{e}_1\mathrm{e}_1^{\mathrm{T}} + \mathrm{e}_2\mathrm{e}_2^{\mathrm{T}}$ describes a plate, and $\mathrm{e}_1\mathrm{e}_1^{\mathrm{T}} + \mathrm{e}_2\mathrm{e}_2^{\mathrm{T}} + \mathrm{e}_3\mathrm{e}_3^{\mathrm{T}}$ describes a ball. Here we are interested in the plate component. $\nu_1 - \nu_3$ is related to the strength of the planar junction. The junction is detected if

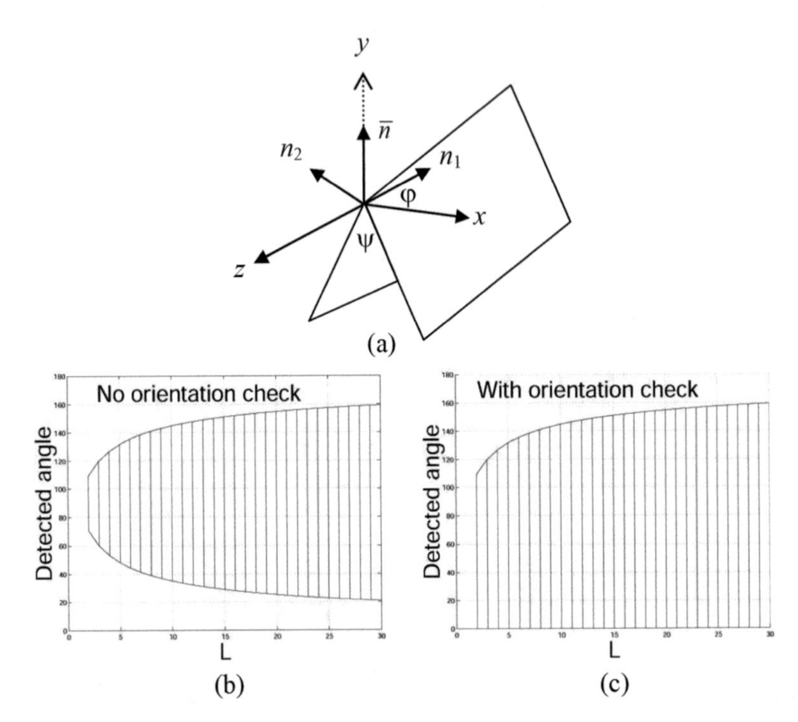

Fig. 11.3 (**a**) A crease edge with angle ψ. (**b**) Angle ψ that can be detected without an orientation check. Edges with sharp angles are missing. (**c**) Angle ψ that can be detected with an orientation check

$$\nu_2 - \nu_3 > \nu_1 - \nu_2 \text{ and } \nu_2 - \nu_3 > \nu_3. \tag{11.48}$$

However, this only works for junctions near 90°. To explain this situation, assume that a crease edge is parallel to the z axis, and two planes generating the edge are symmetric according to the y axis. The normals of the two surfaces are

$$n_1 = (\cos\varphi, \sin\varphi, 0)^{\mathrm{T}} \text{ and } n_2 = (-\cos\varphi, \sin\varphi, 0)^{\mathrm{T}}, \tag{11.49}$$

as shown in Fig. 11.3a.

According to Eq. (11.42), if μ_i is set to 1, the collected tensor at the vertex on the edge is given as

$$T = \begin{bmatrix} M\cos^2\phi & 0 & 0 \\ 0 & M\sin^2\phi & 0 \\ 0 & 0 & 0 \end{bmatrix}. \tag{11.50}$$

And the eigenvalues are simply obtained as

$$\nu_1 = M\cos^2\varphi,\ \nu_2 = M\sin^2\varphi \text{ and } \nu_3 = 0,\ \text{for } 0^\circ < \varphi \le 45^\circ, \tag{11.51}$$

or

$$\nu_1 = M\sin^2\varphi,\ \nu_2 = M\cos^2\varphi,\ \text{and}\ \nu_3 = 0,\ \text{for}\ 45^\circ < \varphi \le 90^\circ. \tag{11.52}$$

From Eq. (11.48), where we have

$$\nu_1 < 2\nu_2\ \text{and}\ \psi = 2\varphi, \tag{11.53}$$

we get a limited range of detectable edge angle, such as

$$70.53^\circ < \psi < 109.47^\circ. \tag{11.54}$$

We propose a method to solve this problem. Initially, it is obvious that if we set the edge detection condition as

$$\nu_1 > L\nu_2\ \text{and}\ \nu_2 > 2\nu_3, \tag{11.55}$$

we can detect crease edges with

$$2\tan^{-1}\frac{1}{\sqrt{L}} < \psi < 2\tan^{-1}\sqrt{L}, \tag{11.56}$$

which is depicted by the shaded region in Fig. 11.3b. Crease edges with large slopes can be detected by appropriately increasing L. However, Eq. (11.56) is still not complete because of a limit on sharp edges. This limit stems from the covariance matrix T that contains no orientation information. In other words, ν and $-\nu$ result in the same T. Based on this observation, we define

$$\overline{n} = \sum_{i=1}^{M} \mu_i n_i', \tag{11.57}$$

as shown in Fig. 11.3a. Observing that $\overline{n}$ tends to align with e_1 if $\psi > 90^\circ$ or to be perpendicular with e_1 if $\psi > 90^\circ$, we use

$$|\overline{n} \cdot \mathrm{e}_1| < \delta \tag{11.58}$$

as an additional edge detection condition, where δ is a positive threshold. In the experiments, $\delta = 0.3$ is used. Condition Eq. (11.55) and the orientation check in Eq. (11.58) provide a reasonable range of detectable edge angles as shown in Fig. 11.3c.

The crease edge strength is defined as

$$s = \begin{cases} 1, & |\overline{n} \cdot \mathrm{e}_1 < \delta| \\ (\nu_1 - \nu_2)/\nu_3, & \text{otherwise} \end{cases}, \tag{11.59}$$

such that $0 \leq s \leq 1$, and Eq. (11.32) is then modified by inserting an additional control factor based on the crease edge strength, such as

$$v^{(k+1)} = v^{(k)} + e^{-5s} ln^{(k)} \tag{11.60}$$

to realize the edge-adaptive smoothing.

The crease edge strength for each vertex is computed only once before deforming the surface. Assume the triangle mesh has N vertices, and there are M triangles inside the geodesic window on average. The computational complexity of crease edge strength is $O(NM \log M)$. The complexity for each iteration of smoothing is $O(N)$. The iteration is stopped if the following condition is satisfied:

$$\left|Z^{(k+1)} - Z^{(k)}\right| / Z^{(k+1)} < \varepsilon, \tag{11.61}$$

where

$$Z^{(k)} = \sum \left\| v^{(k+1)} - v^{(k)} \right\|, \tag{11.62}$$

and ε is a threshold chosen as 0.1 in our experiment.

Figure 11.4 shows the raw data captured by the Perceptron laser range scanner. The size of the original range image is 1024×1024 pixels. The Perceptron scanner is able to scan objects in a range from 2 to 20 m. Besides random noise, measurement accuracy is also sensitive to the surface material.

Figure 11.5 shows the corresponding nonadaptive regularization results that are much smoother than the raw surfaces shown in Fig. 11.4.

Figure 11.6a shows the result of a 3×3 median filtering conducted twice, which does not produce sufficiently smoothed surface. Additional median filtering provides no discernible improvement. For fair comparison, we did not include results with a larger median filtering window because the proposed algorithm is based on operation with a 3×3 window. Figure 11.6b shows the regularization result using the simple 2D Laplacian as the smoothing term. Unstable results along edges are

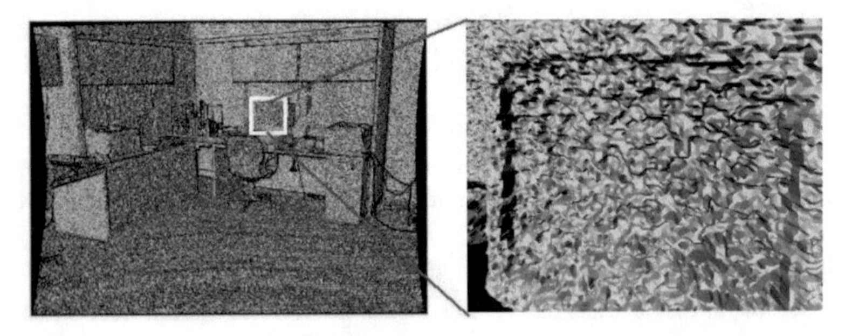

Fig. 11.4 Raw range data (*left*) and zoomed portion (*right*). The image was taken by the Perceptron range scanner. The size of the original range image is 1024×1024. The 3D model has 1,996,958 *triangles*

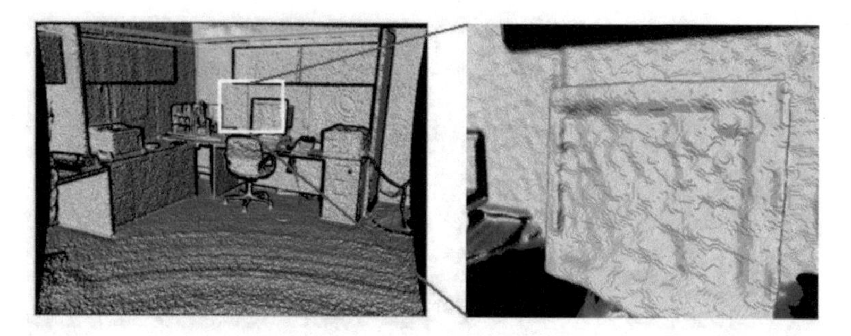

Fig. 11.5 Range data regularization result (*left*) and zoomed portion (*right*). The smoothed image is obtained by 50 iterations of nonadaptive regularization using area decreasing flow

obtained, which coincide with the results reported in [blake87]. The edge map and 50 iterations of nonadaptive and adaptive regularization results are shown in Fig. 11.6c–e, respectively. In the regularization, $w = 10^{-5}$, $\rho = 0.01$, and $\kappa = 0.5$ were selected. Note in Fig. 11.6e, the wires on the cubicle wall behind the monitor are preserved by the adaptive regularization. We note that the adaptive regularization technique gives much better results than the median filtering method, even though their ARAs are comparable. In the experiments Eq. (11.29) makes the minimization more robust, and the regularization factor can be set to a large value to speed up the convergence. One iteration using Eq. (11.29) achieves almost the same smoothing result as 50 iterations using Eq. (11.26).

Figure 11.7 shows nonadaptively smoothed results of the surface mesh for both synthetic and real data. The blocky-looking surfaces in Fig. 11.7a, c are caused by binary reconstruction using the marching cube algorithm [lorensen87]. Binary reconstruction means the voxels status is either empty or occupied. The aliasing artifacts are caused by the discontinuous transition of the status. Figure 11.7b shows the nonadaptively smoothed result of Fig. 11.7a after six iterations with $\lambda = 10^{-6}$. Figure 11.7d shows the nonadaptively smoothed result of Fig. 11.7c after seven iterations with $\lambda = 0.1$. Figure 11.7e shows the raw surface captured by the RIEGL laser mirror scanner LMS-Z210 [riegl00], with 99,199 triangles. The scanner is able to capture range images and color images simultaneously in a range from 2 up to 350 m. The standard deviation of the measurement error is 2.5–5 cm. Figure 11.7f shows the corresponding nonadaptively smoothed result after six iterations with $\lambda = 0.01$, in which noise is effectively suppressed.

Figure 11.8 shows the experimental results of adaptive smoothing on the triangle mesh. Figure 11.8a, b show the raw surface captured by the RIEGL scanner with and without texture. The size of the original range image is 524 × 223 pixels. The building is approximately 50 m away from the scanning position. There are no data obtained behind the building where the distance is greater than the scanner's capturing range. The tower is separated due to self occlusion. We removed the trees in front of the building to highlight the smoothing on the building surface. The 3D model has 139,412 triangles. Crease edge detection on the triangle mesh using

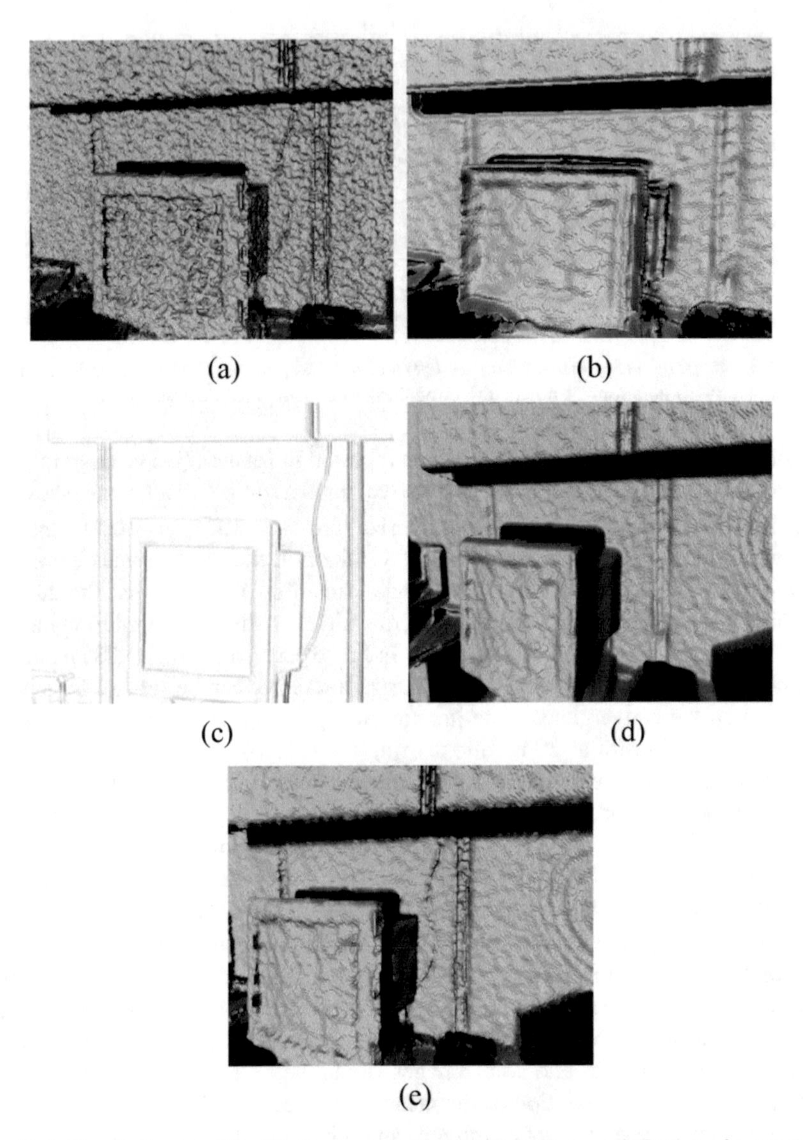

Fig. 11.6 Results from median filtering, nonadaptive, and adaptive regularization of range data. (**a**) Result from three by three median filtering conducted twice. (**b**) Result from regularization using Laplacian smoothing term. Note the instability along edges. (**c**) The edge map. (**d**) Result from nonadaptive regularization. (**e**) Result from adaptive regularization. Note the wire on the wall preserved by the adaptive method

Eqs. (11.55) and (11.58) is shown in Fig. 11.8c, where each vertex on the crease edge is marked by a small sphere. The window frame portion is zoomed and shown in Fig. 11.8d. Figure 11.8e, f show the nonadaptively and adaptively smoothed

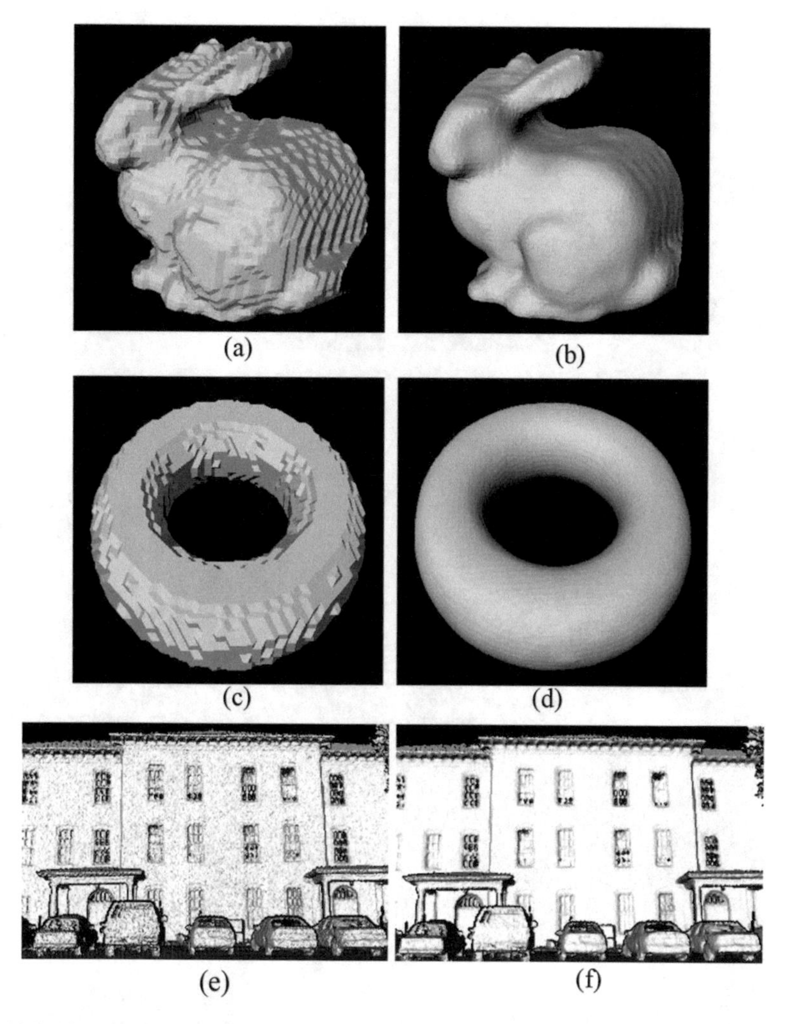

Fig. 11.7 Nonadaptive smoothing of surface mesh. (**a**) Synthetic surface mesh of Stanford bunny model generated by binary reconstruction, 15,665 *triangles*. (b) Seven iteration, nonadaptive smoothed result of (**a**). (**c**) Synthetic surface mesh of a torus model generated by binary reconstruction, 14,604 *triangles*. (**d**) Six iteration, nonadaptive smoothed result of (**c**). (**e**) Raw surface captured by RIEGL laser range scanner, 99,199 *triangles*. (**f**) Six iteration, nonadaptive smoothed result of (**e**)

results, respectively, after five iterations with $\lambda = 0.01$. The geometric details such as window frames are well preserved by the adaptive smoothing.

The area decreasing flow for surface smoothing instead of the mean curvature flow is considered. Despite their mathematical equivalence, area minimization generates a more efficient algorithm for discrete surface smoothing, where the mean curvature is not easy to compute. For a triangle mesh, the optimal magnitude of surface variation under area decreasing flow can be solved uniquely, which is not

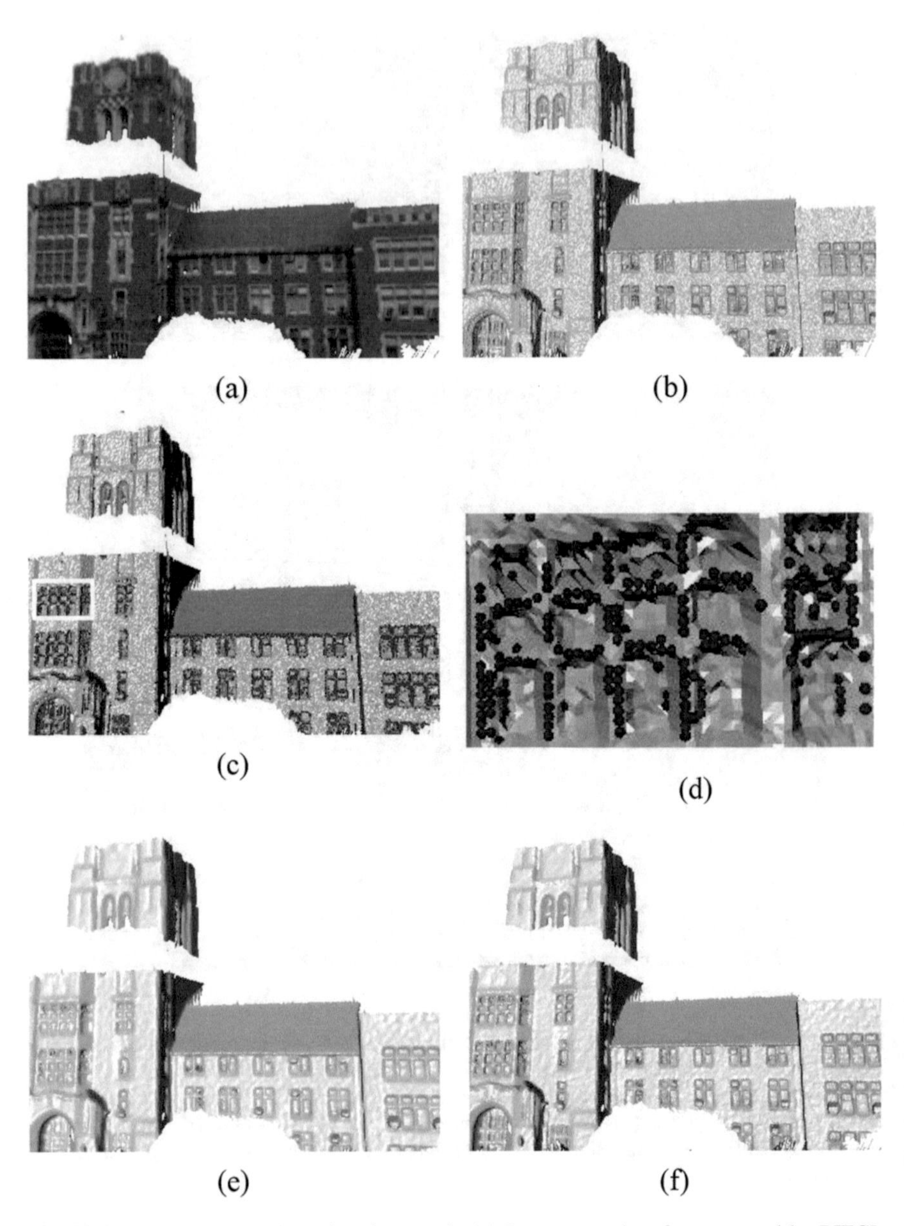

Fig. 11.8 Adaptive smoothing of surface mesh. (**a**) Raw textured surface captured by RIEGL laser range scanner. (**b**) Raw surface without texture, 139,412 *triangles*. (**c**) Crease edge detection using tensor voting approach and orientation check. Vertices on the edges are marked by small spheres. (**d**) Zoomed window frame portion of (**c**). (**e**) Five iterations of nonadaptively smoothed result of (**b**). (**f**) Five iterations of adaptively smoothed result of (**b**). Note the well-preserved window frame structure

possible for mean curvature flow. A method to preserve the crease edges on a triangle mesh by adaptive smoothing is also presented. The edge strength of each vertex is computed by the eigen analysis of the normal vector field in a geodesic window. An orientation check is introduced to guarantee the correct crease edge detection. Experimental results show that the proposed method achieves better performance than existing methods.

References

[blake87] A. Blake, A. Zisserman, *Visual Reconstruction* (MIT Press, Cambridge, MA, 1987)

[stevenson92] R.L. Stevenson, E.J. Delp, Viewpoint invariant recovery of visual surface from sparse data. IEEE Trans. Pattern Anal. Mach. Intell. **14**(9), 897–909 (1992)

[yi95] J.H. Yi, D.M. Chelberg, Discontinuity-preserving and viewpoint invariant reconstruction of visible surface using a first order regularization. IEEE Trans. Pattern Anal. Mach. Intell. **17** (6), 624–629 (1995)

[taubin95] G. Taubin, A Signal Processing Approach to Fair Surface Design, Proc. SIGGRAPH, 351–358 (1995)

[vollmer99] J. Vollmer, R. Mencl, and H. Muller, Improved laplacian smoothing of noisy surface meshes, Computer Graphics Forum, (Proc. Eurographics Conf.), 131–138 (1999)

[ohtake00] Y. Ohtake, A. Belyaev, and I. Bogaevski, Polyhedral surface smoothing with simultaneous mesh regularization, Proc. Geometric Modeling and Processing (2000)

[sethian98] J.A. Sethian, *Level set methods and fast marching methods: evolving interfaces in computational geometry, fluid mechanics, computer vision and material sciences*, 2nd edn. (Cambridge University Press, New York, 1998)

[whitaker98] R.T. Whitaker, A level-set approach to 3D reconstruction from range data. Int. J. Comput. Vis. **29**(3), 203–231 (1998)

[zhao00] H.K. Zhao, S. Osher, B. Merriman, M. Kang, Implicit and non-parametric shape reconstruction from unorganized points using variational level set method. Comput. Vis. Image Underst. **80**, 295–319 (2000)

[desbrun99] M. Desbrun, M. Meyer, P. Schroder, and A.H. Barr, Implicit fairing of irregular meshes using diffusion and curvature flow, Proc. SIGGRAPH, 317–324 (1999)

[docarmo76] M. DoCarmo, *Differential Geometry of Curves and Surfaces* (Prentice Hall, Saddle, River NJ, 1976)

[vaidya98] N.M. Vaidya, K.L. Boyer, Discontinuity-preserving surface reconstruction using stochastic differential equations. Comput. Vis. Image Underst. **72**(3), 257–270 (1998)

[perceptron93] Perceptron Inc., 23855 Research Drive, Farmington Hills, Michigan 48335, LASAR Hardware Manual (1993)

[hoover96] A. Hoover, The space envelope representation for 3D scenes, Ph.D. thesis, Department of Computer Science and Engineering, University of South Florida, 1996

[canny86] J. Canny, A computational approach to edge detection. IEEE Trans. Pattern Anal. Mach. Intell. **8**(6), 679–698 (1986)

[katsaggelos89] A.K. Katsaggelos, Iterative image restoration algorithms. Opt. Eng. **28**(7), 735–748 (1989)

[riegl00] RIEGL Laser Measurement Systems, Laser Mirror Scanner LMS-Z210, Technical documentation and User's Instructions, 2000

[kobbelt98] L. Kobbelt, S. Campagna, J. Vorsatz, and H.P. Seidel, Interactive multi-resolution modeling on arbitrary meshes, Proc. SIGGRAPH, 105–114 (1998) [kimmel98] R. Kimmel, J.A. Sethian, Computing geodesic paths on manifolds. Proc. Natl. Acad. Sci. U.S.A. **95**(15), 8431–8435 (1998)

[sun01] Y. Sun, M.A. Abidi, Surface matching by 3D point's fingerprint. Proc. IEEE Int. Conf. Comput. Vis. **2**, 263–269 (2001)

[medioni00] G. Medioni, M.S. Lee, C.K. Tang, *A Computational Framework for Segmentation and Grouping* (Elsevier, New York, 2000)

[lorensen87] W. E. Lorensen and H.E. Cline, Marching cubes: a high resolution 3D surface construction algorithm, Proc. SIGGRAPH, 163–169 (1987)

Chapter 12
Multimodal Scene Reconstruction Using Genetic Algorithm-Based Optimization

Abstract Many applications require the use of 3D graphics to create models of real environments. These models are usually built from range or depth images. In the scene modeling process, the use of additional 2D digital sensorial information leads to multimodal scene representation, where an image acquired by a 2D sensor is used as a texture map for a geometric model of a scene. In this chapter we present, as an example of optimization, a photo-realistic scene reconstruction procedure using laser range data and color photographs.

The reconstruction system involves the creation of triangle meshes from range images as a scene surface representation, but the main emphasis is made on the registration of laser range and photographic images. Major 3D data acquisition techniques are discussed in Appendix C, and a real range data is acquired by using a light amplitude detection and ranging (LADAR) range scanner.

The proposed multimodal image registration approach uses random distributions of pixels to measure the amount of dependence between two images and estimates the relative pose of one imaging system to the other. The similarity metric used in the proposed automatic registration algorithm is based on the χ^2 measure of dependence, which is presented as an alternative to the standard mutual information criterion. These two criteria belong to the class of information theoretic similarity measures, which quantify the dependence in terms of the information provided by one image about the other. For the maximization of the similarity measure, a robust optimization scheme is needed. To achieve both accurate and robust results, genetic algorithms are investigated in the heuristic manner.

12.1 Introduction

In the past few years, 3D computer graphics (CG) has gained attraction to represent and visualize real-world scenes. At the same time, the fields of virtual and augmented reality based on 3D scene modeling become more influencing, and examples of the corresponding applications include simulation and training for the military and civilians, medical imaging, computer and robot vision, remote sensing, reverse engineering, and digital entertainment, to name a few. Manual design of 3D

M.A. Abidi et al., *Optimization Techniques in Computer Vision*, Advances in Computer Vision and Pattern Recognition, DOI 10.1007/978-3-319-46364-3_12

scene models has been a very tedious and inefficient task and requires the intervention of experts in many professional areas: computer graphics, art, computer programming, etc.

For this reason, automatic 3D model building and fusion of multisensory data is being considered as an alternative to the manual, CG-based scene modeling. For example, in the application of artificial vision-based robot guidance, it is crucial to have accurate geometric reconstruction of the real environment. More specifically, the Pathfinder Project for the exploration of the surface of Mars [stoker98] and the Pioneer robot [maimone] used for the navigation inside the radiation-contaminated Chornobyl plant have shown the possibility of real-time 3D vision and multisensor (or multimodal) data acquisition and interpretation.

The major advantage of multimodal scene reconstruction is that a scene point could be represented in a multidimensional space, so it makes tasks such as scene segmentation and object recognition much easier. An example is the use of a γ-radiation map, which allows robots operating in nuclear plants for the identification of radioactive containers together with the 3D model information. In this case, the use of both shape and radiation enables higher recognition performance.

Multimodal (or multisensor) scene reconstruction is considered as a part of data fusion [abidi92] that encapsulates both geometric and other intensity data. For the visualization of such scene representations, a reconstructed 3D geometric model is rendered, and the associated sensorial 2D data, such as photographic, infrared, X-ray, UV, and γ-radiation, is used as a texture map on top of it as shown in Fig. 12.1.

If reconstruction uses color photographs as textures for scene models, it is called photo-realistic reconstruction [elhakim98, manga99, sequira99]. In many cases scene models created by photo-realistic reconstruction are designed to provide virtual environments that resemble the real world as seen by the human eye. More applications of automatic photo-realistic reconstruction include computer games, digital architecture, and special effects for cinema movies.

Among many approaches for building accurate geometric scene models, there are two major categories: the volumetric and surface representations. Volumetric representation requires the use of 3D occupancy information in order to build a spatial representation based on volume elements, called *voxels*, which are out of the scope of this chapter. On the other hand, we use the surface-based reconstruction approach to build a 3D model of a scene. In the case of the 3D surface-based reconstruction, a model of a scene is created from range (or depth) images by using a range scanner or the stereovision technique. The visualization standards generally require the use of polygon meshes to represent the surface model of a scene. It is, in general, necessary to use multiple range images from different viewpoints for more complete representation of a scene.

In addition to geometric surface reconstruction, photo-realistic, multimodal scene reconstruction requires the use of other 2D data such as color photographs and reflectance images as a texture map for the model. For this purpose, it is crucial to register the range data with the associated image by using correspondence between the same physical structures in the two images. Once these two modalities

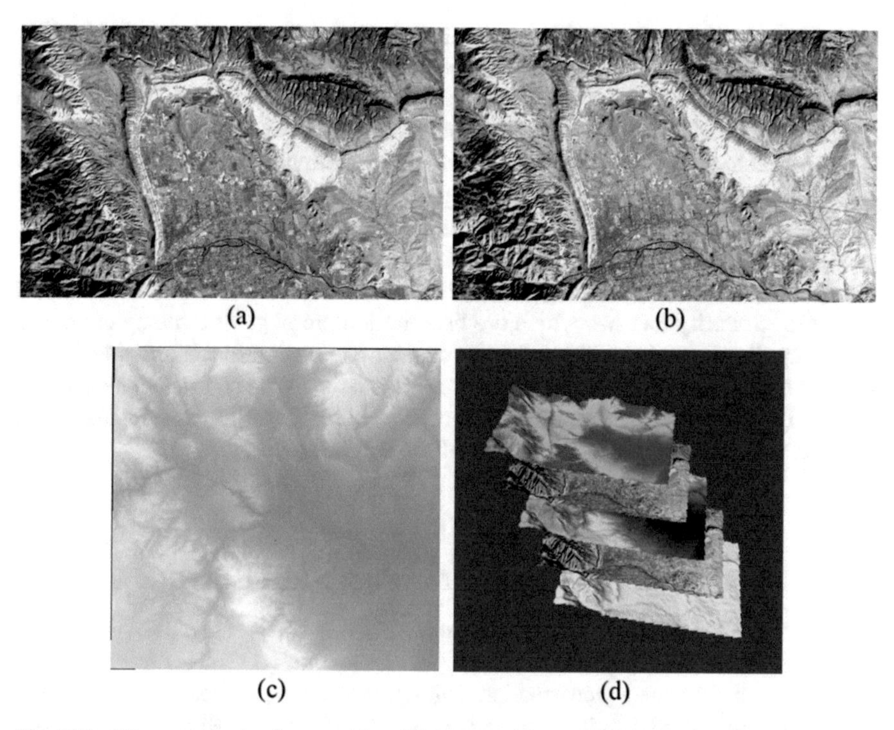

Fig. 12.1 3D reconstruction from multimodal data set: (**a**) and (**b**) show two different reflectance images from LANDSAT, (**c**) the corresponding digital elevation map, and (**d**) the rendering of a model built from the digital elevation map with the reflectance images as textures on top of it

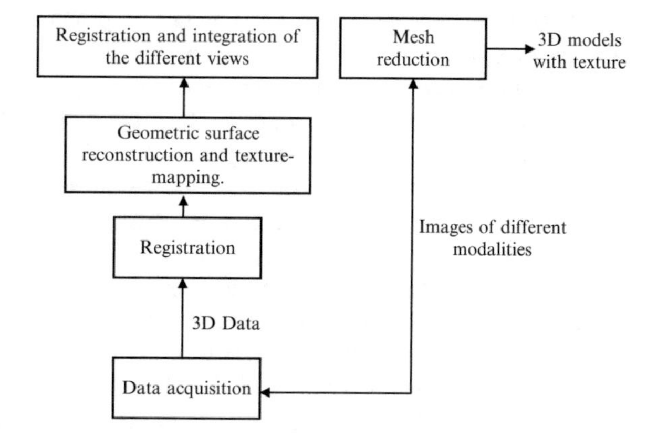

Fig. 12.2 Paradigm for multimodal scene reconstruction

are registered, the image is used as a texture map on top of the model. This step in the reconstruction process requires the determination of occluded areas in the scene, which will not receive a texture because they are visible for the range sensing device but occluded for the other 2D sensor. Other preprocessing and preliminary tasks, such as smoothing the range images and reducing the size of polygon mesh, may be required in order to have higher-quality visualization. Figure 12.2 shows the overall strategy commonly adopted for multimodal 3D reconstruction.

In this work, we mainly focus on the problem of automatic photo-realistic reconstruction from range images, obtained by a laser range finder, which is called *LADAR imagery*, and digital color photographs. The most important step is the registration of different sensorial data related to a real-world scene.

Many techniques have been proposed in the literature to achieve multimodal registration, or alignment. The classic approaches are based on the extraction of corresponding features in the two modalities. In this chapter we explore the more recent and promising techniques based on the use of similarity metrics to quantify the dependence of the two images. Particular emphasis is made on the information theoretic similarity metrics, which are becoming more popular in many computer vision applications. In addition to the standard mutual information metric, we present a similarity measure based on the χ^2 statistics. We use the latter metric in our LADAR-color registration algorithm. A genetic algorithm is adopted as an optimization tool for the maximization of the similarity metric.

12.2 Multisensor Registration for Photo-Realistic Scene Reconstruction

In photo-realistic scene reconstruction, color photographs are used as a texture map for a geometric model built from range images of a scene. Since we are using unregistered data set, two sensors, such as a range sensor and a color camera, must be aligned first. This alignment problem falls into the estimation of the relative pose of one sensor to the other.

Multisensor registration is a fundamental problem in many computer vision applications including medical imaging, remote sensing, and model-based object recognition. Given two image representations of the same object, registration is performed by geometrically transforming one image (the floating image) to the other (the reference image), so that pixels or voxels representing the same physical structures may be superimposed. The laser reflectance image, which is perfectly registered with the range image, is used to align the LADAR data with the color picture. The range and reflectance images present important dissimilarities with a color picture of the scene. The relationship between the brightness values of corresponding pixels is complex and unknown, and aligning 3D to 2D data exhibits additional difficulties due to the occlusion problem.

Most registration approaches can be classified into two categories: (a) *feature-based* and (b) *similarity metric-based* approaches. The feature-based approach uses invariant features, such as edge, contour, and feature point, which exhibit visual similarities between different modalities. This approach can be easily implemented using a single-modality registration technique at the cost of losing a significant amount of texture information in the feature extraction process.

The similarity metric-based approach is also known as the intensity-based approach. This approach uses statistical tools to measure the degree of dependence between the different images. One popular technique in the field of medical imaging is the use of information theoretic mutual information measure.

In this work we investigated the use of similarity measures for automatically aligning LADAR and color images, which presents the advantage of avoiding preliminary feature extraction and takes into account all the available texture information.

12.2.1 Landmark-Based Image Registration

Landmark-based image registration falls into the category of feature-based approaches and is equivalent to the external camera calibration problem, or also known as the space resection problem. The reflectance image was used to match landmark points in the two images, which have a certain degree of similarity as shown in Fig. 12.3. Since the reflectance image is perfectly registered with the range image, we can produce a list of 3D-2D correspondences.

Particular feature points such as corners and intersections of edges were assigned as landmarks. It could be extracted either manually in an interactive fashion or automatically by using, for example, a corner detection algorithm.

This list of corresponding points will allow us to calculate the rigid transformations $\{t, R\}$ which represent translation and rotation, aligning the two imaging systems. A similar approach based on an interactive technique was proposed by Sequira et al. [sequira99] and is used to register color pictures with a model built from range data.

12.2.1.1 Projective Geometry

Projective geometry is a mathematical tool that bridges the 3D position of an object in the real world and the camera plane for image formation. Figure 12.4 shows a

Fig. 12.3 Color and LADAR reflectance images of the "boxes scene" with corresponding landmarks

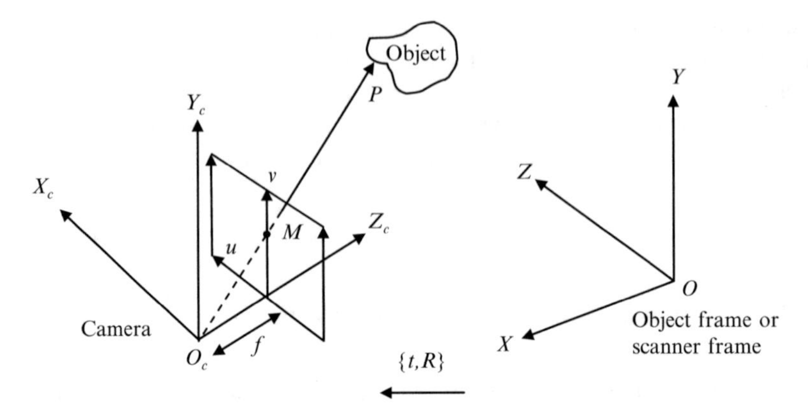

Fig. 12.4 Projective geometry describing a camera frame attached to the imaging systems

model for the image formation in a color camera plane. A more elaborate camera model can be found in [tsai87]. Let P be a 3D point and $(x, y, z)^t$ its coordinates relative to the object coordinates system, and $(x_c, y_c, z_c)^t$ the point coordinates in the camera frame. The u and v axes, respectively, parallel to the X_c and Y_c axes, define the image coordinate system with its origin located at the principal point.

Let $(u, v)^t$ be the coordinate of M, which is the projection of P in the image coordinate frame. We have

$$\lambda \begin{pmatrix} u \\ v \\ f \end{pmatrix} = \begin{pmatrix} x_c \\ y_c \\ z_c \end{pmatrix}, \tag{12.1}$$

where λ represents a scalar value and f the camera's focal length.

The rigid transformation consists of a rotation represented by a 3×3 *orthonormal* matrix $R = \{r_{ij}\}$, and a translation defined by a vector $t = (t_x, t_y, t_z)^t$. They relate the coordinate of P in the two imaging systems as follows:

$$\begin{pmatrix} x_c \\ y_c \\ z_c \end{pmatrix} = R \begin{pmatrix} x \\ y \\ z \end{pmatrix} + t. \tag{12.2}$$

The *collinearity* equations are then obtained as

$$u = f.\frac{r_{11}x + r_{12}y + r_{13}z + t_x}{r_{31}x + r_{32}y + r_{33}z + t_z} \quad \text{and} \quad v = f.\frac{r_{21}x + r_{22}y + r_{23}z + t_y}{r_{31}x + r_{32}y + r_{33}z + t_z}. \tag{12.3}$$

12.2.1.2 Pose Estimation

Let $\{P_i\}$ and $\{M_i(u_i, v_i)\}$, $i = 1, \ldots, N$, be the selected set of corresponding 3D points and their image projections, respectively. Different approaches have been adopted to solve the pose estimation problem from a set of 3D-2D correspondences.

The classical approaches [faugeras93, haralick93, horn86] are to find $\{t, R\}$ that minimizes the least squared distance of the projected model points to the image points, which leads to a nonlinear least squares

$$\varepsilon^2 = \sum \{\hat{u}_i(R,t) - u_i\}^2 + \{\hat{x}_i(R,t) - x\}^2, \tag{12.4}$$

with the orthonormality constraint $R^t R = I$ that is required for the rotation matrix. $\hat{u}_i$ and $\hat{v}_i$ are the calculated coordinates in the image frame of the projection of P_i according to the motion parameters $\{t, R\}$.

Since there is no closed-form solution for this problem, numerical techniques are required to find the pose parameters [horn86]. An alternative approach linearizes the equations with respect to the unknown exterior orientation parameters [faugeras93, ji98]. This type of method is usually very sensitive to noise and hardly guarantees the orthonormality of the rotation matrix. They are commonly used to get initial estimation of the parameters, which will be further refined by a nonlinear technique. In order to improve pose estimation performance, higher-order features such as lines, circles, and ellipses are often used [ji98].

There have been different approaches to point correspondence-based pose estimation. Kanatani used an accurate statistical characterization of the alignment error, along with a technique to minimize the error [kanatani98]. Basri introduced a new type of 3D-2D transformation metric, which penalizes for the deformations applied to the object to produce the observed image [basri96].

To register model and image accurately, we used an iterative method for solving the nonlinear least squares, which requires a good initial estimation. Because of the high level of noise, a linear method cannot guarantee a good initialization. Instead, a stochastic minimization scheme, such as simulated annealing, is used to get a suboptimal solution. This stochastic method requires the discretization of the different pose parameters: rotation angles and translation vector components.

For fine-tuning the parameters, we used the Levenberg-Marquardt method [press92], which is a common technique for solving a nonlinear least square fitting problem.

To avoid keeping track of the orthonormality constraint, we parameterize R as follows:

$$R = R(\phi_x)R(\phi_y)R(\phi_z), \tag{12.5}$$

where ϕ_x, ϕ_y, and ϕ_z respectively represent the rotation angles around the x, y, and z axes. Each rotation matrix is defined as

$$R(\phi_x) = \begin{bmatrix} 1 & 0 & 0 \\ 0 & \cos\phi_x & -\sin\phi_x \\ 0 & \sin\phi_x & \cos\phi_x \end{bmatrix}, \tag{12.6}$$

$$R(\phi_y) = \begin{bmatrix} \cos\phi_y & 0 & -\sin\phi_y \\ 0 & 1 & 0 \\ \sin\phi_y & 0 & \cos\phi_y \end{bmatrix}, \tag{12.7}$$

and

$$R(\phi_z) = \begin{bmatrix} \cos\phi_z & -\sin\phi_z & 0 \\ \sin\phi_z & \cos\phi_z & 0 \\ 0 & 0 & 1 \end{bmatrix}. \tag{12.8}$$

Let $\theta = \left(\phi_x, \phi_y, \phi_z, t_x, t_y, t_z\right)^t$ be the parameter vector. We define the matrix $A = \{\alpha_{kl}\}$, for $k, l = 1, \ldots, 6$, where

$$\alpha_{kl} = \sum_{i=1}^{N} \frac{\partial \hat{u}_i(\theta)}{\partial \theta_k} \frac{\partial \hat{u}_i(\theta)}{\partial \theta_l} + \frac{\partial \hat{v}_i(\theta)}{\partial \theta_k} \frac{\partial \hat{v}_i(\theta)}{\partial \theta_l}. \tag{12.9}$$

A is an approximation of the Hessian matrix, and

$$\beta = \{\beta_k\}, \text{ where } \beta_k = -\frac{1}{2} \frac{\partial \varepsilon^2(\theta)}{\partial \theta_k}, \text{ for } k = 1, \ldots, 6. \tag{12.10}$$

The Levenberg-Marquardt algorithm iteratively solves the following linear system

$$A' \delta\theta = \beta, \tag{12.11}$$

where $\delta\theta^n = \theta^{n+1} - \theta^n$, for the nth iteration. The matrix $A' = \{\alpha'_{kl}\}$ is defined as follows:

$$\alpha'_{ij} = \begin{cases} (1+\lambda)\alpha_{ii}, & i = j \\ \alpha_{ij}, & i \neq j \end{cases}, \tag{12.12}$$

where λ represents a stabilization parameter used to ensure convergence. The algorithm is summarized in the following.

Algorithm 12.1

(Step 1) Evaluate $\varepsilon^2(\theta)$.
(Step 2) Assign a small value to λ.
(Step 3) Solve the linear Eq. (12.11) for $\delta\theta$ and evaluate $\varepsilon^2(\theta + \delta\theta)$.
(Step 4) If $\varepsilon^2(\theta + \delta\theta) \geq \varepsilon^2(\theta)$, increase λ by a substantial factor. Otherwise, decrease λ by some factor, and update the solution as: $\theta \leftarrow \theta + \delta\theta$.
(Step 5) If ε^2 stops decreasing, terminate the algorithm. Otherwise, go back to (step 3).

12.2.2 Similarity Metric-Based Registration

The use of similarity measures for multimodal image registration is based on the maximization of an objective function, which quantifies the pixel (or voxel) similarity of the images to be aligned. These techniques have become popular for registration and recognition. The early similarity-based registration techniques used correlation as a measure. They, in general, require strong assumptions on the relationship between the two images, such as linearity and functionality.

A particular class of similarity measures, based on the information theory, has been proven effective to evaluate the quality of registration. These similarity measures can be applied to a wide range of alignment problems since they do not need impractical assumptions such as functional dependency between the different modalities. The current standard mutual information (MI) criterion belongs to this class and was first introduced independently by Viola [viola97] and Collignon [collignon95]. MI is related to the notion of entropy and quantifies the degree of dependence in terms of the amount of information given by one modality to the other. This criterion was successfully used in aligning both 2D and volumetric images, which have different modalities presenting gross dissimilarities in the medical imaging [collignon95, maes97, viola97], remote sensing [nikou98], and pose estimation [viola97] areas.

Another class of similarity measures which is not related with the information theory has also been applied to the multisensor alignment problem. Woods et al. devised a criterion based on the inter-uniformity hypothesis [woods93], assuming that the dispersion of the joint histogram is minimal in the registered position. They used this measure to register medical images. Irani used global estimation along with locally normalized correlation similarity measure for the alignment of electrooptical and infrared images [irani98]. Roche et al. presented a technique based on standard statistics by assuming functional dependence between the image intensities [roche98]. In this technique, correlation ratio was used as a similarity measure. Nikou et al. introduced a robust similarity metric based on robust *M* estimators and compared it to MI and Woods' criteria [nikou98].

The registration algorithms presented in the literature commonly require the use of optimization techniques in order to find the maximum of the similarity measure.

Viola used a stochastic gradient descent method to maximize MI [viola97]. Maes et al. used Powell's multidimensional technique [maes97]. Nikou et al. applied a stochastic optimization scheme using simulated annealing and iterative conditional modes (ICM) to avoid local maxima [nikou98].

On the other hand, the 3D-2D alignment problem, also known as pose estimation, is widely addressed and plays an important role in many applications such as object recognition and robot self-localization.

For the registration of X-ray or video (2D) with MRI or CT (3D), Lavallée and Szeliski used the occluding contours of the 3D object in the 2D image to find the correct pose [lavallee95]. Feldmar et al. similarly defined a distance between the 3D object surface and the 2D image contour [feldmar97]. They used an extended version of the iterative closest point (ICP) algorithm to minimize the distance [besl92]. In object recognition, feature-based techniques are still widely used [hausler99, ji98], and hybrid approaches, using both feature and intensity information, are also used in different applications such as automatic target recognition [stevens97].

12.2.3 Similarity Measures for Automatic Multimodal Alignment

The pixel-based or voxel-based registration techniques require a function to measure the similarity of the images intensities. For this purpose, no feature extraction is required but gray-level values of the two images are used.

In the case of images of the same modality, correlation measures have been extensively used. They assume linear dependence of the two images. These measures are not able to handle multimodal images. For solving the problem of dissimilar image registration, different measures have been devised based on the statistical comparison of gray levels.

Let r and f denote the reference and the transformed (or floating) images, M the similarity measure, and T_θ the transformation that aligns r and f with the motion parameter θ. In the case of 3D-2D alignment, T_θ is the rigid transformation followed by a perspective projection. The aligning problem is to find the transformation parameter $\hat{\theta}$ that maximizes the similarity measure

$$\hat{\theta} = \underset{\theta}{\operatorname{argmax}}\,(M(\theta)), \tag{12.13}$$

where $M(\theta) = M(f(x), r(T_\theta(x)))$, and x represents a variable representing the coordinates of the pixels in the floating image.

The similarity measures are based on the notion of statistical dependence. Some measures assume functional dependence [irani98, roche98] or use corresponding uniform areas in the images [woods93].

In the following sections, we focus on the information theoretic similarity metrics and we present the mutual information criterion and the χ^2 information metric, the latter is used in our automatic LADAR-color registration algorithm.

12.2.4 Information Theoretic Metrics

In this section pixel intensities in the images to be aligned are considered as random variables. Let us consider P and Q the two probability distributions corresponding to the gray levels of the two images of different modalities, such as $P = \{p_1, p_2, \ldots\}$ and $Q = \{q_1, q_2, \ldots\}$.

Information theory defines a class of dependence measure of two probability distributions, known as *divergences* [vajda89]. An example is the widely used Kullback-Leibler divergence,

$$I(P||Q) = \sum_i q_i \ln\left(\frac{p_i}{q_i}\right). \tag{12.14}$$

This divergence belongs to a class called *f*-divergences compatible with a *metric* on the space of probability distributions, which are defined as

$$I_f(P||Q) = \sum_i q_i f\left(\frac{p_i}{q_i}\right). \tag{12.15}$$

Examples of *f*-divergences include $I\alpha$-divergence defined as

$$I\alpha = \frac{1}{\alpha(\alpha-1)}\left[\sum_i \frac{p_i^{\alpha}}{q_i^{\alpha-1}} - 1\right], \tag{12.16}$$

and χ^2-divergence defined as

$$\chi^2 = \sum_i \frac{(p_i - q_i)^2}{q_i}. \tag{12.17}$$

To quantify rigorously the dependence, in terms of information, of two probability distributions A and B, the notion of *f*-information is derived from *f*-divergence. The corresponding measure of dependence can be denoted as

$$I_f(P||P_1 \times P_2), \tag{12.18}$$

where P represents the set of joint probability distributions $P(A, B)$, and $P_1 \times P_2$ the set of joint probability distributions $P(A)P(B)$ under the assumption that A and B are mutually independent.

An example of the corresponding *f*-information is $I\alpha$-information defined as

$$I\alpha(P||P_1 \times P_2) = \frac{1}{\alpha(\alpha-1)}\left[\sum_i \frac{{p_{ij}}^{\alpha}}{(p_{i.}p_{.j})^{\alpha-1}} - 1\right], \tag{12.19}$$

where $p_{ij} = P(i,j)$, $p_{i.} = \sum_j p_{ij}$, and $p_{.j} = \sum_i p_{ij}$, for $\alpha = 1$. The *mutual information* metric can be derived as

$$I_1(P||P_1 \times P_2) = \sum_{i,j} p_{ij} \ln\left(\frac{p_{ij}}{p_{i.}p_{.j}}\right). \tag{12.20}$$

Another example is the χ^2-information as

$$\chi^2(P||P_1 \times P_2) = \sum_{i,j} \frac{(p_{ij} - p_{i.}p_{.j})^2}{p_{i.}p_{.j}}. \tag{12.21}$$

In the following two subsections, we present the use of mutual information for multimodal image alignment and discuss the χ^2-information metric as a robust similarity measure.

12.2.5 *Mutual Information Criterion*

Mutual information is a very popular measure for similarity and proved efficient in medical imaging and other multimodal alignment applications [collignon95, maes97, nikou98, viola97]. Like the other information theoretic metrics, mutual information quantifies the dependence between two distributions *X* and *Y* by measuring the distance, in the space of probabilities, between the joint probability distribution $p_{XY}(x, y)$ and the distribution associated with the case of complete independence $p_X(x)p_Y(y)$. This distance is the Kullback-Leibler measure:

$$I(X,Y) = \sum_{x,y} p_{XY}(x,y)\ln\left(\frac{p_{XY}(x,y)}{p_X(x)\,p_Y(y)}\right). \tag{12.22}$$

Mutual information can be expressed in terms of entropy as

$$I(X,Y) = H(X) + H(Y) - H(X,Y) = H(X) - H(X|Y) = H(Y) - H(Y|X), \tag{12.23}$$

where $H(X)$ and $H(Y)$ respectively represent the entropy of X and Y, $H(X,Y)$ their joint entropy, and $H(X|Y)$ and $H(Y|X)$ the conditional entropy of X given Y and that of Y given X, respectively. More specifically, we have that

$$H(X) = -\sum_{x} p_X(x)\log(p_X(x)), \tag{12.24}$$

$$H(X,Y) = -\sum_{x,y} p_{XY}(x,y)\log(p_{XY}(x,y)), \tag{12.25}$$

and

$$H(X|Y) = -\sum_{x,y} p_{XY}(x,y)\log\left(\frac{p_{XY}(x,y)}{p_Y(y)}\right). \tag{12.26}$$

In the case of image registration, x and y represent the pixels' intensities, and X and Y the associated random variables.

To calculate the previously introduced quantities, we assume that the random variables are defined in the discrete space and that the marginal and joint probabilities could then be estimated by normalizing each image histogram and their 2D histogram as

$$p_x = \frac{N_x}{N} \text{ and } p_{xy} = \frac{N_{xy}}{N}, \tag{12.27}$$

where N_x and N_y respectively represent the numbers of pixels corresponding to the gray levels x and y, N the total number of pixels in the overlapped area, and N_{xy} the number of the co-occurrence of x and y in the joint histogram for a given overlap. In another approach, variables are considered continuous and are estimated using the Parzen windowing technique [mcgarry96].

Mutual information works well in many cases of multimodal registration, particularly for medical images or other sensorial data presenting important variation in the texture. Unfortunately, this measure is sensitive to subsampling and false matching in the case of data presenting large patches of uniform texture, which is generally the case of large indoor or outdoor scenes. For this reason we present another measure of dependence based on the χ^2 statistic.

12.2.6 χ^2-Information Criterion

The χ^2-information metric, also known as the χ^2 measure of dependence [kass97, press92], is used mainly for characterizing the significance of an association between two distributions. Like mutual information, the variables, which represent gray levels in this case, are considered nominal. The χ^2 statistic is more suitable for comparing the significance of different associations than the mutual information criterion, which is a measure of strength of an association already known to be significant.

Let n_{xy} be the number of co-occurrences of x and y in the case of total independence of the two distributions as

$$n_{xy} = \frac{N_x N_y}{N}, \tag{12.28}$$

and the χ^2-information metric can be expressed as

$$\chi^2 = \sum_{x,y} \frac{\left(N_{xy} - n_{xy}\right)}{n_{xy}}. \tag{12.29}$$

To compare different associations over varying overlapped areas, we use the contingency coefficient

$$C = \sqrt{\frac{\chi^2}{\chi^2 + N}}, \tag{12.30}$$

which lies between zero and one.

Figure 12.5 shows the variation of MI and contingency coefficient versus the rotation parameter, and Fig. 12.6 shows the same plots versus the translation parameter. Both plots are obtained from the "boxes" scene by the use of the landmark-based technique. We can see that in some cases, both metrics exhibit more than one peak. In the case of mutual information, the global maximum does not always correspond to the correct registration parameters (see plots for the rotations around the x-axis and translations along the y-axis). This has been observed in previous works [roche98], while for the contingency coefficient associated with the χ^2-information metric, we find a peak that is the global maximum, corresponding to the "correct" parameters.

These metrics become more sensitive to rotations out of the XY plan and translations resulting in partial overlap between the two images, when we project the reflectance image to the color image plan after applying the rigid transformations.

12.3 LADAR-Color Registration and Integration

12.3.1 Optimization Technique for Registration Algorithms

Since the χ^2-information metric seems more robust than mutual information based on the qualitative evaluation, we chose this similarity measure in our algorithm for automatic LADAR-color registration. The reflectance image is considered as the floating image, while the gray-scaled color image will be the reference image. The range map, which is perfectly registered with the reflectance image, will provide the 3D position corresponding to every pixel. For solving the similar problem, Viola

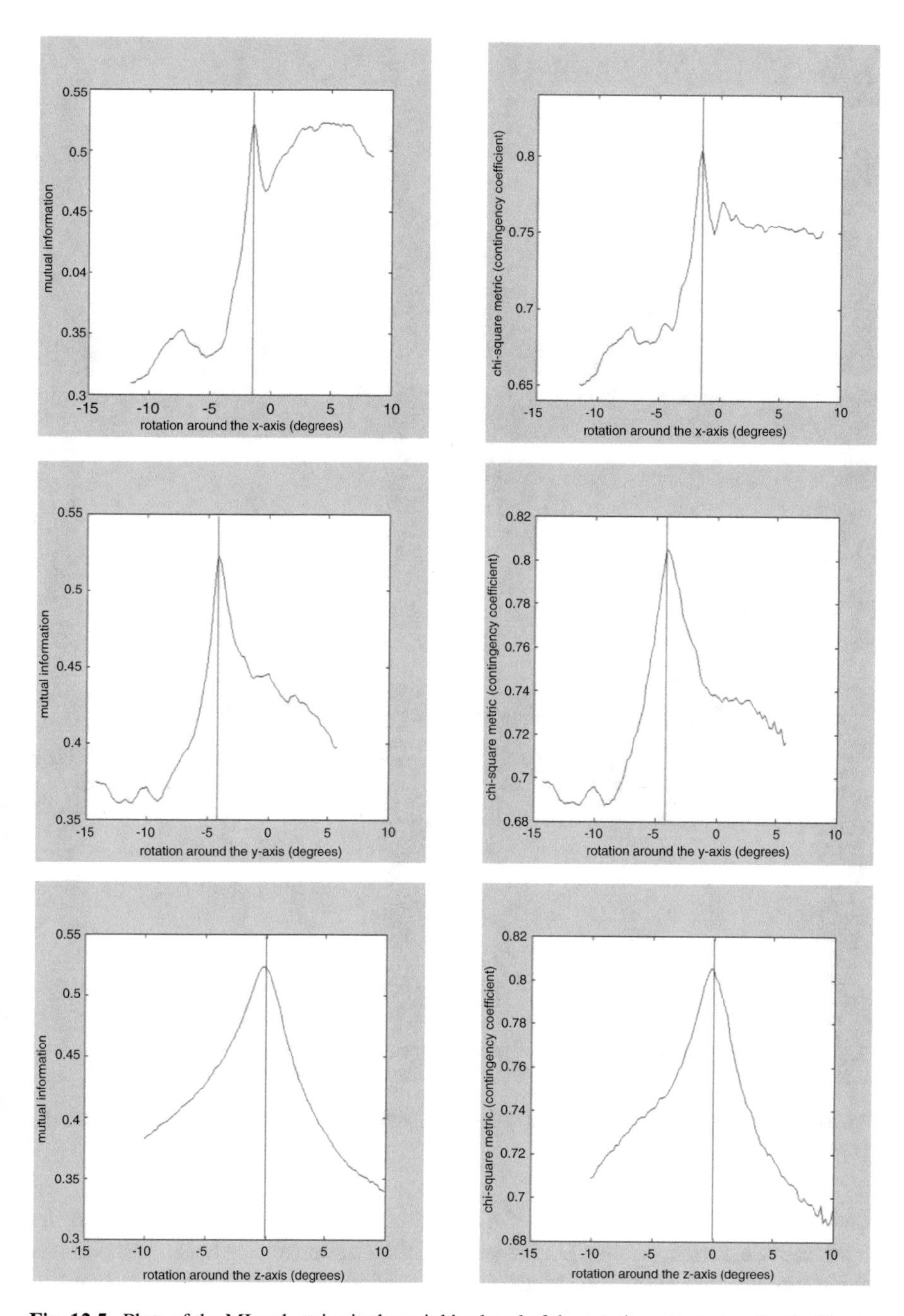

Fig. 12.5 Plots of the MI and -etrics in the neighborhood of the rotation parameters for the "boxes scene"

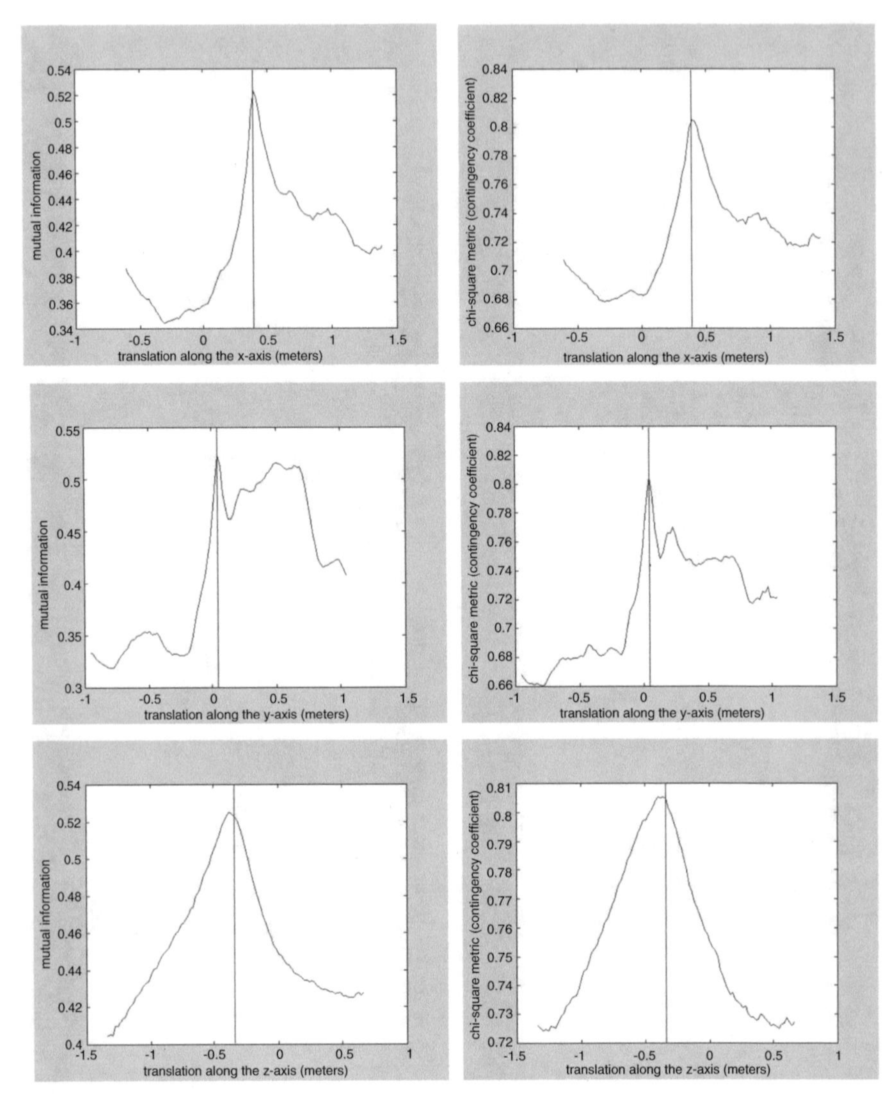

Fig. 12.6 Plots of the MI and -etrics in the neighborhood of the translation parameters for the "boxes scene"

used the normals to the vertices of a model built from a range map to align the model with a corresponding 2D image with the relationship between the normals and image shading [viola97]. Since the laser reflectance image provides intensity values characterizing the surface vertices, we do not need to use normals.

The registration problem is to find the pose parameters by measuring the similarity of the projected reflectance image in the camera plane with the color image. We want to find the rigid transformations $\{t, R\}$ for which the χ^2 metric (or the corresponding contingency coefficient) is maximized. The optimum pose parameter can be expressed as

$$\hat{\theta} = \underset{\theta}{\operatorname{argmax}}\, C(f(x), r(T_\theta(x))). \tag{12.31}$$

where $\theta = \left(\phi_x, \phi_y, \phi_z, t_x, t_y, t_z\right)^t$ represents the parameter vector containing the three rotation angles and the three translations, and $T_\theta(x)$ projects a 3D surface point to the image plane after rigid transformations in the 3D space. The image point will not necessarily fall onto the 2D grid, which requires the use of interpolation to determine $r(T_\theta(x))$. We used bilinear interpolation in our algorithm.

A similarity measure-based image registration algorithm always requires the use of an appropriate optimization step. In general, similarity measures, including the χ^2 metric, are highly nonlinear and present many local extrema.

Most existing registration algorithms use deterministic optimization techniques. For the maximization of mutual information, Viola used a stochastic gradient descent approach, relying on the noise introduced by the evaluation of the gradient over small samples to escape local hills in the mutual information landscape [viola97]. Maes et al. [maes97] applied Powell's multidimensional direction set method for multimodal volumetric medical image registration. Gradient approaches are known to be very sensitive to local maxima and require a good initial guess of the parameters. Stochastic approaches have been used to avoid this problem. Nikou et al. applied a simulated annealing optimization technique for multimodal registration using the robust inter-uniformity metric [nikou98].

In this work we investigated the use of two heuristic optimization techniques to get suboptimal solution and robust alignment of the images. Genetic algorithms [beasly93, beasly93b, brooks96] are used for the rough maximization, and the additional local search [battiti94, glover97, vaessens98] allows further refinement of the registration parameters.

12.3.2 Genetic Algorithms

Genetic algorithms are a computational paradigm belonging to the class of optimization techniques known as evolutionary computation. These algorithms have been successfully implemented to solve for many difficult optimization problems, including sensor calibration and many computer vision problems. As shown in Fig. 12.7, the solutions of a given problem are coded as a string, called *chromosomes*, which convey information of the pose parameters. In the proposed implementation, chromosomes consist of the six binary-coded parameters.

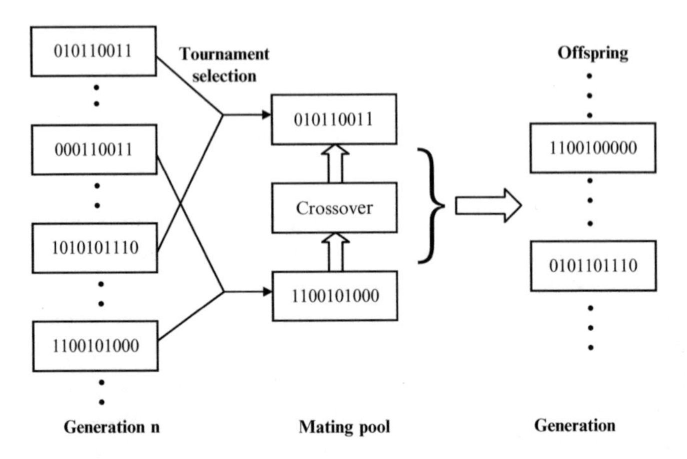

Fig. 12.7 Genetic algorithm scheme: in the tournament selection, two parents are selected at random, and we select the one with the best fitness function. In the mating pool the parents exchange bits (crossover) and are subject to mutation. They then become members of the new generation

A parameter θ_i is coded in N bits, such as

$$\theta_i = \theta_{\min} + \frac{C_i}{2^N - 1}(\theta_{\max} - \theta_{\min}), \tag{12.32}$$

where $\theta_i \in [\theta_{\min}, \theta_{\max}]$ represents a specified range for the parameter θ_i and C_i the numerical value of the corresponding chromosome coded in N bits.

A large set of these strings forms a *gene pool*, and the quality of these strings is evaluated using a *fitness function* that is the registration metric in our case. A new *generation* is created using a *selection* process, which mimics the process of natural selection. Among various strategies for selecting the new generation, we used the tournament approach, which consists in picking at random two or more strings and choosing the best with some probability close to the unity. The selected elements are placed in a *mating pool*. Mixing the elements of the two randomly chosen chromosomes (or parents) in the mating pool creates a new generation (or offspring) which has the same number of individuals or strings. This step is called *crossover*. We used a uniform crossover that exchanges bits at random positions of the *chromosomes*, as shown in Fig. 12.8.

A certain amount of mutations occur in the system with a fixed probability to allow for some exploration of the search space. Mutation is implemented by flipping a bit at a random position of the chromosome as shown in Fig. 12.9.

This process continues for some number of fixed generations or when the average fitness of the population reaches a satisfactory level. We then select the solution with the best fitness function.

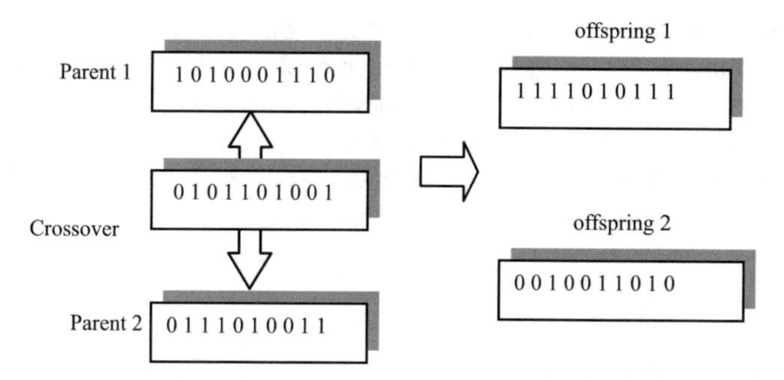

Fig. 12.8 Uniform crossover for binary-coded parameters

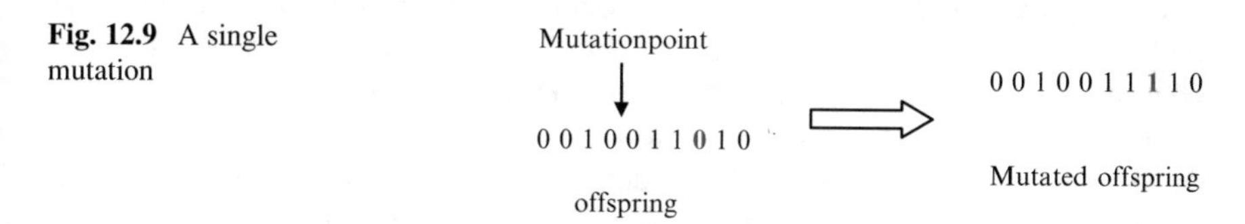

Fig. 12.9 A single mutation

12.3.3 Local Search: Tabu Search

We used a local search technique for further refinement of the solution obtained by the genetic algorithm. We use Tabu search for local improvement of the solution for a small number of iterations.

Tabu search is a local search technique, which attempts to avoid falling into local extrema by keeping a history of the search and forbidding the most recently visited nodes in the search space. The forbidden moves are placed in a Tabu list; this approach will allow a search algorithm to climb out of shallow local minima (or maxima).

This heuristic requires defining incremental moves in the parameter space to explore the neighborhood of a given node. At the kth iteration, the pose parameter θ_i is updated as

$$\theta_i^{k+1} = \theta_i^k + \Delta\theta_i. \tag{12.33}$$

The length of step depends on the resolution of the 3D model and the distance from sensor to objects. After each move the search decides to visit the next node in the direction of the best-fitted function unless it is on the Tabu list. Every visited node is placed in this list which is updated in a cyclical fashion.

The combined genetic algorithm and Tabu search can be used for parameter estimation, where the resulting global extremum of an objective function is hidden between multiple local extrema. They were in particular applied to sensor alignment and calibration. The population size, mutation parameters, and the size of the Tabu list can be determined heuristically. The speed of convergence depends on the complexity of the similarity metric, which is calculated over the two images. To accelerate convergence, we can use subsampling and evaluate the metric over smaller samples rather than over all the pixels of the images at the cost of accuracy.

The algorithm also requires a step of occlusion culling. In other words, for a given set of parameters, many points of the model created from the range map are not visible in the color image plan because of occlusion, so they should not be included when we compute the similarity measure.

We can extend the algorithm to the uncalibrated case and search also for the focal length f or other camera model parameters.

12.3.4 Texture Mapping

Once the pose parameters are found, we should generate the texture map, which is a color image used by the 3D-rendering engine to wrap the reconstructed 3D model. This can be done by re-projecting the color image to the range scanner's reference frame. This operation normally associates each 3D point with a texture element and identifies occluded and non-visible regions from the color camera's point of view. The occlusion problem results in sharing the same texture element for multiple model points when projected in the color image plane. One solution for this occlusion problem is using a Z-buffering operation, which associates a depth value to each color pixel as shown in Fig. 12.10.

If another 3D point attempts to use the color pixel that has already been used, its depth value is checked. If the depth value of the new point is less than that of the old one, the texture element associated with the 3D point is labeled as occluded (turned black in our textures), and the new point will receive the color pixel as texture.

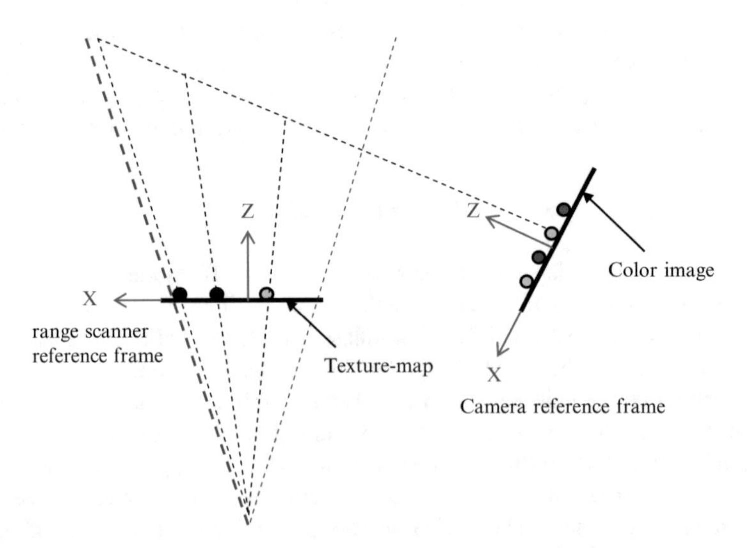

Fig. 12.10 Texture map generation and occlusion culling using Z-buffering. Pixels are represented by *filled circles*. Note that three points of a surface share the same picture element and that the occluded points will receive *black* texture

Fig. 12.11 Color, range, and reflectance images of the "boxes" used in the reconstruction experiments

Fig. 12.12 Color, range, and reflectance images of the "lab" used in the reconstruction experiments

12.4 Experimental Results

The two data sets of Figs. 12.11 and 12.12 are used to test the registration algorithm and to build 3D photo-realistic models. Color images of size 768×511 and the LADAR reflectance-range images of size 1024×1024 are obtained using a Kodak DCS460 digital camera and a Perceptron P5000 laser range finder, respectively. Using these data sets, we test the registration algorithm, and the obtained parameters are used later for texture generation. The final step in the process is the application of the texture maps and the visualization of the photo-realistic models.

12.4.1 Registration Results

To test the registration algorithm, reference results are built using an interactive landmark-based registration technique. Corresponding landmarks are manually extracted, and the pose estimation technique is applied to find the parameters. To maximize the fitness function, which quantifies a contingency coefficient associated with the χ^2-information similarity metric, the genetic algorithm is set with the parameters given in Table 12.1. Care must be taken to fix the tournament size appropriately to avoid either premature or too slow convergence.

Table 12.1 Optimal parameters for the genetic algorithm used in the registration algorithm for the two data sets

Parameters	Boxes scene	Lab scene
Population size	50	60
Tournament size	2	3
Mutation probability	0.03	0.03

Table 12.2 LADAR-color registration results for the "boxes" scene and comparison with reference results built by registering landmarks

Parameters	Registration results	Reference	Difference
$\varphi_x(°)$	−1.5942	−1.4975	0.0967
$\varphi_y(°)$	−6.1375	−4.1943	1.9432
$\varphi_z(°)$	−0.0061	0.0586	0.0525
t_x(m)	0.5551	0.3908	0.1643
t_y(m)	0.0610	0.0462	0.0148
t_z(m)	−0.4132	−0.3377	0.0755

Table 12.3 LADAR-color registration results for the "lab" scene and comparison with reference results built by registering landmarks

Parameters	Registration results	Reference	Difference
$\varphi_x(°)$	−10.8137	−12.3721	1.5584
$\varphi_y(°)$	2.9318	4.0797	1.1479
$\varphi_z(°)$	1.3352	0.8583	0.4769
t_x(m)	−0.4981	−0.5953	0.0972
t_y(m)	−0.1246	−0.0015	0.1231
t_z(m)	0.1715	0.1251	0.0464

Contextual information of the scene could be used to limit the range of the parameters and thus increase the accuracy of the algorithm. Tables 12.2 and 12.3 show comparisons between the results of the registration algorithm and the reference. For the "boxes" scene, the average error for a set of 18 landmarks after registration is 1.93 pixels, while for the "lab" scene, it is 2.77 pixels. The registration results could be improved so that the resulting resolution may reach up to sub-pixel accuracy by the use of a more elaborate camera model.

The algorithm is terminated after the average fitness stops varying significantly or the best solution found so far is not improved any more. The average fitness function of the population in Figs. 12.13 and 12.14 shows some oscillations near the convergence. This is due to the effect of mutation. Algorithms converged after about 60 generations for both data sets.

The algorithm was implemented in C++ and was relatively expensive in terms of computation. More specifically, it takes about 45 min in average in our experiments to have convergence on an SGI R10000 processor of 200 MHz. Such high computational load results from the large number of function evaluations. For example,

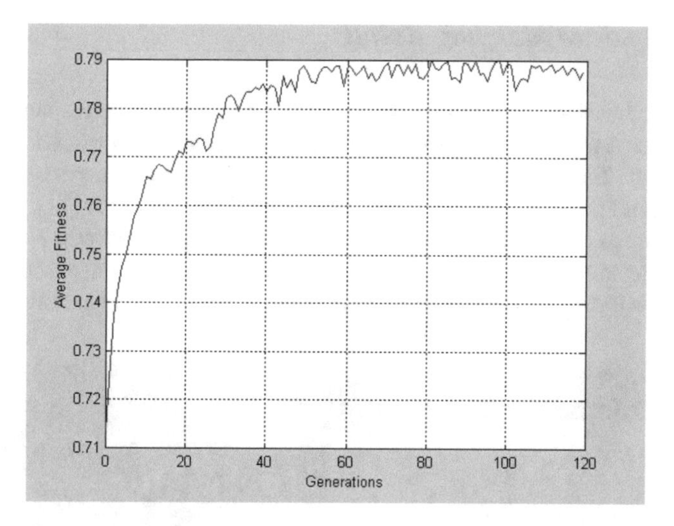

Fig. 12.13 Plot of the run of the genetic algorithm used in the registration of the "boxes scene" data set

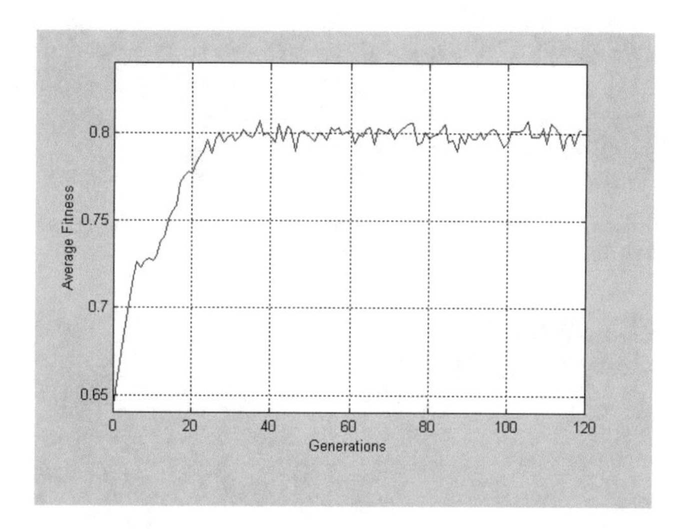

Fig. 12.14 Plot of the run of the genetic algorithm for the registration of the "lab scene" data set

over 3000 evaluations were performed in our experiments. A Z-buffering operation, required for elimination of occluded pixels from fitness computation, is also computationally expensive. It was found that omitting the occlusion culling step might increase the speed of convergence but significantly deteriorates the accuracy of the results.

The obtained pose parameters are refined using Tabu search for a limited number of iterations, and the size of the Tabu list is considered infinite.

12.4.2 Texture Mapping Results

The final parameters are used to generate the texture maps corresponding to the range scanner's point of view. The occluded areas are, as mentioned earlier, labeled in black color. Figures 12.15 and 12.16 show the texture maps generated using the manual landmark-based technique and the automatic technique using the similarity measure, respectively. Manual results are considered as ground truth. We can notice that there are some inaccuracies in the occluded areas for the automatic technique, which is due to the errors in pose parameters estimation. We also note that only a

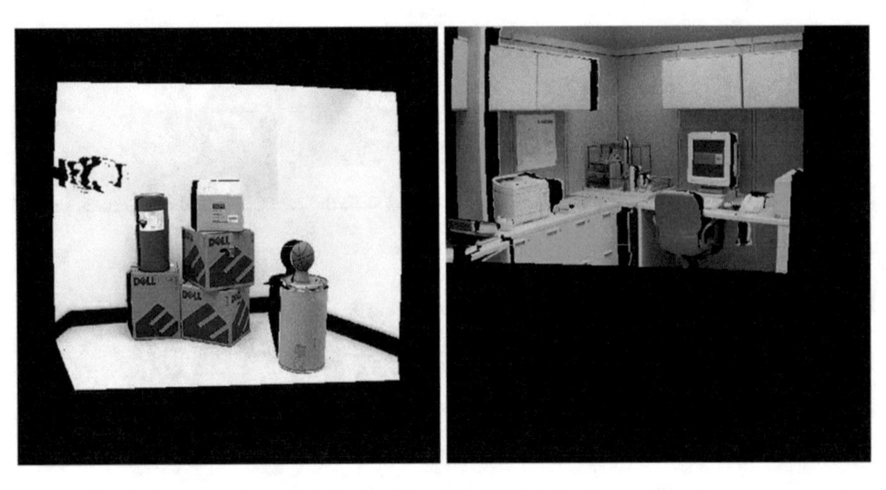

Fig. 12.15 Texture maps created for the "boxes" and "lab" scenes using the pose parameters obtained by the manual landmarks registration technique

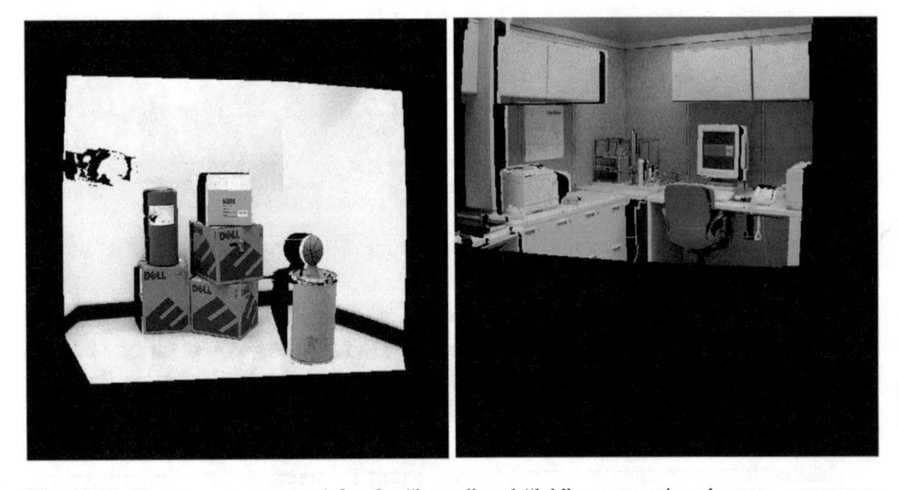

Fig. 12.16 Texture maps created for the "boxes" and "lab" scenes using the pose parameters obtained by the automatic registration technique

part of the model will receive a texture since the field of view of the scanner is larger than the color camera's field of view. Thus, in order to wrap entirely the model with a single texture map, multiple photographs from different angles must be merged and blended.

12.4.3 Reconstruction Results

The models are rendered using OpenInventor software, and triangle meshes are formed. The generated texture maps are applied on top of the geometric models. Given a practical size of the range images, these models become very large and require a significant amount of computing resources to visualize and manipulate. In our experiment, for example, 2,093,058 triangles were used. A mesh reduction step could be useful to ease rendering the models because of significant reduction of memory space. A mesh reduction system is built by using both geometry and color information in the reduction process. For practical applications such as robot vision and navigation, real-time reconstruction is required; however, it is still difficult for such large data sets and should be left as a future research topic. Figure 12.17 shows two snapshots of the reconstructed models with a uniform texture, for which OpenInventor, a 3D rendering engine, uses vertex normals to calculate the shading and the appearance of the scene from a given viewpoint of a virtual camera model. Thus, the 3D model could be viewed from different angles. Figures 12.18 and 12.19 show snapshots from different viewpoints of reconstructed photo-realistic models. Since we are rendering a surface mesh created from a single-range image, the quality of scene reconstruction is optimal when the model is rendered from the

Fig. 12.17 Snapshots of the reconstructed models of the "boxes" and "lab" scenes. Here, the models are rendered with uniform textures

Fig. 12.18 Snapshots of different views of the 3D-reconstructed "boxes" scene rendered with texture

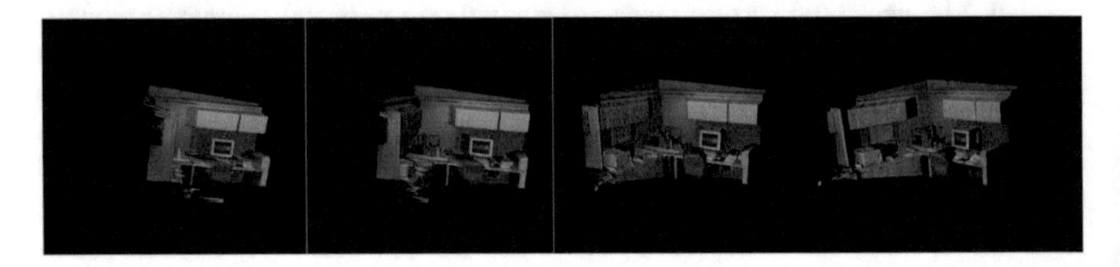

Fig. 12.19 Snapshots of different views of the 3D-reconstructed "lab" scene rendered with texture

scanner's point of view. To obtain a more complete model, we need to merge multiple triangle meshes. In this context, registration and integration of multiple range images is a very active research field and is crucial particularly for robot vision and 3D scene modeling.

Automatic multimodal and, in particular, photo-realistic reconstruction from unregistered sensorial data has proven to be an attractive and feasible way to build 3D scene models.

A fundamental step is the registration of the different modalities. For this purpose, information theoretic similarity measures are a powerful tool which is theoretically rigorous and allows a simple implementation. The quantification of the images' dependence in terms of information avoids the more restrictive classical assumptions on the relationship of the different modalities. This approach presents the advantage of using the raw data without preprocessing. More specifically, no feature extraction is required.

In this work, we have presented the approach to 3D surface reconstruction of a real scene from unregistered LADAR and color images. The first step was the creation, from range images, of a triangle mesh representing the model's skeleton. But the main focus was on the multimodal registration issue.

One important contribution was to show that the χ^2-information metric performs generally well in the registration task and could compete with the standard mutual information criterion. This proves that other dependence measures based on statistical comparison of random distributions could be applied to the multimodal registration problem as well.

For dealing with the landscape of the similarity measure with multiple local maxima, we used genetic algorithms as an optimization method, which makes the registration algorithm remain simple and robust with little need for operator intervention at the cost of increased computational overhead.

One of the possible continuations of this work is to fuse different textured surface meshes, of a given scene, to have a more complete 3D representation. Indeed, recent works [johnson98, maimone] demonstrated that the association of color with geometry eases the tasks of 3D registration and integration of surfaces reconstructed from different sensorial views. It is also important to improve the computational performance of the automatic registration algorithm and the overall photo-realistic reconstruction pipeline in order to reach real-time or quasi real-time performances required in applications such as robot navigation and remote monitoring through artificial vision. Another interesting use of these models is automatic object recognition, where the presence of both color and shape aspects play an important role in providing more information on the scene features and objects.

References

[abidi92] M.A. Abidi, R.C. Gonzalez, *Data Fusion in Robotics and Machine Intelligence* (Academic, Boston, MA, 1992)

[basri96] R. Basri, D. Weinshall, Distance metric between 3D models and 2D images for recognition and classification. IEEE Trans. Pattern Anal. Mach. Intell. **18**(4), 465–470 (1996)

[battiti94] R. Battiti, G. Tecchiolli, The reactive Tabu search. ORSA J. Comput. **6**(2), 126–140 (1994)

[beasly93] D. Beasly, D.S. Bull, R.R. Martin, An overview of genetic algorithms: part 1, fundamentals. Univ. Comput. **15**(2), 58–69 (1993)

[beasly93b] D. Beasly, D.S. Bull, R.R. Martin, An overview of genetic algorithms: part 2, research topics. Univ. Comput. **15**(4), 170–181 (1993)

[besl92] P.J. Besl, N.D. McKay, A method for registration of 3D shapes. IEEE Trans. Pattern Anal. Mach. Intell. **14**, 239–256 (1992)

[brooks96] R.R. Brooks, S.S. Iyenger, J.C. Chen, Automatic correlation and calibration of noisy sensor readings using elite genetic algorithms. Artif. Intell. **84**, 339–354 (1996)

[collignon95] A. Collignon, F. Maes, D. Delaere, D. Vandermeulen, P. Suetens, and G. Marchal, Automated multimodality medical image registration using information theory, in *Proceeding of the 14th International Conference Information Processing in Medical Imaging; Computational Imaging and Vision 3* (1995), pp. 263–274

[faugeras93] O. Faugeras, *Three-Dimensional Computer Vision* (The MIT Press, Cambridge, MA, 1993)

[feldmar97] J. Feldmar, N. Ayache, F. Betting, 3D-2D registration of free-form curves and surfaces. Comput. Vis. Image Underst. **65**(3), 403–424 (1997)

[glover97] F. Glover, M. Laguna, *Robotics, Tabu Search* (Kluwer's Academic Publishers, Boston, 1997)

[haralick93] R.M. Haralick, L.G. Shapiro, *Computer and Robot Vision* (Addison-Wesley Publishing Company, Michigan, 1993)

[hausler99] G. Haüsler, D. Ritter, Feature-based object recognition and localization in 3D space using a single video image. Comput. Vis. Image Underst. **73**(1), 64–81 (1999)

[horn86] B.K.P. Horn, *Robot Vision* (The MIT Press, Cambridge, MA, 1986)

[irani98] M. Irani and P. Anandan, Robust Multi-sensor Image Alignment, Technical Report, The Weizmann Institute of Science, 1998

[ji98] Q. Ji, M.S. Costa, R.M. Haralick, and L.G. Shapiro, An Integrated Linear Technique for Pose Estimation from Different Geometric Features, Technical Report, University of Washington, Department of Electrical and Computer Engineering, 1998

[johnson98] A.E. Johnson and S.B. Kang, Registration and integration of textured 3D data, in *Proceedings of the International Conference of Recent Advances in 3D Digital Imaging and Modeling* (1998), pp. 331–338

[kanatani98] K. Kanatani, N. Ohta, Optimal robot self-localization and reliability evaluation, in *Proceedings of the European Conference Computer Vision, 1998* (Springer, Berlin, 1998), pp. 796–808

[kass97] R.E. Kass, P.W. Vos, M.S.J. Tsao, *Geometrical Foundations of Asymptotic Inference* (John Wiley & Sons, New York, 1997)

[lavallee95] S. Lavallée, R. Szeliski, Recovering the position and orientation of free-form objects from image contours using 3D distance maps. IEEE Trans. Pattern Anal. Mach. Intell. **17**(4), 378–390 (1995)

[lohmann98] G. Lohmann, *Volumetric Image Analysis* (John Wiley & Sons and B. G. Teubner Publishers, New York, 1998)

[manga99] A.P. Mangan, Photo-Realistic Surface Reconstruction, Master's Thesis, University of Tennessee, Knoxville, 1999

[maes97] F. Maes, A. Collignon, D. Vandermeulen, G. Marchal, P. Suetens, Multi-modality image registration by maximization of mutual information. IEEE Trans. Med. Imaging **16**(2), 187–198 (1997)

[maimone] M. Maimone, L. Matthies, J. Osborn, E. Rollins, J. Teza, and S. Thayer, A Photo-Realistic Mapping System for Extreme Nuclear Environments: Chornobyl, IEEE/RSJ International Conference on Intelligent Robotic Systems (1998)

[maybank] S. Maybank, *Theory of Reconstruction from Image Motion* (Springer, Berlin, 1993)

[mcgarry96] D.P. McGarry, T.R. Jackson, M.B. Plantec, N.F. Kassell, and J. Hunter Downs, III, Registration of Functional Magnetic Resonance Imagery using Mutual Information, Technical Report, University of Virginia, Charlottesville, Neuro-surgical Visualization Laboratory, 1996

[nikou98] C. Nikou et al., Robust registration of dissimilar single and multi-modal images, in *Proceedings of the European Conference Computer Vision, 1998* (Springer, Berlin, 1998), pp. 51–63

[press92] W.H. Press, B.P. Flannery, S.A. Teukolsky, W.T. Vetterling, *Numerical Recipes in C*, 2nd edn. (Cambridge University Press, New York, 1992)

[roche98] A. Roche, G. Malandin, X. Pennec, and N. Ayache, Multi-modal Image Registration by Maximization of the Correlation Ratio, Research Report, Institut National de Recherche en informatique et en Automatique, No. 3378, 1998

[rogers91] D.F. Rogers, R.A. Earnshaw, *State of the Art in Computer Graphics: Visualization and Modeling* (Springer, Berlin, 1991)

[sequira99] V. Sequira, E. Wolfart, J.G.M. Gonclaves, D. Hogg, Automated reconstruction of 3D models from real environments. ISPRS J. Photogramm. Remote Sens. **54**, 1–22 (1999)

[stoker98] C. Stoker, T. Blackmon, J. Hagen, B. Kanefsky, D. Rasmussen, K. Schwehr, M. Sims, E. Zbinden, *MARSMAP: An Interactive Virtual Reality Model of the Pathfinder Landing Site* (NASA Ames Research Center, Moffett Field, CA, 1998)

[stevens97] M.R. Stevens, J.R. Beveridge, Precise matching of 3D target models to multi-sensor data. IEEE Trans. Med. Imaging **6**(1), 126–142 (1997)

[szeliski96] R. Szeliski, Video mosaics for virtual environments, IEEE Comput. Graph. Appl., 22–30 (1996)

[tsai87] R.Y. Tsai, A versatile camera calibration technique for high-accuracy 3D machine vision metrology using off-the-shelf TV cameras and lenses. IEEE J. Robot. Autom. **3**(4), 323–344 (1987)

[tucceryan95] M. Tuceryan, D.S. Greer, R.T. Whitaker, D.E. Breen, C. Crampton, E. Rose, K.H. Ahlers, Calibration requirements and procedures for a monitor-based augmented reality system. IEEE Trans. Vis. Comput. Graph. **1**, 255–273 (1995)

[vaessens98] R.J.M. Vaessens, E.H.L. Aarts, J.K. Lenstra, A local search template. Comput. Oper. Res. **25**(11), 969–979 (1998)

[vajda89] I. Vajda, *Theory of Statistical Inference and Information* (Kluwer, Boston, MA, 1989)

[viola97] P. Viola, W. Wells III, Alignment by maximization of mutual information. Proc. IEEE Int. Conf. Comput. Vis. **24**(2), 137–154 (1997)

[woods93] R.P. Woods, J.C. Mazziotta, S.R. Cherry, MRI-PET registration with automated algorithm. J. Comput. Assist. Tomogr. **17**(4), 536–546 (1993)

Appendix A
Matrix-Vector Representation for Signal Transformation

A set of numbers can be used to represent discrete signals. These numbers carry a certain amount of information and are subject to change by various kinds of transformations, called systems. For example, a one-dimensional linear time-invariant system can be expressed by its corresponding impulse response. The output of the system is then determined by the convolution of the impulse response and the input signal. Convolution equations, in general, are too complicated to efficiently express related theories and algorithms.

Analysis and representation of signal transformations can be substantially simplified by using matrix-vector representation, where a vector and a matrix, respectively, represent the corresponding signal and transformation.

A.1 One-Dimensional Signals and Systems

Suppose a one-dimensional system has input signal $x(n)$, $n = 0, 1, \ldots, N-1$ and impulse response $h(n)$. The output of the system can be expressed as the one-dimensional convolution:

$$y(n) = \sum_{q=0}^{N-1} h(n-q)x(q), \quad \text{for } n = 0, 1, \ldots, N-1. \tag{A.1}$$

By simply rewriting Eq. (A.1), we have

$$\begin{aligned}
y(0) &= h(0)x(0) + h(-1)x(1) + h(-2)x(2) + \cdots \\
y(1) &= h(1)x(0) + h(0)x(1) + h(-1)x(2) + \cdots \\
&\vdots \\
y(N-1) &= h(N-1)x(0) + h(N-2)x(1) + h(N-3)x(2) + \cdots.
\end{aligned} \tag{A.2}$$

M.A. Abidi et al., *Optimization Techniques in Computer Vision*, Advances in Computer Vision and Pattern Recognition, DOI 10.1007/978-3-319-46364-3

If we express both input and output signals as $N \times 1$ vectors, such as

$$x = [x(0) \quad x(1) \quad \cdots \quad x(N-1)]^{\mathrm{T}} \text{ and } y = [y(0) \quad y(1) \quad \cdots \quad y(N-1)]^{\mathrm{T}}, \tag{A.3}$$

then the output vector is obtained by the following matrix-vector multiplication

$$y = Hx, \tag{A.4}$$

where

$$H = \begin{bmatrix} h(0) & h(-1) & h(-2) & \cdots & h(-N+1) \\ h(1) & h(0) & h(-1) & \cdots & h(-N+2) \\ h(2) & h(1) & h(0) & \cdots & h(-N+3) \\ \vdots & \vdots & \vdots & \ddots & \vdots \\ h(N-1) & h(N-2) & h(N-3) & \cdots & h(0) \end{bmatrix}. \tag{A.5}$$

We note that H is a Toeplitz matrix having constant elements along the main diagonal and sub-diagonals.

If two convolving sequences are periodic with period N, their circular convolution is also periodic. In case, $h(-n) = h(N-n)$, which results in the circulant matrix and can be expressed as

$$H = \begin{bmatrix} h(0) & h(N-1) & h(N-2) & \cdots & h(1) \\ h(1) & h(0) & h(N-1) & \cdots & h(2) \\ h(2) & h(1) & h(0) & \cdots & h(3) \\ \vdots & \vdots & \vdots & \ddots & \vdots \\ h(N-1) & h(N-2) & h(N-3) & \cdots & h(0) \end{bmatrix}. \tag{A.6}$$

The first column of H is the same as that of the vector $h = [h(0) \quad h(1) \quad \cdots \quad h(N-1)]^{\mathrm{T}}$, and the second column is the same as that of the rotated version of h indexed by one element, such as $[h(N-1) \quad h(0) \quad \cdots \quad h(N-2)]^{\mathrm{T}}$. The remaining columns are determined in the same manner.

Example A.1: One-Dimensional Shift-Invariant Filtering and the Circulant Matrix Consider a discrete sequence $\{1 \quad 2 \quad 3 \quad 4 \quad 5 \quad 4 \quad 3 \quad 2 \quad 1\}$. Suppose that the corresponding noisy observation is given as $x = [1.10 \quad 1.80 \quad 3.10 \quad 4.20 \quad 5.10 \quad 3.70 \quad 3.20 \quad 2.10 \quad 0.70]^{\mathrm{T}}$. One simple way to remove the noise is to replace each observed sample by the average of the neighboring samples. If we use an averaging filter that replaces each sample by the average of two neighboring samples, plus the sample itself, we have the output $y = [1.20 \quad 2.00 \quad 3.03 \quad 4.03 \quad 5.00 \quad 4.00 \quad 3.00 \quad 2.00 \quad 1.30]^T$, where the first and the last samples have been computed under the assumption that the input sequence is periodic with period 9, because they are located at a boundary and do not have enough neighboring samples for convolution with the impulse response. The averaging process can be expressed as a one-dimensional time-invariant system whose impulse response is

$$h(n) = \frac{1}{3}\{\delta(n+1) + \delta(n) + \delta(n-1)\}. \quad \text{(A.7)}$$

We can make the corresponding circulant matrix by using the impulse response as

$$H = \frac{1}{3}\begin{bmatrix} 1 & 1 & 0 & \cdots & 1 \\ 1 & 1 & 1 & \cdots & 0 \\ 0 & 1 & 1 & \ddots & 0 \\ \vdots & \vdots & \ddots & \ddots & \vdots \\ 1 & 0 & 0 & \cdots & 1 \end{bmatrix}. \quad \text{(A.8)}$$

It is straightforward to prove that $y = Hx$.

A.2 Two-Dimensional Signals and Systems

In the previous section we obtained the matrix-vector expression of one-dimensional convolution by mapping an input signal to a vector and the impulse response to a Toeplitz or circulant matrix. In a similar manner, we can also represent two-dimensional convolution as a matrix-vector expression by mapping an input two-dimensional array into a row-ordered vector and the two-dimensional impulse response into a doubly block circulant matrix.

A.2.1 Row-Ordered Vector

Two-dimensional rectangular arrays or matrices usually represent image data. Representing two-dimensional image processing systems, however, becomes too complicated to be analyzed if we use two-dimensional matrices for the input and output signals. Based on the idea that both vectors and matrices can represent the same data, only in different formats, we can represent two-dimensional image data by using a row-ordered vector.

Let the following two-dimensional $M \times N$ array represent an image

$$X = \begin{bmatrix} x(0,0) & x(0,1) & \cdots & x(0,N-1) \\ x(1,0) & x(1,1) & \cdots & x(1,N-1) \\ \vdots & \vdots & \ddots & \vdots \\ x(M-1,0) & x(M-1,1) & \cdots & x(M-1,N-1) \end{bmatrix}, \tag{A.9}$$

which can also be represented by the row-ordered $MN \times 1$ vector, such as

$$x = \left[x(0,0)\ x(0,1)\ \cdots\ x(0,N-1) | x(1,0)\ \cdots\ x(1,N-1) | \cdots | x(M-1,0)\ \cdots\ x(M-1,N-1) \right]^{\mathrm{T}}. \tag{A.10}$$

A.2.2 Block Matrices

A space-invariant two-dimensional system is characterized by a two-dimensional impulse response. The output of the system is determined by two-dimensional convolution, expressed as

$$y(m,n) = \sum_{p=0}^{M-1} \sum_{q=0}^{N-1} h(m-p, n-q) x(p,q), \tag{A.11}$$

where $y(m,n)$, $h(m,n)$, and $x(m,n)$, respectively, represent the two-dimensional output, the impulse response, and the input signals.

Like the one-dimensional case, two-dimensional convolution can also be expressed by matrix-vector multiplication.

Example A.2: Two-Dimensional Space-Invariant Filtering and the Block Circulant Matrix Suppose that an $N \times N$ image $x(m,n)$ is filtered by the two-dimensional low-pass filter with impulse response:

$$h(m,n) = \frac{1}{16} \left\{ \begin{array}{l} \delta(m+1,n+1) + 2\delta(m+1,n) + \delta(m+1,n-1) \\ +2\delta(m,n+1) + 4\delta(m,n) + 2\delta(m,n-1) \\ +\delta(m-1,n+1) + 2\delta(m-1,n) + \delta(m-1,n-1) \end{array} \right\}. \tag{A.12}$$

The output is obtained by two-dimensional convolution as given in Eq. (*A.11*). We can also express the two-dimensional convolution by multiplying the block matrix and the row-ordered vector. If we assume that both the impulse response and the input signal are periodic with period $N \times N$, it is straightforward to prove that the matrix-vector multiplication

$$y = Hx, \tag{A.13}$$

is equivalent to the two-dimensional convolution, where the row-ordered vector x is obtained as in Eq. (*A.10*), and the block matrix is obtained as

$$H = \frac{1}{16} \begin{bmatrix} H_0 & H_{-1} & 0 & \cdots & H_1 \\ H_1 & H_0 & H_{-1} & \cdots & 0 \\ 0 & H_1 & H_0 & \cdots & 0 \\ \vdots & \vdots & \vdots & \ddots & \vdots \\ H_{-1} & 0 & 0 & \cdots & H_0 \end{bmatrix}. \tag{A.14}$$

Each element in H is again a matrix defined as

$$H_0 = \begin{bmatrix} 4 & 2 & 0 & \cdots & 2 \\ 2 & 4 & 2 & \cdots & 0 \\ 0 & 2 & 4 & \cdots & 0 \\ \vdots & \vdots & \vdots & \ddots & \vdots \\ 2 & 0 & 0 & \cdots & 4 \end{bmatrix}, \text{ and } H_1 = H_{-1} = \begin{bmatrix} 2 & 1 & 0 & \cdots & 1 \\ 1 & 2 & 1 & \cdots & 0 \\ 0 & 1 & 2 & \cdots & 0 \\ \vdots & \vdots & \vdots & \ddots & \vdots \\ 1 & 0 & 0 & \cdots & 2 \end{bmatrix}. \tag{A.15}$$

Any matrix A whose elements are matrices is called a block matrix, such as

$$A = \begin{bmatrix} A_{0,0} & A_{0,1} & \cdots & A_{0,N-1} \\ A_{1,0} & A_{1,1} & \cdots & A_{1,N-1} \\ \vdots & \vdots & \ddots & \vdots \\ A_{M-1,0} & A_{M-1,1} & \cdots & A_{M-1,N-1} \end{bmatrix}, \tag{A.16}$$

where $A_{i,j}$ represents a $p \times q$ matrix.

More specifically, the matrix A is called an $m \times n$ block matrix of basic dimension $p \times q$. If the block structure is circulant, that is, $A_{i,j} = A_{i \bmod M, j \bmod N}$, A is called *block circulant*. If each $A_{i,j}$ is a circulant matrix, A is called a *circulant block* matrix. Finally, if A is both block circulant and circulant block, A is called *doubly block circulant*.

A.2.3 Kronecker Products

If A and B are $M_1 \times M_2$ and $N_1 \times N_2$ matrices, respectively, their Kronecker product is defined as

$$A \otimes B \equiv \begin{bmatrix} a(0,0)B & \cdots & a(0,M_2-1)B \\ \vdots & \ddots & \vdots \\ a(M_1-1,0)B & \cdots & a(M_1-1,M_2-1)B \end{bmatrix}, \tag{A.17}$$

which is an $M_1 \times M_2$ block matrix of basic dimension $N_1 \times N_2$. Kronecker products are useful in generating high-order matrices from low-order matrices.

For an $N \times N$ image, X, a separable operation, where A operates on the columns of X and B operates on the rows of the result, can be expressed as

$$Y \equiv AXB^{\mathrm{T}} \tag{A.18}$$

or

$$y(m,n) = \sum_{p=0}^{N-1}\sum_{q=0}^{N-1} a(m,p)x(p,q)b(n,q). \tag{A.19}$$

In addition, Eqs. (A.18) and (A.19) are equivalent to

$$y = (A \otimes B)x, \tag{A.20}$$

where y and x, respectively, represent the row-ordered vectors of $y(m,n)$ and $x(m,n)$.

Example A.3: Two-Dimensional Extension Using the Kronecker Product Consider an $N \times N$ image, denoted by the two-dimensional matrix X as shown in Eq. (*A.9*). If we apply a one-dimensional averaging filter, given in Eq. (A.7), on each column of X, and apply the same filter on each row of the resulting matrix, then we can obtain a two-dimensional averaging filtered image. Let $y(m,n)$ be the output of the two-dimensional averaging filter and $h(m,n)$ the element of matrix H, which represents the circulant matrix for the one-dimensional system given in Eq. (A.8). Then we have that

$$y(m,n) = \sum_{p=0}^{N-1}\sum_{q=0}^{N-1} h(m,p)x(p,q)h(n,q), \tag{A.21}$$

which is equivalent to

$$Y = HXH^{\mathrm{T}}. \tag{A.22}$$

Let x and y respectively represent row-ordered vectors for $x(m,n)$ and $y(m,n)$, as shown in Eq. (*A.10*). Then Eq. (A.22) is equivalent to the following matrix-vector multiplication:

$$y = (H \otimes H)x. \tag{A.23}$$

Appendix B
Discrete Fourier Transform

Continuous-time Fourier transform methods are well known for analyzing frequency characteristics of continuous-time signals. In addition, the inverse transform provides perfect original signal reconstruction. More specifically, an arbitrary frequency component in the signal can be extracted by forming the inner product of the signal with the corresponding sinusoidal basis function. Due to the orthogonality property of sinusoidal basis functions, each frequency component, which is called the Fourier transform coefficient, exclusively contains the desired frequency component. At the same time, the inverse transform can be performed using the exact procedure of the forward transform, except that the complex conjugated basis function is used.

Fourier transforms provide a very powerful tool for the mathematical analysis and synthesis of signals. For continuous signals, the work can be performed using paper and pencil calculations. However, in image processing (and many other fields) where signals are digitized and the resulting arrays are large scale, we find that the computations become digital, and thus the discrete Fourier transform (DFT) is used. The advantage here is that the DFT work can easily be performed with the digital computer using well-known DFT methods along with FFT algorithms.

In this appendix, the basic material needed for applying the DFT and IDFT to discrete signals is presented.

B.1 One-Dimensional Discrete Fourier Transform

The unitary DFT of a sequence $\{u(n),\ \ n = 0,\ \ \ldots,\ \ N-1\}$ is defined as

$$v(k) = \frac{1}{\sqrt{N}} \sum_{n=0}^{N-1} u(n) {W_N}^{kn},\ \ k = 0, 1, \ldots, N-1, \tag{B.1}$$

M.A. Abidi et al., *Optimization Techniques in Computer Vision*, Advances in Computer Vision and Pattern Recognition, DOI 10.1007/978-3-319-46364-3

where

$$W_N = \exp\left(-j\frac{2\pi}{N}\right). \tag{B.2}$$

Let the matrix F_1 be defined as follows:

$$F_1 = \frac{1}{\sqrt{N}}\begin{bmatrix} {W_N}^0 & {W_N}^0 & {W_N}^0 & \cdots & {W_N}^0 \\ {W_N}^0 & {W_N}^1 & {W_N}^2 & \cdots & {W_N}^{N-1} \\ {W_N}^0 & {W_N}^2 & {W_N}^4 & \cdots & {W_N}^{2(N-1)} \\ \vdots & \vdots & \vdots & \ddots & \vdots \\ {W_N}^0 & {W_N}^{N-1} & {W_N}^{2(N-1)} & \cdots & {W_N}^{(N-1)^2} \end{bmatrix}. \tag{B.3}$$

Then, Eq. (B.1) can be expressed in matrix-vector representation as

$$v = F_1 u, \tag{B.4}$$

where v and u represent vectors whose elements take values of the discrete signals $v(k)$ and $u(n)$, respectively.

The inverse transform is given by

$$u(n) = \frac{1}{\sqrt{N}}\sum_{n=0}^{N-1} v(k){W_N}^{-kn},\ n = 0, 1, \ldots, N-1. \tag{B.5}$$

Because the complex exponential function ${W_N}^{nk}$ is orthogonal and F_1 is symmetric, we have that

$$F_1^{-1} = F_1^{*\mathrm{T}} = F_1^*, \tag{B.6}$$

where the superscript * represents the conjugate of a complex matrix. According to Eq. (B.6), the inverse transform can be expressed in vector-matrix form as

$$u = F_1^* v. \tag{B.7}$$

The DFT of Eq. (B.1), and the inverse expression of Eq. (B.5), form a transform pair. This transform pair has the property of being unitary because the matrix F_1 is unitary. In other words, the conjugate transpose of F_1 is equal to its inverse, as expressed in Eq. (B.6). In many applications, differently scaled DFTs and their inverses may be used. In such cases, most properties, which will be summarized in the next section, hold with proper adjustment of a scaling factor.

B.2 Properties of the DFT

B.2.1 Periodicity

The extensions of the DFT and its inverse transform are periodic with period N. In other words, for every k,

$$v(k+N) = v(k), \tag{B.8}$$

and for every n,

$$u(n+N) = u(N). \tag{B.9}$$

B.2.2 Conjugate Symmetry

The DFT of a real sequence is conjugate symmetric about $N/2$, where we assume N is an even number. By applying the periodicity of the complex exponential, such as $W_N{}^N = \exp\left(-j\frac{2\pi}{N}\cdot N\right) = 1$, we obtain

$$v^*(N-k) = v(k). \tag{B.10}$$

From Eq. (B.10), we see that, for $k = 0, \ldots, \frac{N}{2} - 1$,

$$v\left(\frac{N}{2} - k\right) = v^*\left(\frac{N}{2} + k\right), \tag{B.11}$$

and

$$\left|v\left(\frac{N}{2} - k\right)\right| = \left|v\left(\frac{N}{2} + k\right)\right|. \tag{B.12}$$

According to the conjugate symmetry property, only $N/2$ DFT coefficients completely determine the frequency characteristics of a real sequence of length N. More specifically, the N real sequence

$$v(0), \mathrm{Re}\{v(1)\}, \mathrm{Im}\{v(1)\}, \ldots, \mathrm{Re}\{v(N/2-1)\}, \mathrm{Im}\{v(N/2-1)\}, v(N/2) \tag{B.13}$$

completely defines the DFT of the real sequence. It is clear from Eq. (B.1) that $v(0)$ is real and from Eq. (B.11) that $v(N/2)$ is real.

B.2.3 Relationships Between DFT Basis and Circulant Matrices

The basis vectors of the DFT are the orthonormal eigenvectors of any circulant matrix. The eigenvalues of a circulant matrix are the DFT of its first column. Based on these properties, the DFT is used to diagonalize any circulant matrix.

To prove that the basis vectors of the DFT are the orthonormal eigenvectors of any circulant matrix, we must show that

$$H\phi_k = \lambda_k \phi_k, \tag{B.14}$$

where H represents a circulant matrix ϕ_k, the k-th column of the matrix $F_1^{*\mathrm{T}} = F_1^*$, and λ_k the corresponding eigenvalue. Since H is circulant, it can be represented as

$$H = \begin{bmatrix} h_0 & h_{N-1} & h_{N-2} & \cdots & h_1 \\ h_1 & h_0 & h_{N-1} & \cdots & h_2 \\ h_2 & h_1 & h_0 & \cdots & h_3 \\ \vdots & \vdots & \vdots & \ddots & \vdots \\ h_{N-1} & h_{N-2} & h_{N-3} & \cdots & h_0 \end{bmatrix}. \tag{B.15}$$

The k-th column of the matrix $F_1^{*\mathrm{T}} = F_1^*$ is represented as

$$\phi_k = \frac{1}{\sqrt{N}} \begin{bmatrix} W_N{}^0 & W_N{}^{-k} & \ldots & W_N{}^{-(N-1)k} \end{bmatrix}^{\mathrm{T}}. \tag{B.16}$$

From Eqs. (B.15) and (B.16), the m-th element of $H\phi_k$ in (B.14) is obtained as

$$e_m{}^{\mathrm{T}}[H\phi_k] = \frac{1}{\sqrt{N}} \sum_{n=0}^{N-1} h(m-n) W_N{}^{-kn}, \tag{B.17}$$

where e_m represents the m-th unit vector, for example, $e_1 = \begin{bmatrix} 1 & 0 & \ldots & 0 \end{bmatrix}^T$. By changing variables and using the periodicity of the complex exponential function W_N, we can rewrite Eq. (B.17) as

$$e_m{}^{\mathrm{T}}[H\phi_k] = \left(\sum_{l=0}^{N-1} h(l) W_N{}^{kl} \right) \left(\frac{1}{\sqrt{N}} W_N{}^{-km} \right), \tag{B.18}$$

which results in Eq. (B.14). The eigenvalues of the circulant matrix H are defined as

$$\lambda_k = \sum_{l=0}^{N-1} h(l) W_N{}^{kl}, \ \text{for } k = 0, 1, \ldots, N-1. \tag{B.19}$$

From Eq. (B.19), we see that the eigenvalues of a circulant matrix are the DFT of its first column.

Furthermore, since Eq. (B.14) holds for $k = 0, 1, \ldots, N-1$, we can write

$$H[\phi_0 \quad \phi_1 \quad \cdots \quad \phi_{N-1}] = [\phi_0 \quad \phi_1 \quad \cdots \quad \phi_{N-1}]\Lambda, \tag{B.20}$$

where $\Lambda = \mathrm{diag}\{\lambda_0 \quad \lambda_1 \quad \cdots \quad \lambda_{N-1}\}$. We note that $[\phi_0 \quad \phi_1 \quad \cdots \quad \phi_{N-1}] = F_1^*$. Then expression Eq. (B.20) reduces to

$$HF_1^* = F_1^*\Lambda. \tag{B.21}$$

By multiplying F_1 to the left hand side of Eq. (B.21), we obtain

$$F_1 H F_1^* = \Lambda. \tag{B.22}$$

Equation (B.22) shows that any circulant matrix can be diagonalized by multiplying the DFT matrix and its conjugate on the left and the right-hand sides, respectively.

B.3 Two-Dimensional Discrete Fourier Transform

The two-dimensional DFT of an $N \times N$ sequence $u(m, n)$ is defined as

$$v(k,l) = \frac{1}{N}\sum_{m=0}^{N-1}\sum_{n=0}^{N-1} u(m,n) W_N^{km} W_N^{nl}, \quad \text{for} \quad k,l = 0, 1, \ldots, N-1, \tag{B.23}$$

and the inverse transform is

$$u(m,n) = \frac{1}{N}\sum_{k=0}^{N-1}\sum_{l=0}^{N-1} v(k,l) W_N^{-km} W_N^{-nl}, \quad \text{for} \quad m,n = 0, 1, \ldots, N-1. \tag{B.24}$$

There are two major categories of the two-dimensional DFT applications, such as (a) two-dimensional spatial frequency analysis and filtering and (b) diagonalization of block circulant matrices for efficient computation of two-dimensional convolution.

B.3.1 Basis Images of the Two-Dimensional DFT

In order to analyze spatial frequency characteristics of a $N \times N$ two-dimensional signal, we consider N^2 basis images of the same size $N \times N$. If each basis image contains unique two-dimensional spatial frequency components exclusively, we can compute the desired frequency component in the given image by finding the inner product of the given image and the corresponding basis image. If we form N^2 basis images by taking the outer product of two vectors that are permutations from $\{\phi_k,\ k = 0,\ \ldots,\ N-1\}$, defined in Eq. (B.16), each inner product of the given image and the corresponding basis image is equal to the two-dimensional DFT.

In matrix notation, Eqs. (B.23) and (B.24), respectively, become

$$V = F_1 U F_1 \tag{B.25}$$

and

$$U = F_1^* V F_1^*, \tag{B.26}$$

where U and V represent $N \times N$ matrices whose elements are mapped to two-dimensional signals $u(m,n)$ and $v(k,l)$, respectively.

The two-dimensional DFT basis images are given by

$$B_{kl} = \phi_k {\phi_l}^{\mathrm{T}}, \quad \text{for} \quad k,l = 0,\ \ldots,\ N-1. \tag{B.27}$$

We note that Eq. (B.25) is mathematically equivalent to the set of inner products of the given image and basis images defined in Eq. (B.27).

Given two-dimensional DFT coefficients, which are elements of V, the two-dimensional signal can be reconstructed by summation of all basis images weighted by the given DFT coefficients. This reconstruction process is mathematically equivalent to Eq. (B.26).

B.3.2 Diagonalization of Block Circulant Matrices

Let u and v be the lexicographically ordered vectors for two-dimensional signals $u(m,n)$ and $v(k,l)$, respectively. The two-dimensional DFT matrix is defined as

$$F = F_1 \otimes F_1, \tag{B.28}$$

where $\otimes$ represents the Kronecker product of matrices and F_1 represents the one-dimensional DFT matrix. According to the property of the Kronecker product, we see that the two-dimensional DFT matrix is also symmetric, that is,

$$F^{\mathrm{T}} = (F_1 \otimes F_1)^{\mathrm{T}} = F_1^{\mathrm{T}} \otimes F_1^{\mathrm{T}} = F_1 \otimes F_1 = F. \tag{B.29}$$

The two-dimensional DFT is written in matrix-vector form as

$$v = Fu. \tag{B.30}$$

In order to investigate the orthogonality of the two-dimensional DFT matrix, we have

$$\begin{aligned} F^{*\mathrm{T}}F = F^*F = (F_1 \otimes F_1)^*(F_1 \otimes F_1) = \left(F_1^* \otimes F_1^*\right)(F_1 \otimes F_1) \\ = \left(F_1^*F_1 \otimes F_1^*F_1\right) = I_N \otimes I_N = I_{N^2 \times N^2}. \end{aligned} \tag{B.31}$$

From Eq. (B.31), we know that

$$F^{*\mathrm{T}} = F^* = F^{-1}, \tag{B.32}$$

which yields the following inverse transform

$$u = F^*v. \tag{B.33}$$

We note that the matrix-vector notation of the two-dimensional DFT in Eq. (B.30) is equivalent to both Eqs. (B.23) and (B.25) and that Eq. (B.33) is equivalent to both Eqs. (B.24) and (B.26).

Consider the two-dimensional circular convolution

$$y(m,n) = \sum_{p=0}^{N-1}\sum_{q=0}^{N-1} h(m-p, n-q)_C x(p,q), \tag{B.34}$$

where

$$h(m,n)_C = h(m \bmod N, n \bmod N). \tag{B.35}$$

The reason for using the circular convolution in image processing is to deal with boundary problems. In other words, when processing an image with a convolution kernel, boundary pixels do not have a sufficient number of neighboring pixels. In order to compensate for this pixel shortage, we may use one of the following: (1) assign zero values to nonexistent neighbors, or (2) replicate the value of the outermost existing pixel to its nonexistent neighbors, or (3) suppose that the input image is periodic with a period that is the same as the size of the image.

Although none of the three methods give us the ideal solution for boundary problems, the third method that assumes two-dimensional periodicity allows us to use circular convolution, which can be diagonalized by using a two-dimensional DFT.

Given (p,q), we can obtain the two-dimensional DFT of $h(m-p,n-q)_C$ as

$$\mathrm{DFT}\{h(m-p,n-q)_C\} = \sum_{m=0}^{N-1}\sum_{n=0}^{N-1} h(m-p,n-q)_C {W_N}^{mk+nl}. \tag{B.36}$$

Writing $i = m-p$, $j = n-q$, and using Eq. (B.35), we can rewrite Eq. (B.36) as

$$\begin{aligned} {W_N}^{pk+ql} \sum_{i=-p}^{N-1-p} \sum_{i=-q}^{N-1-q} h(i,j)_C {W_N}^{ik+jl} &= {W_N}^{pk+ql} \sum_{m=0}^{N-1}\sum_{n=0}^{N-1} h(m,n) {W_N}^{mk+nl} \\ &= {W_N}^{pk+ql} \mathrm{DFT}\{h(m,n)\}. \end{aligned} \tag{B.37}$$

Since Eq. (B.36) holds for $p,q = 0,\ldots,N-1$, the right side of Eq. (B.36) can be expressed as

$$(F_1 \otimes F_1) h_{p,q}, \tag{B.38}$$

where $h_{p,q}$, represents the $N^2 \times 1$ row-ordered vector obtained from the rotated version of the $N \times N$ matrix $\{h(m,n)\}$ by (p,q). By equating Eqs. (B.38) and (B.37), for $p,q = 0,\ldots,N-1$, we obtain

$$(F_1 \otimes F_1) H = D(F_1 \otimes F_1), \tag{B.39}$$

where H represents the $N^2 \times N^2$ doubly block circulant matrix, and D the diagonal matrix whose N^2 diagonal elements are equal to the two-dimensional DFT of $h(m,n)$.

Equation (B.39) can be rewritten as

$$FH = DF \;\text{ or }\; FHF^{-1} = FHF^* = D, \tag{B.40}$$

which shows that a doubly block circulant matrix is diagonalized by the two-dimensional unitary DFT.

Furthermore, if a double block circulant matrix is used to represent the two-dimensional circular convolution, such as

$$y = Hx, \tag{B.41}$$

then the two-dimensional DFT of the output can be obtained by multiplying the DFT matrix as

$$Fy = FHx = FHF^* Fx = D \cdot Fx. \tag{B.42}$$

Equation (B.42) can be rewritten as

$$\mathrm{DFT}\{y(m,n)\} = \mathrm{DFT}\{h(m,n)\} \cdot \mathrm{DFT}\{x(m,n)\}. \tag{B.43}$$

After having the DFT of the output, its inverse transform can easily be obtained by multiplying the conjugate of the two-dimensional DFT matrix.

Appendix C
3D Data Acquisition and Geometric Surface Reconstruction

C.1 Introduction

The first step in building the 3D model of a real scene is the acquisition of raw data. For this purpose, a common approach needs to acquire depth information from a given point of view. Two major depth acquisition methods include:

- Stereovision, which uses classic photography from two or more viewpoints in order to retrieve the third dimension and build depth maps [ayache97, faugeras93, horn86].
- Direct range acquisition, which directly uses range finding devices, also called 3D scanners. The time-of-flight and active laser triangulation methods fall into this category.

For the experimental input data, a time-of-flight laser range finder (LRF), Perceptron LASAR P5000 [dorum95, perceptron93], was used. The depth is retrieved by computing the phase shift between an outgoing laser beam and its returned (bounced back) signal. This kind of imaging is also known as light amplitude detection and ranging (LADAR). Perceptron is an azimuth-elevation scanner, which uses two kinds of rotating mirrors as shown in Fig. C.1. A faceted mirror controls the horizontal displacement of the laser beam and a second planar mirror controls the vertical deflection. The LRF is then able to scan a scene *point by point*, generating a range map in which a pixel represents the distance value of a scene point from the scanner. The scanner has some limitations like a maximum distance beyond which it could differentiate the range and the horizontal and vertical fields of view.

In addition to the range image, the LRF outputs a reflectance image based on the intensity of the returned laser beam as shown in Fig. C.2. This image is perfectly registered with the range image with pixel by pixel correspondence and will be useful later in the registration of LADAR and color data.

M.A. Abidi et al., *Optimization Techniques in Computer Vision*, Advances in Computer Vision and Pattern Recognition, DOI 10.1007/978-3-319-46364-3

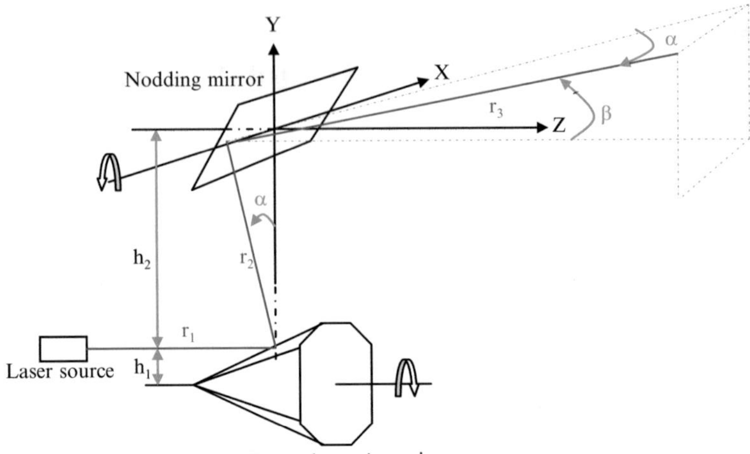

Each pixel in a Perceptron range image, denoted by $r(i,j)$, $0 \leq i < R$ (rows) and $0 \leq j < C$ (columns), is converted to Cartesian coordinates as follows:

$$x(i,j) = dx + r_3.\sin\alpha$$

$$y(i,j) = dy + r_3.\cos\alpha.\sin\beta$$

$$z(i,j) = dz + r_3.\cos\alpha.\cos\beta$$

$$r_1 = (dx - h_2)/\delta$$

$$r_2 = \left(\sqrt{(dx)^2 + (h_2 + dy)^2}\right)/\delta$$

$$r_3 = (r(i,j) + r_0 - (r_1 + r_2)).\delta$$

$$\beta = \frac{(\frac{R-1}{2} - i)}{R}.V$$

$$\alpha = \frac{(\frac{C-1}{2} - j)}{C}.H$$

$$dz = (h_2 + dy).\tan\alpha$$

$$dy = dz.\tan(\theta + \frac{\beta}{2})$$

$$dz = -h_1.(1 - \cos\alpha)/\tan\gamma$$

H and V respectively represent the horizontal and vertical fields of view; r_0 the standoff distance (length of the laser beam at the point where $r = 0$), γ the slope of the facets of the rotating mirror with respect to the z-axis, and θ the angle of the nodding mirror with respect to the z-axis when $\beta = 0$.

Fig. C.1 Principle of 3D image acquisition using Perceptron [dorum95, perceptron93]

It is important to notice that the range map is a partial 3D representation since it allows only for surface reconstruction from a given point of view. In order to build a more complete representation of an object, we need more than one range image; thus, this kind of data is known as $2\frac{1}{2}$-dimensional data.

C.2 From Image Pixels to 3D Points

To reconstruct the 3D Cartesian coordinates of the scanned scene points from the range image, we use the Perceptron LRF imaging model [dorum95], which includes the different parameters allowing the retrieval of scene points position in the LRF coordinates frame. Figure C.1 shows the coordinate systems attached to the scanner

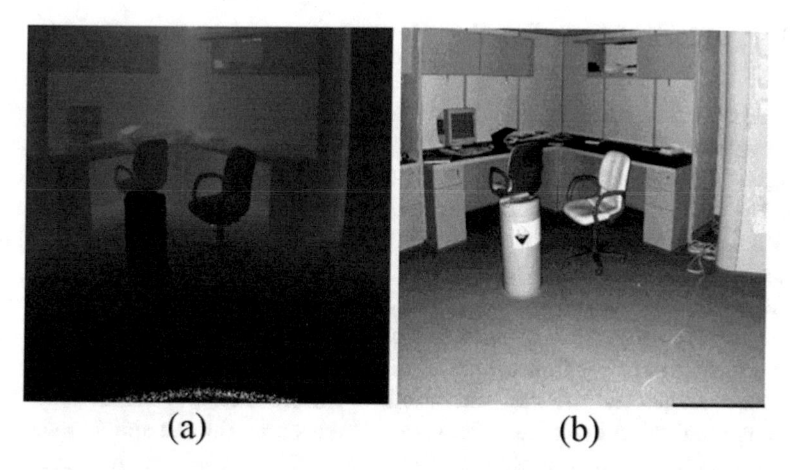

Fig. C.2 Two perfectly registered outputs of Perceptron: (**a**) range and (**b**) reflectance images

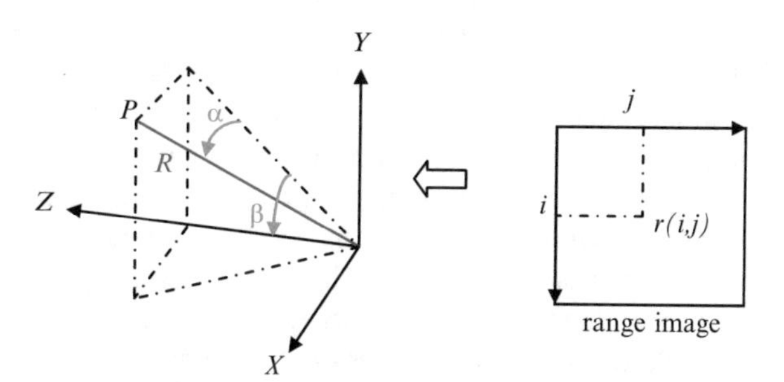

Fig. C.3 Spherical model for the reconstruction of 3D points from a range image

and the different model parameters with the equations allowing for the calculation of the x, y, and z coordinates of a given scene point.

A simple spherical model is shown in Fig. C.3, based on approximation of the complete model. A pixel at the (i,j)-th position in a range image has as intensity value $r(i,j)$. The coordinates of a point are calculated from the pixel value as

$$R = r_0 + \frac{r(i,j)}{\delta}, \tag{C.1}$$

where r_0 represents a standoff distance, or an *offset*, and δ represents the range resolution. Both r_0 and δ can be obtained through calibration. The azimuth and elevation angles α and β are given by

$$\alpha = \frac{\left(\frac{C-1}{2} - j\right)}{C} \cdot H \quad \text{and} \quad \beta = \frac{\left(\frac{R-1}{2} - i\right)}{R} \cdot V, \tag{C.2}$$

where H and V, respectively, represent the horizontal and vertical fields of view of the same scanner. The x, y, and z coordinates are finally calculated as

$$x(i,j) = R\sin\alpha, y(i,j) = R\cos\alpha\sin\beta, \text{and}\, z(i,j) = R\cos\alpha\cos\beta. \tag{C.3}$$

C.3 Surface Reconstruction

The purpose of surface reconstruction is to fit surfaces to the 3D point's cloud reconstructed from the range image. The visualization standards require the use of polygons as basic elements composing the object surface. The most simple and common polygon used is the triangle; thus, the process of creation of surface model is also called triangulation. Different techniques were used for triangulation [elhakim98, sequira99].

Figure C.4 shows a simple approach of creating a triangle mesh from a range image, which is considered as a 2D grid. The first step is the creation of a rectangular grid. Every four neighbors in the grid correspond to four neighbors in the range image. The quadratic mesh is then transformed into a triangle mesh by dividing the different rectangles into two triangles according to simple rules.

The resulting triangle mesh built from range images could be visualized using different software standards, such as VRML and OpenInventor of SGI.

Some other hardware specifications could be required to get optimal visualization performances, particularly in the case of large models, such as those we are dealing with in this work. The number of model triangles is $(C - 1) \times (R - 1)$; for example, a model built from a 1000×1000 range image contains around two million triangles. Hence, it is, in many cases, suitable to reduce the number of triangles in the mesh [gourley98]. This is done in such a way as to keep a maximum number of triangles in the scene areas with a high level of detail and a minimum number of triangles in areas with large flat surfaces.

Other steps in model building from range images include 3D segmentation, smoothing, and other preprocessing. In addition to these, the 3D models are usually rendered with textures on top of them. The 3D visualization engine also requires the calculation of surface or vertex normals, as shown in Fig. C.5, in order to determine the shading, using the lighting and camera models in a given viewpoint.

Figure C.6 shows an example of a model reconstructed from range images. The model is rendered using OpenInventor, and we can see the rendered model with a uniform texture and also the triangle mesh forming the skeleton of the scene model.

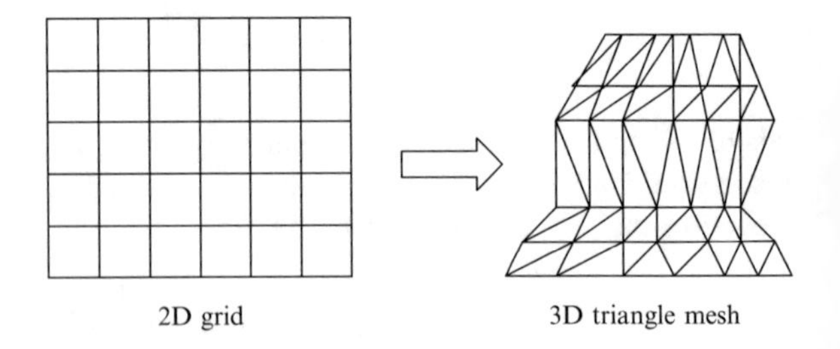

Fig. C.4 3D triangle mesh reconstructed from a 2D grid representing range data

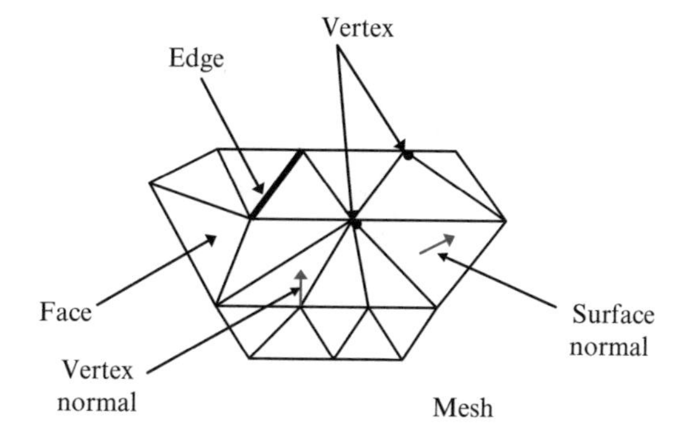

Fig. C.5 Terminology used for the different features of a triangle mesh

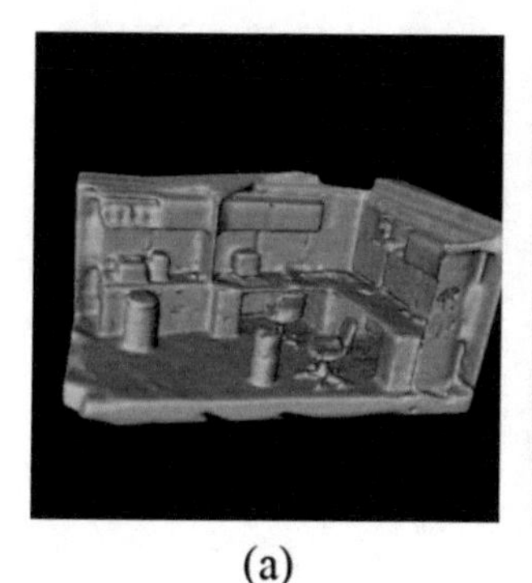

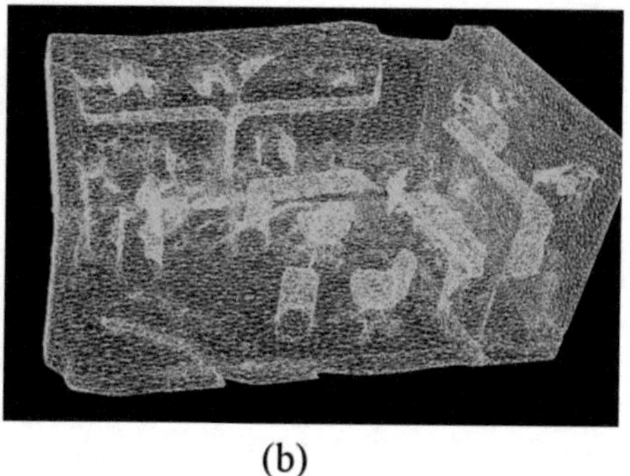

(a) (b)

Fig. C.6 A 3D model built from the previous range image with other views: (**a**) the model is rendered with a uniform texture and (**b**) a rendering of the wireframe

References

[ayache97] N. Ayache, *Artificial Vision for Mobile Robots: Stereo-Vision and Multi-Sensory Perception* (MIT Press, Cambridge, MA, 1997)

[dorum95] O.H. Dörum, A. Hoover, J.P. Jones, Calibration and Control for Range Imaging in Mobile Robot Navigation, in *Research in Computer and Robot Vision*, ed. by C. Archibald, P. Kwok (World Scientific, Singapore, 1995), pp. 1–18

[elhakim98] S.F. El-Hakim, C. Brenner, G. Roth, An Approach to Creating Virtual Environments using Range and Texture. ISPRS J. Photogramm. Remote Sens. **53**, 379–391 (1998)

[faugeras93] O. Faugeras, *Three-Dimensional Computer Vision* (MIT Press, Cambridge, MA, 1993)

[gourley98] C. Gourley, Pattern vector based reduction of large multi-modal datasets for fixed rate interactivity during the visualization of multi-resolution models, Ph.D. Thesis, University of Tennessee, Knoxville, 1998

[horn86] B.K.P. Horn, *Robot Vision* (MIT Press, Cambridge, MA, 1986)

[perceptron93] Perceptron Inc, *LASAR Hardware Manual*, 23855 Research Drive, Farmington Hills, Michigan 48335, 1993.

[sequira99] V. Sequira, E. Wolfart, J.G.M. Gonclaves, D. Hogg, Automated Reconstruction of 3D Models from Real Environments. ISPRS J. Photogramm. Remote Sens. **54**, 1–22 (1999)

Appendix D
Mathematical Appendix

D.1 Functional Analysis

D.1.1 Real Linear Vector Spaces

Real linear vector space is a set V of elements (objects) x, y, $z\ldots$ for which operations of addition and multiplication by real numbers are defined satisfying the following nine axioms:

1. $x + y \in V$.
2. $\alpha \cdot x \in V, \alpha \in R$.
3. $x + y = y + x$.
4. $x + (y + z) = (x + y) + z$.
5. $x + y = x + z \quad \text{iff} \quad y = z$.
6. $\alpha(x + y) = \alpha \cdot x + \alpha \cdot y$.
7. $(\alpha + \beta)x = \alpha \cdot x + \beta \cdot x, \beta \in R$.
8. $\alpha \cdot (\beta \cdot x) = (\alpha \cdot \beta) \cdot x$.
9. $1 \cdot x = x \quad \text{where} \quad \alpha, \beta \in R$.

Examples

1. The real numbers themselves with the ordinary operations of arithmetic $-R$.
2. The set of ordered real N-tuples $(x_1, x_2, \ldots, x_N)$, or N-dimensional vectors $-R^N$.
3. The set of all functions continuously differentiable to order n on the real interval $[a, b] - C^n[a, b]$.

M.A. Abidi et al., *Optimization Techniques in Computer Vision*, Advances in Computer Vision and Pattern Recognition, DOI 10.1007/978-3-319-46364-3

D.1.2 Normed Vector Spaces

Real linear vector space equipped with the measure of the size of its elements which satisfies the conditions:

1. $\|x\| = 0$ iff $x = 0$
2. $\|\alpha \cdot x\| = |\alpha| \cdot \|x\|$
3. $\|x + y\| \leq \|x\| + \|y\|$

is called normed vector space. A norm of an element is a real number.

Examples

1. Absolute value of a real number $|x|$ is a norm.
2. Euclidean norm of a vector $\|x\|_2 = \sqrt{\left(x_1^2 + x_2^2 + \ldots + x_N^2\right)}$.
3. Functional norms for continuously differentiable functions $x(t)$, $a \leq t \leq b$:

   (a) $\|x\|_\infty = \sup\limits_{a \leq t \leq b} |x(t)|$.

   (b) $\|x(t)\|_1 = \int\limits_a^b |x(t)| dt$.

   (c) $\|x(t)\|_2 = \left\{ \int\limits_a^b |x(t)|^2 dt \right\}^{1/2}$.

   (d) $\|x(t)\|_p = \left\{ \int\limits_a^b |x(t)|^p dt \right\}^{1/p}$.

D.1.3 Convergence, Cauchy Sequences, Completeness

An infinite sequence of elements x_1, x_2, x_3,... in a normed vector space is said to converge to an element y if as $k \to \infty$, $\|x_k - y\| \to 0$. As the sequence converges, elements of the sequence tend to get closer and closer.

$$\|x_n - x_m\| \to 0 \text{ as } n, m \to \infty. \tag{D.1}$$

Such sequences are called Cauchy sequences. Every convergent sequence is Cauchy. If every Cauchy sequence converges to a limit that is an element of the original vector space, then such space is called complete. Not every normed linear space is complete. Complete normed linear spaces are called Banach spaces. The sequence may be convergent under one norm but not under another; hence, a normed vector space may be complete under one norm but not under another. Convergence under norm a. is the most important and is called uniform convergence.

D.1.4 Euclidean Spaces and Hilbert Spaces

As a norm is an abstraction of the size of an element in a real linear vector space, the inner (scalar, dot) product is a generalization of the angle between two elements in a real linear vector space. The inner product is a rule $(\cdot)$ which assigns to any elements of a real linear vector space a real number. This rule should have the following four properties:

1. $(x, y) = (y, x)$.
2. $(\alpha \cdot x, y) = \alpha \cdot (x, y)$.
3. $(x + y, z) = (x, z) + (y, z)$.
4. $(x, x) > 0$, if $x \neq 0$.

The inner product induces a natural norm

$$\|x\| = (x, x)^{1/2}. \tag{D.2}$$

A linear vector space equipped with an inner product and an induced norm is called Euclidean space or pre-Hilbert. A Euclidean space which is complete under an induced inner product norm is called Hilbert space.

Examples

1. Vector scalar product in R^N, $(x, y) = \sum_{i=1}^{N} x_i \cdot y_i$ with the induced norm

$$\|x\| = (x, x)^{1/2} = \left[\sum_{i=1}^{N} x_i^2\right]^{1/2}.$$

2. Scalar product in $C^n\ [a, b]$, $(x(t), y(t)) = \int_a^b x(t) \cdot y(t) dt$ with the induced norm

$$\|x(t)\| = \left[\int_a^b [x(t)]^2\right]^{1/2}.$$

D.1.5 Approximations in Hilbert Spaces, Fourier Series

Two vectors x, y in a Hilbert space are said to be orthogonal if $(x, y) = 0$. A system of elements $e_1, e_2, \ldots, e_N \ldots$ is called an orthonormal system of elements in a Hilbert space if

$$(e_i, e_j) = \begin{cases} 1 \text{ if } i = j \\ 0 \text{ if } i \neq j \end{cases}. \tag{D.3}$$

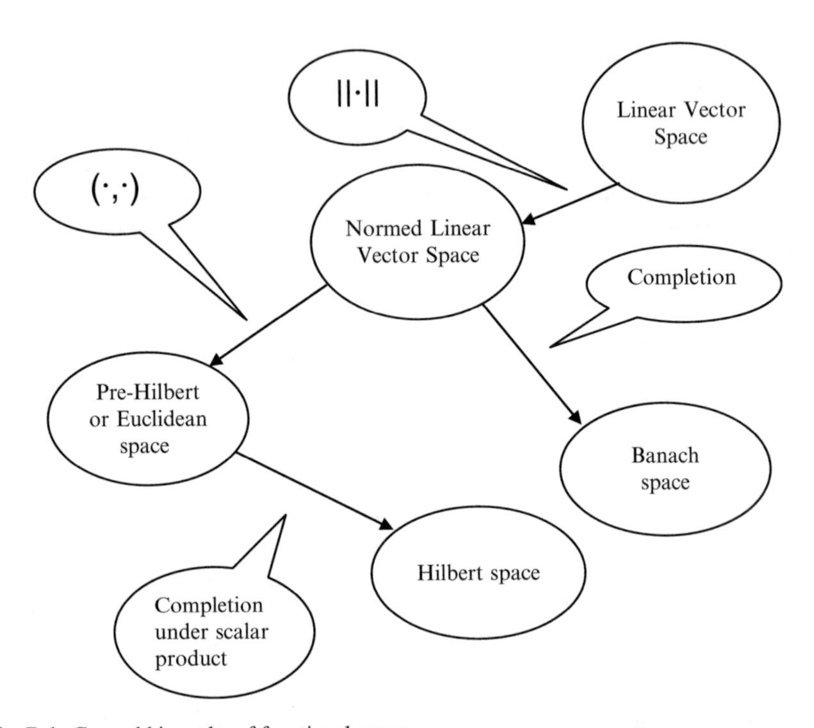

Fig. D.1 General hierarchy of functional spaces

An element x of a Hilbert space can be represented as a linear combination of e_i as $x = \sum_{i=1}^{\infty} (x, e_i) \cdot e_i$, where (x, e_i) are called Fourier coefficients of x and the series itself is called the Fourier expansion of x.

The general hierarchy of functional spaces is presented in Fig. D.1.

D.1.6 Operators and Their Norms

Let V and U be two vector spaces. Then transformation A (mapping, rule) which assigns to each element in V a unique element in $U, A : V \rightarrow U$ is called an operator. An operator A is called linear if

$$A(\alpha \cdot x + \beta \cdot y) = \alpha \cdot A \cdot x + \beta \cdot A \cdot y, \tag{D.4}$$

for all x and $y \in V$ and for all real scalars α and β.

Examples

1. Differential operators, $\frac{d}{dt}x(t), \frac{d^2}{dt^2}x(t), \ldots, \frac{d^n}{dt^n}x(t)$.
2. Indefinite integral operator $\int x(t)dt$.
3. Convolution or Fredholm integral operator $y(t) = \int_a^b K(t,\tau) \cdot x(\tau)d\tau$.

A type 3 operator is of the utmost importance in image processing.

Normally, if a mapping carries a function to a function, it is called an operator; if it carries a function to a number, it is called functional; and if it carries a number to a number, it is called a function.

Operator A is called bounded if there is a constant K such that

$$||Ax|| \leq K||x||, x \in V. \tag{D.5}$$

The norm of the operator is then defined as

$$||A|| = \sup_{x \neq 0} \frac{||Ax||}{||x||}. \tag{D.6}$$

D.1.7 Vector and Matrix Norms

Let x be an N-dimensional vector $x = (x_1, x_2, \ldots, x_N)$. Then the following norms can be defined:

1. $||x||_1 = \sum_{i=1}^{N} |x_i|$.
2. $||x||_2 = \left[\sum_{i=1}^{N} x_i^2\right]^{1/2}$.
3. $||x||_p = \left[\sum_{i=1}^{N} x_i^p\right]^{1/p}$.
4. $||x||_W = (x \cdot W \cdot x)^{1/2}$.
5. $||x||_\infty = \max_i |x_i|$.

Let A be an $(m \times n)$ matrix. Then following norms can be defined:

1. $||A||_1 = \max_j \left[\sum_{i=1}^{m} |a_{i,j}|\right]$.
2. $||A||_\infty = \max_i \left[\sum_{j=1}^{n} |a_{i,j}|\right]$.

3. $\|A\|_2 = \left[\max \text{eigenvalue of } \left(A^T A\right)\right]^{1/2}$—spectral norm.
4. $\|A\|_F = \left[\sum_{i=1}^{m}\sum_{j=1}^{n} a_{i,j}^2\right]^{1/2}$—Frobenius norm.
5. $\|A\|_E = \max_{\|x\|=1}\left[\frac{\|A\cdot x\|}{\|x\|}\right]$—Euclidean norm.

D.1.8 SVD, Eigenvalues and Eigenvectors, Condition Number

Every $(m \times n)$ matrix A can be factored into $A = U \cdot \Sigma \cdot V^T$, where $U^T \cdot U = V^T \cdot V = I$ and $\Sigma = \text{diag}(\sigma_1, \sigma_2, \ldots, \sigma_n)$.

$$A = \underbrace{\begin{bmatrix} u_1, u_2, \ldots, u_m \end{bmatrix}}_{m\times m} \cdot \underbrace{\begin{bmatrix} \sigma_1 & & \\ & \ddots & \\ & & \sigma_n \end{bmatrix}}_{m\times n} \cdot \underbrace{\begin{bmatrix} v_1, v_2, \ldots, v_n \end{bmatrix}}_{n\times n}^T. \tag{D.7}$$

Columns of matrices U and V are called left and right singular vectors of A, and vector $(\sigma_1, \sigma_2, \ldots, \sigma_n)$ is called a vector of singular values which appear in nonincreasing order. The condition number of a matrix A is with respect to a norm is defined as

$$\text{cond}(A) = \|A\| \cdot \left\|A^{-1}\right\|. \tag{D.8}$$

If a spectral norm is used, then the condition number is given by a ratio

$$\text{cond}(A) = \frac{\sigma_{\max}}{\sigma_{\min}}, \tag{D.9}$$

where $\sigma_{\max}$ and $\sigma_{\min}$ are maximum and minimum singular values of matrix A.

For a nonsingular system of linear equations $Ax = b$, the condition number defines a possible relative change Δx in the solution x due to a given relative change Δb in the right-hand side vector b, namely,

$$\frac{\|\Delta x\|}{\|x\|} \le \text{cond}(A) \cdot \frac{\|\Delta b\|}{\|b\|}. \tag{D.10}$$

Hence, if cond(A) is large then the relative change in the solution could be very large even for small changes in the right-hand side of the equation.

D.2 Matrix Algebra and the Kronecker Product

For the matrices A, B, and C, the following theorems are valid:

1. $A+B=B+A$.
2. $(A+B)+C=A+(B+C)$.
3. $(AB)C=A(BC)$.
4. $C(A+B)=CA+CB$.
5. $\alpha(A+B)=\alpha A+\alpha B$, $\alpha\in R$.
6. $(A^{\mathrm{T}})^{\mathrm{T}}=A$.
7. $(A+B)^{\mathrm{T}}=A^{\mathrm{T}}+B^{\mathrm{T}}$.
8. $(AB)^{\mathrm{T}}=B^{\mathrm{T}}A^{\mathrm{T}}$.
9. $(ABC)^{\mathrm{T}}=C^{\mathrm{T}}B^{\mathrm{T}}A^{\mathrm{T}}$.
10. $(AB)^{-1}=B^{-1}A^{-1}$.
11. $(ABC)^{-1}=C^{-1}B^{-1}A^{-1}$.
12. $(A^{-1})^{-1}=A$.
13. $(A^{\mathrm{T}})^{-1}=(A^{-1})^{\mathrm{T}}$.

The Kronecker product of two matrices A, $(m\times n)$, and B, $(k\times l)$, is the block matrix

$$A\otimes B=\begin{bmatrix} a_{11}\cdot B & \cdots & a_{1n}\cdot B \\ \vdots & \ddots & \vdots \\ a_{m1}\cdot B & \cdots & a_{mn}\cdot B \end{bmatrix}, (mk\times ml), \tag{D.11}$$

with the following properties:

1. $(A\otimes B)^{\mathrm{T}}=A^{\mathrm{T}}\otimes B^{\mathrm{T}}$.
2. $(A\otimes B)^{-1}=A^{-1}\otimes B^{-1}$.
3. $(A\otimes B)\otimes C=A\otimes(B\otimes C)$.
4. $(A+B)\otimes C=A\otimes C+B\otimes C$.
5. $(A\otimes B)=(U_A\otimes U_B)\cdot(\Sigma_A\otimes\Sigma_B)\cdot(V_A\otimes V_B)^{\mathrm{T}}$.

Operation vec(A) is defined as

$$\mathrm{vec}(A)=[a_{11},\ldots,a_{m1},a_{12},\ldots,a_{m2},\ldots,a_{1n},\ldots,a_{mn}]^{\mathrm{T}}, \tag{D.12}$$

where vec(A) is an $(mn\times 1)$ vector—the stacked columns of A. Let

$$A\cdot X=B \tag{D.13}$$

represent a matrix equation with A, $(m\times n)$; X, $(n\times k)$; and B, $(m\times k)$. Then Eq. (*D.13*) can be written in equivalent form as

$$(I\otimes A)\cdot\mathrm{vec}(X)=\mathrm{vec}(B), \tag{D.14}$$

and the solution to this equation can be expressed as

$$\operatorname{vec}(X) = (I \otimes A)^{-1} \cdot \operatorname{vec}(B). \tag{D.15}$$

The following important relationship exists between the Kronecker product and the operation vec:

$$(A \otimes B) \cdot \operatorname{vec}(X) = \operatorname{vec}\left(BXA^{\mathrm{T}}\right). \tag{D.16}$$

This relationship is extensively used in image processing.

The convolution of an image with separable point spread function can be represented as

$$(K_1 \otimes K_2) \cdot \operatorname{vec}(X) = \operatorname{vec}(B), \tag{D.17}$$

where X is the true image, K_1 and K_2 are separable kernels, and B is the blurred image.

D.2.1 Derivatives and Trace

Let $x = (x_1, x_2, \ldots, x_n)$ and $y = (y_1, y_2, \ldots, y_n)$ be two vectors and A an $(m \times n)$ matrix. Then

1. $\frac{\partial}{\partial x}(x^{\mathrm{T}}y) = y$.
2. $\frac{\partial}{\partial x}(x^{\mathrm{T}}x) = 2x$.
3. $\frac{\partial}{\partial x}(x^{\mathrm{T}}Ay) = Ay$.
4. $\frac{\partial}{\partial x}(y^{\mathrm{T}}Ax) = A^{\mathrm{T}}y$.
5. $\frac{\partial}{\partial x}(x^{\mathrm{T}}Ax) = \left(A + A^{\mathrm{T}}\right)x$.
6. $\frac{\partial}{\partial A}(x^{\mathrm{T}}Ay) = xy^{\mathrm{T}}$.
7. $\frac{\partial}{\partial A}(x^{\mathrm{T}}Ax) = 2xx^{\mathrm{T}} - \operatorname{diag}(xx^{\mathrm{T}})$.

Assume A and B are $(n \times n)$ matrices, and α and β are two scalars. Then

$$\operatorname{trace}(A) = \sum_{i=1}^{n} a_{ii} \tag{D.18}$$

with the following properties:

1. $\operatorname{tr}(AB) = \operatorname{tr}(BA)$.
2. $\operatorname{tr}(A) = \sum_{i=1}^{n} \sigma_i$.

3. $\mathrm{tr}(\alpha A + \beta B) = \alpha \mathrm{tr}(A) + \beta \mathrm{tr}(B)$.
4. $\mathrm{tr}\left(A^{\mathrm{T}}\right) = \mathrm{tr}(A)$.

D.3 Probability and Statistics

The random variable is a real-valued function which is defined on the space of random events (outcomes of experiments). The same space of random events may define an infinite number of random variables.

Examples

1. In the die experiment, we can assign the six outcomes their corresponding values, thus $f(1)=1, f(2)=2,\ldots, f(6)=6$.
2. In the same experiment we can assign number $10 \cdot i$ to the i-th outcome; thus $f(1)=10, f(2)=20,\ldots, f(6)=60$.
3. Finally, in the same experiment we can assign number 1 to every even outcome and 0 to every odd outcome; thus $f(1)=f(3)=f(5)=0$ and $f(2)=f(4)=f(6)=1$.

All these three random variables are defined on the same sample space.

A random variable can assume discrete, continuous, or both values. The probability distribution function of a random variable x is the function defined for every X such that

$$F_x(X) = P(x \leq X). \tag{D.19}$$

The derivative of a probability distribution function is called a probability density function and is denoted by

$$f(x) = \frac{dF(x)}{dx}. \tag{D.20}$$

The Gaussian or normal probability density function can be written as

$$f(x) = \frac{1}{\sigma\sqrt{2\pi}} \mathrm{e}^{\frac{-(x-\mu)^2}{2\sigma^2}}, \tag{D.21}$$

where σ and μ are two parameters which define the shape of the distribution.

The expected or mean value of a continuous random variable x is by definition

$$E(x) = \int_{-\infty}^{\infty} x f(x) dx. \tag{D.22}$$

Sometimes this value is called mathematical expectation. It has the following important properties:

1. $E(c) = c$, where c is constant.
2. $E(cx) = cE(x)$.
3. $E(x + y) = E(x) + E(y)$.
4. $|E(x)| \leq E(|x|)$.

Another very important numerical characteristic of a random variable is variance which in a continuous case is defined as

$$\mathrm{Var}(x) = \int_{-\infty}^{\infty} (x - E(x))^2 f(x) dx, \tag{D.23}$$

with the following properties:

1. $\mathrm{Var}(c) = 0$.
2. $\mathrm{Var}(cx) = c^2 \mathrm{Var}(x)$.
3. $\mathrm{Var}(x + y) = \mathrm{Var}(x) + \mathrm{Var}(y)$ if x and y are independent random variables.

Two random variables are called independent if their joint probability density function can be factored into their individual probability densities as in

$$f_{x,y}(X, Y) = f_x(X) f_y(Y). \tag{D.24}$$

By definition, probability theory is concerned with the statements that can be made about a random variable if its probability density function or probability distribution function is known.

Theory of estimation or statistical theory is concerned with the statements that can be made about a probability density function with only a limited amount of samples drawn from that probability density function. If the general form of underlying distribution is known, and the problem is to evaluate parameters from the available data, then the problem of estimation is called parametric. Otherwise, the estimation problem is called nonparametric. Having a set of experimental observations $x_1, x_2, \ldots, x_n$ drawn from an underlying probability density function $f(x|\theta)$, the goal of statistical inference is to obtain a reliable estimate $\hat{\theta}$ of the parameter θ. Any function of observations $f(x_1, x_2, \ldots, x_n)$ is called statistics or estimator.

An estimator $\hat{\theta}$ of the parameter θ is called unbiased if

$$E\left[\hat{\theta}\right] = \theta; \tag{D.25}$$

that is, the mathematical expectation of $\hat{\theta}$ equals the true parameter θ.

An estimator $\hat{\theta}$ of the parameter θ is called consistent if

$$\lim_{n \to \infty} P\left(\left|\hat{\theta} - \theta\right| > \varepsilon\right) = 0. \tag{D.26}$$

In other words an estimator is converging in probability to the true parameter as the number of samples grows.

An estimator $\hat{\theta}$ of the parameter θ is called efficient if it has the smallest variance within a class of estimators

$$\mathrm{Var}\left(\hat{\theta}\right) \leq \mathrm{Var}\left(\tilde{\theta}\right). \tag{D.27}$$

Let $x_1, x_2, \ldots, x_n$ be a sample of independent observations drawn from an unknown parametric family of probability density functions $f(x|\theta)$. Then by definition the likelihood function is

$$L(x_1, x_2, \ldots, x_n|\theta) = f(x_1|\theta) f(x_2|\theta) \ldots f(x_n|\theta). \tag{D.28}$$

The likelihood function represents an "inverse" probability since it is a function of the parameter.

The most important principle of statistical inference is the maximum likelihood principle which can be stated as follows: Find an estimate for θ such that it maximizes the likelihood of observing the data that actually had been observed. Or, in other words, find the parameter values that make the observed data most likely.

D.3.1 Maximum Likelihood and Least Squares

Suppose we have a set of measurements $(y_1,x_1),(y_2,x_2),\ldots,(y_n, x_n)$, and we believe that the independent variable x and dependent variable y are linked through the following linear relationships:

$$y = xb + \varepsilon, \tag{D.29}$$

with $\varepsilon \sim N(0, \sigma)$ being normally distributed noise. The functional relationship can be written as

$$\varepsilon = y - xb. \tag{D.30}$$

The functional form of the relation is known; however, the parameter b needs to be estimated from the given data sample. The likelihood function for the observed error terms is

$$L(\varepsilon_1, \varepsilon_2, \ldots, \varepsilon_n|b, \sigma^2) = \prod_{i=1}^{n} \frac{1}{\sigma\sqrt{2\pi}} \mathrm{e}^{\frac{-\varepsilon_i^2}{2\sigma^2}} = \prod_{i=1}^{n} \frac{1}{\sigma\sqrt{2\pi}} \mathrm{e}^{\frac{-(y_i - x_i b)^2}{2\sigma^2}}. \tag{D.31}$$

It is computationally more convenient to work with sums rather than with products, and, since logarithm is a monotonic transformation and densities are nonnegative, the $\log\left[L\left(\varepsilon_1, \varepsilon_2, \ldots, \varepsilon_n \middle| b, \sigma^2\right)\right]$ is usually considered and maximized. Taking the logarithm of the likelihood function and maximizing it, we obtain

$$\max_b \left(\ln\left(L\left(\varepsilon_1, \varepsilon_2, \ldots, \varepsilon_n \middle| b, \sigma^2\right)\right)\right) = -n\ln\sigma - \frac{n}{2}\ln 2\pi - \frac{1}{2\sigma^2}\sum_{i=1}^{n}(y - x_i b)^2. \tag{D.32}$$

Since the first two terms are not dependent on the parameter of maximization b, the last expression amounts to maximization of

$$\max\left(-\frac{1}{2\sigma^2}\sum_{i=1}^{n}(y - x_i b)^2\right) \tag{D.33}$$

or

$$\min\left(\frac{1}{2\sigma^2}\sum_{i=1}^{n}(y - x_i b)^2\right), \tag{D.34}$$

which is exactly the least squares cost function. Thus, the least squares solution to a linear estimation problem is equivalent to the maximum likelihood solution under the assumption of Gaussian uncorrelated noise distribution.

The variance of maximum likelihood estimator

$$\mathrm{var}\left(\hat{\theta}\right) \sim \left(E\left[-\left(\frac{\partial L\left(\hat{\theta}\right)}{\partial \hat{\theta}}\right)^2\right]\right)^{-1}, \tag{D.35}$$

where matrix

$$J = E\left[-\left(\frac{\partial L\left(\hat{\theta}\right)}{\partial \hat{\theta}}\right)^2\right] \tag{D.36}$$

is called the Fisher information matrix. The variance of any unbiased estimator is lower bounded as

$$\mathrm{var}\left(\hat{\theta}\right) \geq J^{-1}, \tag{D.37}$$

which is the celebrated Cramer-Rao inequality.

D.3.2 Bias-Variance

Though the maximum likelihood estimator has the smallest variance among unbiased estimators, it is not an estimator with the smallest variance in general. If we are willing to introduce bias into an estimator, then in general we can obtain an estimator with a much smaller variance then the maximum likelihood estimator. This is called the bias-variance decomposition of the mean squared error and can be derived as follows:

$$\begin{aligned} E_D\left[\hat{\theta} - \theta\right]^2 &= E_D\left[\hat{\theta} - E_D(\hat{\theta}) + E_D(\hat{\theta}) - \theta\right]^2 \\ &= E_D\left[(\hat{\theta} - E_D(\hat{\theta}))^2 + (E_D(\hat{\theta}) - \theta)^2 + 2(\hat{\theta} - E_D(\hat{\theta}))(E_D(\hat{\theta}) - \theta)\right] \\ &= E_D\left[(\hat{\theta} - E_D(\hat{\theta}))^2\right] + E_D\left[(E_D(\hat{\theta}) - \theta)^2\right] \\ &\quad + 2E_D\left[(\hat{\theta} - E_D(\hat{\theta}))(E_D(\hat{\theta}) - \theta)\right] \\ &= E_D\left[(\hat{\theta} - E_D(\hat{\theta}))^2\right] + \left[(E_D(\hat{\theta}) - \theta)\right]^2 \\ &= \text{Var} + \text{Bias}^2, \end{aligned} \tag{D.38}$$

where D denotes that averaging is performed over the available data set. To arrive at the final formula, we consider the following:

$$2E_D\left[(\hat{\theta} - E_D(\hat{\theta}))(E_D(\hat{\theta}) - \theta)\right] = 0, \tag{D.39}$$

due to the fact that

$$E_D\left[(\hat{\theta} - E_D(\hat{\theta})\right] = E_D(\hat{\theta}) - E_D(E_D(\hat{\theta})) = E_D(\hat{\theta}) - E_D(\hat{\theta}) = 0. \tag{D.40}$$

The last transformation is made because $E_D(E_D(\hat{\theta})) = (E_D(\hat{\theta}))$, since the mathematical expectation of a mathematical expectation is just a mathematical expectation. For the same reason

$$E_D\left[(E_D(\hat{\theta}) - \theta)^2\right] = \left[E_D(\hat{\theta}) - \theta\right]^2. \tag{D.41}$$

The bias-variance decomposition shows that the mean squared error between the estimate and the true parameter consists of two parts—variance of the estimate plus squared bias of the estimate. One term can be increased or decreased at the expense of the other term; hence, we can trade a little bit of bias for smaller variance. The idea of regularization exploits this fact as almost any regularized solution is biased but has a smaller variance than the maximum likelihood solution.

D.3.3 Bayes Theorem

Maximum likelihood historically is not the oldest method to "invert" probabilities. In this respect the Bayesian approach came first. The Bayesian approach makes use of Bayes theorem to recover unknown parameters or models from the data.

$$P(\text{Model}/\text{Data}) = \frac{P(\text{Data}/\text{Model})P(\text{Model})}{P(\text{Data})}. \tag{D.42}$$

Sometimes the Bayes formula is written in the other form:

$$\text{Posterior} = \frac{\text{Likelihood} \cdot \text{prior}}{\text{Evidence}}. \tag{D.43}$$

Notice that in addition to the likelihood, the Bayesian approach requires a priori probability of the model to be specified which is the most vexing and controversial issue of Bayesian inference.

D.3.4 Bayesian Interpretation of Regularization

Let us reformulate our maximum likelihood example using Bayesian interpretation. Suppose that prior distribution of the parameter b is Gaussian with zero mean and unknown standard deviation σ_b as in

$$p(b) = \frac{1}{\sqrt{2\pi}\sigma_b} e^{-\frac{b^2}{2\sigma_b^2}}. \tag{D.44}$$

Then, combining likelihood and prior distribution using Bayes theorem, we obtain

$$P(b|\varepsilon_1, \varepsilon_2, \ldots, \varepsilon_n) \propto \left[\prod_{i=1}^{n} \frac{1}{\sigma_\varepsilon \sqrt{2\pi}} e^{\frac{-(y_i - x_i b)^2}{2\sigma_\varepsilon^2}}\right] \cdot \left[\frac{1}{\sqrt{2\pi}\sigma_b} e^{-\frac{b^2}{2\sigma_b^2}}\right]. \tag{D.45}$$

Taking the logarithm of both sides and performing arithmetical manipulations, we obtain

$$\ln\big(P\big(b|\varepsilon_1, \varepsilon_2, \ldots, \varepsilon_n\big)\big) \propto -\frac{1}{2\sigma_\varepsilon^2} \sum_{i=1}^{n} (y - x_i b)^2 - \frac{1}{2\sigma_b^2} b^2. \tag{D.46}$$

Maximization of the last expression amounts to minimization of

$$E(b) = \sum_{i=1}^{n} (y - x_i b)^2 + \frac{\sigma_\varepsilon^2}{\sigma_b^2} b^2 = \sum_{i=1}^{n} (y - x_i b)^2 + \lambda^2 b^2 \tag{D.47}$$

with

$$\lambda^2 = \frac{\sigma_\varepsilon^2}{\sigma_b^2}, \tag{D.48}$$

which is exactly the expression for zero-order Tikhonov regularization. Thus Tikhonov regularization can be interpreted as Bayesian inference with Gaussian likelihood and Gaussian prior on the parameters. In statistics, this solution is known as the maximum penalized likelihood or MPL solution.

D.4 Multivariable Analysis

Let function $F(x_1, x_2, \ldots, x_n)$ be a scalar real-valued function of a real vector x. The gradient of x is defined as a column vector:

$$\nabla F(x) = \left[\frac{\partial F}{\partial x_1}, \frac{\partial F}{\partial x_2}, \ldots, \frac{\partial F}{\partial x_n}\right]^{\mathrm{T}}. \tag{D.49}$$

For the same function $F(x_1, x_2, \ldots, x_n)$, the Hessian is defined as a symmetric $(n \times n)$ matrix of second derivatives:

$$\nabla^2 F(x) = \begin{bmatrix} \dfrac{\partial^2 F}{\partial x_1^2} & \dfrac{\partial^2 F}{\partial x_1 \partial x_2} & \cdots & \dfrac{\partial^2 F}{\partial x_1 \partial x_n} \\ \dfrac{\partial^2 F}{\partial x_2 \partial x_1} & \dfrac{\partial^2 F}{\partial x_2^2} & \cdots & \dfrac{\partial^2 F}{\partial x_2 \partial x_n} \\ \vdots & \vdots & \cdots & \vdots \\ \dfrac{\partial^2 F}{\partial x_n \partial x_1} & \dfrac{\partial^2 F}{\partial x_n \partial x_2} & \cdots & \dfrac{\partial^2 F}{\partial x_n^2} \end{bmatrix}. \tag{D.50}$$

The Hessian matrix is widely used in linear as well as nonlinear optimization. The Hessian also appears in multivariable Taylor series expansion of $F(x)$ around the point x_0 as

$$\begin{aligned} F(x_0 + \Delta x_0) &= F(x_0) + \Delta x^{\mathrm{T}} \nabla F(x_0) \\ &\quad + \frac{1}{2} \Delta x^{\mathrm{T}} \nabla^2 F(x_0) \Delta x + \text{higher order terms}. \end{aligned} \tag{D.51}$$

D.5 Convolution and Fourier Transform

The following integral is called the convolution of two functions:

$$g(x) = \int_{-\infty}^{\infty} K(x - y)f(y)dy, \tag{D.52}$$

where $K(x - y)$ is known under different names in different fields such as the kernel function in the theory of integral equations, the impulse response function in engineering, the point spread function in imaging, Green's function in physics, and the fundamental solution in mathematics.

The Fourier transform of function $f(x)$ is a function depending on frequency w:

$$\hat{f}(w) = \int_{-\infty}^{\infty} f(x)e^{-iwx}dx. \tag{D.53}$$

The fundamental relations linking convolution of two functions and their Fourier transforms is as follows:

$$g(x) = \int_{-\infty}^{\infty} K(x - y)f(y)dy. \tag{D.54}$$

$$\hat{g}(x) = \hat{K}(w)\hat{f}(w). \tag{D.55}$$

D.6 The Trace Result

Consider a random m-vector b normally distributed as $\sqrt{n}b \sim N(0, \Sigma)$. The expected value of $(b^{\mathrm{T}}Ab)$ is given by

$$E\left(b^{\mathrm{T}}Ab\right) = \frac{1}{n}\mathrm{trace}(A\Sigma). \tag{D.56}$$

Indeed, the expected value can be calculated using the properties of the expectation operator as

$$
\begin{aligned}
E\big(b^{\mathrm{T}}Ab\big) &= E\left(\begin{pmatrix} b_1 & \cdots & b_m \end{pmatrix} \begin{pmatrix} a_{11} & \cdots & a_{1m} \\ \vdots & & \vdots \\ a_{m1} & \cdots & a_{mm} \end{pmatrix} \begin{pmatrix} b_1 \\ \vdots \\ b_m \end{pmatrix} \right) \\
&= E\left(\sum_{i=1}^{m} b_1 b_i a_{i1} + \sum_{i=1}^{m} b_2 b_i a_{i2} + \cdots + \sum_{i=1}^{m} b_m b_i a_{im} \right) \\
&= \sum_{k=1}^{m} \sum_{i=1}^{m} a_{ik} E(b_k b_i) = \frac{1}{n} \sum_{k=1}^{m} \sum_{i=1}^{m} a_{ik} \sigma_{ki} = \frac{1}{n} \mathrm{trace}(A\Sigma).
\end{aligned}
\tag{D.57}
$$

This proves the result.

Index

M.A. Abidi et al., *Optimization Techniques in Computer Vision*, Advances in Computer Vision and Pattern Recognition, DOI 10.1007/978-3-319-46364-3

Printed in the United States
By Bookmasters